Microanalysis of Solids

Edited by

B.G. Yacobi

EMTECH
Toronto, Ontario, Canada

D.B. Holt

Imperial College
London, England

and

L.L. Kazmerski

National Renewable Energy Laboratory
Golden, Colorado

Plenum Press • New York and London

0589221

CHEMISTRY

Library of Congress Cataloging-in-Publication Data

Microanalysis of solids / edited by B.G. Yacobi, D.B. Holt, and L.L.
  Kazmerski.
        p.   cm.
    Includes bibliographical references and index.
    ISBN 0-306-44433-X
    1. Microchemistry.   2. Solids--Analysis.   I. Yacobi, B. G.
  II. Holt, D. B.   III. Kazmerski, Lawrence L.
  QD79.M5M53   1994
  543'.081--dc20   93-40984
                                                             CIP

ISBN 0-306-44433-X

© 1994 Plenum Press, New York
A Division of Plenum Publishing Corporation
233 Spring Street, New York, N.Y. 10013

Printed in the United States of America

Contributors

S. E. Asher, National Renewable Energy Laboratory, Golden, Colorado 80401

G. A. D. Briggs, Department of Materials, University of Oxford, Oxford OX1 3PH, England

R. F. Cohn, 51-5 Jacqueline Road, Waltham, Massachusetts 02154

A. J. Garratt-Reed, Center for Materials Science and Engineering, Massachusetts Institute of Technology, Cambridge, Massachusetts 02139

D. A. Grigg, Digital Instruments, Santa Barbara, California 93117

D. B. Holt, Department of Materials, Imperial College, London SW7 2BP, England

J. J. Hren, Department of Materials Science and Engineering, North Carolina State University, Raleigh, North Carolina 27695

L. L. Kazmerski, National Renewable Energy Laboratory, Golden, Colorado 80401

J. Li, Department of Materials Science and Engineering, Cornell University, Ithaca, New York 14853. *Present address:* Intel Corporation, Santa Clara, California 95052

J. Liu, Department of Materials Science and Engineering, North Carolina State University, Raleigh, North Carolina 27695

A. G. Michette, Department of Physics, King's College, London WC2R 2LS, England

P. Mutti, Dipartimento di Ingegneria Nucleare Politecnico di Milano, Milano, Italy 20133

A. J. Nelson, National Renewable Energy Laboratory, Golden, Colorado 80401

R. W. Odom, Charles Evans & Associates, Redwood City, California 94063

A. W. Potts, Department of Physics, King's College, London WC2R 2LS, England

F. Radicati di Brozolo, Charles Evans & Associates, Redwood City, California 94063

P. Revesz, Department of Materials Science and Engineering, Cornell University, Ithaca, New York 14853

P. E. Russell, Department of Materials Science and Engineering, North Carolina State University, Raleigh, North Carolina 27695

T. Wilson, Department of Engineering Science, University of Oxford, Oxford OX1 3PJ, England

B. G. Yacobi, EMTECH, Toronto, Ontario, Canada

Preface

The main objective of this book is to systematically describe the basic principles of the most widely used techniques for the analysis of physical, structural, and compositional properties of solids with a spatial resolution of approximately 1 μm or less.

Many books and reviews on a wide variety of microanalysis techniques have appeared in recent years, and the purpose of this book is not to replace them. Rather, the motivation for combining the descriptions of various microanalysis techniques in one comprehensive volume is the need for a reference source to help identify microanalysis techniques, and their capabilities, for obtaining particular information on solid-state materials.

In principle, there are several possible ways to group the various microanalysis techniques. They can be distinguished by the means of excitation, or the emitted species, or whether they are surface or bulk-sensitive techniques, or on the basis of the information obtained. We have chosen to group them according to the means of excitation. Thus, the major parts of the book are: Electron Beam Techniques, Ion Beam Techniques, Photon Beam Techniques, Acoustic Wave Excitation, and Tunneling of Electrons and Scanning Probe Microscopies.

We hope that this book will be useful to students (final year undergraduates and graduates) and researchers, such as physicists, material scientists, electrical engineers, and chemists, working in a wide variety of fields in solid-state sciences.

B. G. Yacobi
D. B. Holt
L. L. Kazmerski

Toronto, London, and Denver

Contents

PART I. INTRODUCTION

1. An Introduction to Microanalysis of Solids
B. G. Yacobi and D. B. Holt

PART II. ELECTRON BEAM TECHNIQUES

2. Scanning Electron Microscopy
B. G. Yacobi and D. B. Holt

3. Transmission Electron Microscopy
A. J. Garratt-Reed

4. Auger Electron Spectroscopy
L. L. Kazmerski

PART III. ION BEAM TECHNIQUES

5. Secondary Ion Mass Spectrometry
S. E. Asher

PART IV. PHOTON BEAM TECHNIQUES

10. Laser Ionization Mass Spectrometry
R. W. Odom and F. Radicati di Brozolo

11. Microellipsometry
R. F. Cohn

PART V. ACOUSTIC WAVE EXCITATION

12. Scanning Acoustic Microscopy
P. Mutti and G. A. D. Briggs

PART VI. TUNNELING OF ELECTRONS AND SCANNING PROBE MICROSCOPIES

13. Field Emission, Field Ion Microscopy, and the Atom Probe
J. J. Hren and J. Liu

14. Scanning Probe Microscopy
D. A. Grigg and P. E. Russell

Part I

Introduction

1

An Introduction to Microanalysis of Solids

B. G. Yacobi and D. B. Holt

1.1. Introduction

The continuing advances in the development of submicron structures and devices require a better understanding of materials and device properties with correspondingly high spatial resolution. In general, in order to understand a wide variety of macroscopic phenomena in solid-state materials, it is essential to elucidate their properties on a microscopic level. With the recent development of scanning tunneling microscopy (STM) and other scanning probe techniques, it has become possible to examine (and even manipulate) individual atoms on surfaces and to analyze materials properties with nanometer-scale spatial resolution. Recent advances of both theory and experiment also led to a new field, *nanotechnology* (or *nanoengineering*). All these catalyzed further developments of the existing microanalysis techniques and the emergence of new methods.

The emphasis in this book is on the description of the principles of the techniques that provide analysis of a variety of physical, structural, and compositional properties with spatial resolutions on the order of 1 μm or less.

B. G. Yacobi • EMTECH, Toronto, Ontario, Canada.   *D. B. Holt* • Department of Materials, Imperial College, London SW7 2BP, England.

Microanalysis of Solids, edited by B. G. Yacobi *et al.* Plenum Press, New York, 1994.

In the rapidly developing field of materials microanalysis, it is possible that we have omitted descriptions of some existing techniques or of those recent emerging techniques that await wide applications. And it is certain that new microanalysis techniques will be developed in the future. Thus, this book should not be considered as a definitive source of all of the available microanalysis techniques, and especially not of those with a spatial resolution greater than several microns.

In selecting the specific technique for the analysis of a particular problem, several questions arise: What type of information is obtainable? What is the sensitivity? What is the depth of analysis (i.e., is it surface, subsurface, or bulk analysis)? What is the spatial resolution? What is the depth resolution? Is the analysis quantifiable? What are the data acquisition and processing times? Is the method destructive or nondestructive? (In general, great caution should be exercised with regard to the term *nondestructive,* since bombardment-induced defects can often be introduced in the material with no observable effect on its physical integrity.)

Various microanalysis techniques provide different information related to the material's physical, structural, or compositional properties. In an ideal situation, all of these will be needed for a more complete understanding of the material. Therefore, the tendency in the microanalysis of solids is toward more complete analyses using several tools. Only then, a relatively complete picture emerges, in which the different types of information obtained complement each other in elucidating the relationship between the synthesis and the properties of a wide range of materials. Some techniques, such as electron microscopy, can combine several complementary microanalysis modes in the same instrument that can provide local information on structural, compositional, and physical properties of the material.

The basic principles of the microanalysis of solids are based on the effects produced by perturbing materials using means, such as irradiating the solid with various excitation beams (probes), or applying fields to solid surfaces. Excitation sources can be, for example, electrons, ions, photons, positrons, neutrons, and acoustic waves. Important information about the material is obtained from the effects (or responses) produced by the interaction of the excitation probe with the material. The signals, which provide the information employed in the microanalysis, from a material under bombardment usually take the form of backscattered and secondary particles and photons, diffracted waves, or a variety of processes occurring in the bulk of a solid. A multitude of effects, or signals, can be produced in such interactions between a particular probe (e.g., an electron beam) and the solid. Thus, a variety of materials properties can be determined in an instrument with only one excitation source, provided appropriate detectors are employed.

The concept of scanning microscopy in general is based on the serial formation of an image point by point, which can be realized by either (1) electronically scanning a beam (e.g., the electron beam in a scanning electron microscope), (2) mechanically scanning a probe (e.g., the tip in a scanning tunneling microscope), or (3) mechanically scanning the specimen while keeping the probe stationary.

It is interesting to follow the progress in the development of microscopy techniques from optical microscopy, through electron microscopy and other irradiation-probe techniques, and, currently, to tunneling microscopy and related methods. In this development, each new method did not replace the others, but complemented them. The scaling factor in these microscopies is important, since they provide different applications in the general field of microanalysis. Also, the ambient (vacuum) needed, the magnification range, the cost, sample preparation time, sample size, irradiation-induced effects (damage), and the ease of operation and throughput are important factors in considering a particular analysis technique.

The classification of microanalysis techniques described in this book is based on the scheme presented by Goodhew and Castle (1983). The basic classification is in terms of the primary excitation source used. As a result of that excitation, a secondary effect is produced that is analyzed by a detector monitoring a specific variable. Thus, according to this scheme, each technique can be specified by three parameters: (a) *primary excitation source,* (b) *secondary effect observed,* and (c) *the variable monitored.* Specific examples of (a) the primary excitation source and (b) the secondary effect observed are *electron* (denoted e), *light* (1), *atom* (a), *ion* (i), *x ray* (x), *sound* (s), *heat* (h), and (electrostatic) *field* (f). Examples of variables monitored are *intensity* (I), *energy* (E) (or wavelength), *mass* (m), *time* (t), *temperature* (T), *angle* (θ), *position* (xy), and *phase* (ϕ). Thus, for the secondary electron mode in a scanning electron microscope (SEM) the notation is (e_1, e_2, I_{xy}). For the transmitted electrons in a transmission electron microscope (TEM) the notation is (e_1, e_1, I_{xy}); for electron diffraction it is (e_1, e_1, θ); for energy-dispersive x-ray spectrometry (EDS) it is (e, x, E); for secondary ion mass spectrometry (SIMS) it is (i_1, i_2, m); for optical microscopy it is (l_1, l_1, I_{xy}); and for field-ion microscopy (FIM) it is (f, i, I_{xy}).

As mentioned above, important characteristics of a specific analysis technique are (1) the type of information that is obtainable, (2) the sensitivity, (3) the spatial resolution, and so forth. Some techniques, such as scanning electron microscopy, are multimode systems, i.e., they may be operated in several modes, providing different, and often complementary, information on physical, structural, and compositional properties of solid-state materials and devices. The mechanisms of signal formation are different for these modes, and consequently, their spatial resolution, sensitivity, quantifiability, and other characteristics vary significantly. For example, among the seven modes of scanning electron microscopy that provide information about a wide range of properties of solids, the spatial resolution may vary from about 30 Å for the secondary electron mode that provides information, for example, on specimen topography, to about 1 μm for the cathodoluminescence mode that provides information on electronic properties of the material.

1.2. Electron Beam Techniques

Electron irradiation of a solid generally results in a variety of signals that can provide important information on structural, compositional, and electronic properties of the material. One of the advantages of employing an electron beam is the ease of obtaining finely focused probes by using electromagnetic lenses. The major signals produced by the interaction of the primary (i.e., incident or beam) electrons with a solid are illustrated in Fig. 1.1. Thus, when primary electrons are backscattered from the solid with little or no energy loss, they produce *backscattered* electrons; and when the primary electrons lose energy, they can produce *secondary* electrons. The dissipation of the energy of primary electrons in a variety of electronic excitations may lead to (1) the emission of Auger electrons, (2) to the emission of characteristic x-ray photons, (3) to the generation of electron–hole pairs that may lead to cathodoluminescence (CL), i.e., the emission of photons in the ultraviolet, visible, and infrared spectral ranges, or that may lead to charge-collection (CC) signals in devices, and (4) to thermal effects, including *thermal wave* (*electron acoustic*) signals when the beam is chopped. The formation of these signals is the basis for the various modes in an SEM. The signals produced in thin specimens by transmitted electrons, which may be scattered elastically (i.e., with no energy loss) or inelastically (i.e., with energy loss), provide various modes in a TEM. The Auger electrons provide the basis for a scanning Auger electron microscope (SAEM).

All of these electron beam techniques and modes have different sensitivities, spatial and depth resolutions, and quantifiabilities. The depths from which signals are generated in each of the above modes depend on the specifics of the excitation conditions, the properties of the materials, and the mechanisms of the particular signal formation (see Chapter 2).

SEM modes are routinely used in the microcharacterization of a wide variety of properties of materials. From a few modes of operation in the earlier years, there are now at least seven SEM modes that provide microcharacterization of structural, compositional, and electronic properties of a wide range of materials. The *secondary electrons* provide the most routinely used secondary electron image mode for characterizing topographic features of solid

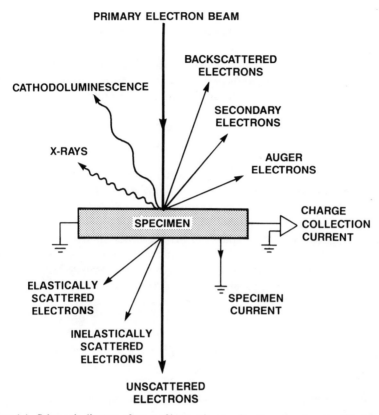

Figure 1.1. Schematic diagram of types of interaction (or signal) produced as a result of electron beam interaction with a solid.

surfaces. In addition, local electric and magnetic fields present in the materials can be inferred from the secondary electron images, and voltage contrast can be used in the analysis of electronic devices. *Backscattered electrons* provide atomic number contrast, i.e., qualitative information on compositional uni-formity. Under appropriate operating conditions, backscattered electrons can also provide crystallographic information relating to electron channeling. *Characteristic x rays* emitted as a result of electronic transitions between inner-core levels provide information about the particular chemical element and, thus, the materials composition. The CL mode can provide contactless and nondestructive characterization of a wide range of electronic properties of a variety of luminescent materials. The detection of impurity concentrations down to limits of 10^{14} at/cm^3 can be attained. In the CC mode, electrical contacts are applied to the specimen and an electrical signal (current or voltage) is monitored in the external circuit. Major applications of this mode include characterization of various semiconductor device structures, microcharacterization of electrically active defects in semiconductors, and quality control and failure analysis of electronic devices. The *electron acoustic mode* uses the chopped electron beam to produce intermittent heating, and thermal expansion and contraction, which result in the propagation of ultrasonic waves that can be detected by piezoelectric transducers. The technique is useful in the characterization of various subsurface defects in solid-state materials. And, finally, scanning deep-level transient spectroscopy (SDLTS), which is based on the capture and thermal release of carriers at traps, allows one to determine the energy levels and the spatial distribution of deep states in semiconductors. This method complements CL spectroscopy for the assessment of nonradiative centers in the semiconductor (see Chapter 2).

Transmitted electrons, which can be detected and analyzed in thin specimens (on the order of 1000 Å and less), provide important information related to the crystal structure and defects in the material. This mode is mainly used in TEMs and scanning transmission electron microscopes (STEMs). In the latter case, the signal can be amplified and processed by detectors that are incorporated in the STEMs. The basic information on the crystalline structure is obtained by using *electron diffraction,* which can be performed with a high spatial resolution by using the *convergent-beam diffraction* technique. Modern instruments have a resolution better than 2 Å that allows atomic resolution imaging in many cases. For electron energies greater than about 100 keV (i.e., in TEMs and STEMs), energy losses of the transmitted electrons are characteristic of the elements present, and thus, electron energy-loss spectroscopy (EELS) can provide chemical and structural information. In modern electron microscope instruments, it is also possible to use *reflected electrons* for reflection high-energy electron diffraction (RHEED), reflection electron microscopy (REM), and scanning reflection electron microscopy (SREM) (e.g., see Cowley, 1989). The latter techniques are useful in the analysis of surfaces, and they are capable of providing images of atom-high surface steps.

Auger electrons emitted from about the top 10 Å of the material have energies characteristic of the elements of the material, and they can provide analysis of surface composition variations with nanometer-scale resolution. Thus, Auger electron spectroscopy (AES) and SAEM provide important surface analytical techniques in ultrahigh-vacuum electron probe instruments. AES is based on the detection of the electron that has been ejected because of the rearrangement of core electrons in the atom as the result of primary electron beam bombardment with energies of up to about 5 keV. Thus, binding energies of core electrons in the atom can be deduced and the chemical elements can be identified. In principle, all elements above helium can be detected with detection limits in the range from about 0.1 to 3 at. %. The analysis is quantifiable, and a spatial resolution of about 500 Å and less can be attained. AES can also be combined with the slow removal of the outermost atomic layers of the solid by sputtering, and thus it can produce depth profiling of chemical constituents with a resolution of about 10 Å (see Chapter 4).

It should be noted that, in comparison with other analytical techniques, the electron beam techniques provide a very broad range of information on a wide variety of materials properties, and, thus, they constitute some of the most important tools in the microanalysis of solids. Some of the disadvantages of the electron beam techniques, compared with photon beam techniques, are the requirements for a vacuum environment and the sample preparation (i.e., thinning for the TEM specimens), potential beam-induced damage in samples, and the charging of insulating specimens (i.e., the need for coating of the specimen with a conductive layer). It should be noted, however, that the latter can be avoided by using a low-voltage (about 1 kV) electron beam excitation, which results in the total electron emission yield being equal to unity.

1.3. Ion Beam Techniques

The ion beam techniques may be distinguished on the basis of the excitation beam energies used, i.e., those with ion beam energies in the kiloelectron volt range and those in the megaelectron volt range. (Ion beam techniques are outlined in Chapters 5 and 6.)

SIMS employs ion beam sputtering of the sample, and thus depth profiling is obtained. Secondary ions emitted as a result of the primary ion (with energies up to 20 keV) bombardment are analyzed in a mass spectrometer for their identification. Quantitative analysis requires using standards. A spatial resolution of less than 1 μm and a depth resolution down to about 3 nm can be obtained. In SIMS measurements, all elements, including hydrogen, can be detected. SIMS imaging with a spatial resolution of less than 1 μm is also possible. The main advantages of this technique are (1) its high sensitivity to low concentrations of all elements with detection limits in the range from about 10^{14} to 10^{16} at/cm^3, (2) an excellent depth resolution, and (3) the capability of obtaining secondary ion images.

High-energy ion beam techniques, which are sometimes termed *nuclear microscopy* (or *proton microscopy,* or *nuclear microprobe*) techniques, employ high-energy (MeV) ion beams, such as protons or He ions, for excitation of the solid. Among these techniques are, for example, Rutherford backscattering (RBS), proton-induced x-ray emission (PIXE), nuclear reaction analysis (NRA), and scanning transmission ion microscopy (STIM). [For recent reviews and examples of applications, see, for example, Grime and Watt (1990), Jamieson *et al.* (1990), Cookson (1991), and references therein.] The spatial resolutions in these cases are on the order of 1 μm (and down to 0.1 μm for the STIM). One of the main advantages of these techniques is their high depth resolution on the order of 200 Å.

RBS is a depth profiling technique, but, in contrast to SIMS, it is non-destructive and quantitative analysis can be performed on the basis of first principles, without the need for standards. In this case, some of the ions of the incident high-energy beam are elastically backscattered. The measurement of the backscattered ion energy provides information about the mass and the depth of the target nucleus (i.e., the composition and depth profiles) of elements in the material. An additional feature of RBS, the channeling phenomenon, can also provide important information about the crystallinity of the specimen and the location of the impurity atoms in the crystal structure (i.e., whether they occupy lattice or interstitial sites).

In the case of PIXE, collision of a high-energy proton with an atom results in emission of the characteristic x ray. Detection of these x rays and measurement of their energies provides a means of microanalysis with excellent sensitivities for detection of trace elements with concentrations down to the parts-per-million level for elements with $z > 10$.

NRA is based on the detection of neutrons, γ rays, and charged particles, which are produced as the result of the nuclear reactions induced by the incident ions. This technique can be used for the measurement of elemental distributions. One of the applications of NRA is in the microanalysis of complex metallurgical systems, in which high-sensitivity measurement of the light elements is required.

In STIM, the energy losses of the individual ions are employed to provide images of thin specimens (as the beam is scanned across the sample, the energy loss variations are equivalent to the corresponding density variations). In this application, very low beam currents can be used, and thus a spatial resolution of less than 0.1 μm can be obtained. Because of the relatively large mass of the ions used as a probe, i.e., because of their greater penetration power, the main advantage of STIM is in its ability to analyze thicker specimens, such as semiconductor microcircuits.

1.4. Photon Beam Techniques

Optical characterization techniques are based on the interaction of electromagnetic radiation with the solid leading to a wide variety of useful signals produced as a result of absorption, reflection, emission, scattering, and so forth, which provide important information on various materials properties. The availability of a wide variety of both photon sources and optical instruments provides a wide range of possible configurations for the microanalysis of the solid.

The general advantages of this class of techniques, relative to those using charged particle (electron, ion) excitation beams, include the absence of charging of insulating materials (i.e., no need for coating of the specimen with a conductive layer) and the ability to examine specimens in air (i.e., no need for a vacuum).

The most versatile and early tool in this category of analytical methods is *optical microscopy,* which is an indispensable technique in all fields of research. The advantages include the versatility of optical examination and the simplicity of sample preparation. Optical microscopy is still one of the most powerful and efficient tools in the microcharacterization of a very wide range of materials. Recent advances in both the hardware and quantitation of optical microscopy have enhanced its power even further. Among these advances are the development of *scanning optical microscopy,* and especially of *confocal scanning optical microscopy,* and also the availability of computer-based image analysis systems. In general, the scanning optical microscope employs laser illumination, the collection of the transmitted (or reflected) light by means of a photodetector, and the display of the resulting video image on a cathode-ray tube (CRT). The confocal configuration improves both the resolution and the depth of field. In a scanning optical microscope, only one point (column) of the sample is illuminated at a time and the light signal is collected from just one (confocal) object point in that column, thus improving contrast and depth resolution. Either the sample or the beam is scanned, so that a raster image can be displayed on a CRT display and/or stored in a computer for image processing. One of the applications of the scanning optical microscope is in the microcharacterization of electronic devices and various defects in semiconductors using the optical beam-induced current (OBIC), which should be compared with the analogous electron beam-induced current (EBIC) technique of the (CC) mode available in the SEM. Another useful application of the scanning optical microscope is in *scanning photoluminescence mapping,* which is analogous to the CL mode in the SEM. Some advantages of the scanning optical microscopy methods, relative to the analogous SEM modes, are that there is no vacuum requirement, no specimen charging, and no electron beam damage (see Chapter 7).

In recent years, *x-ray microscopy* has been increasingly advanced to offer a wide variety of applications in the microanalysis of materials (e.g., Michette, 1988). The main reasons for this progress are the development of bright synchrotron sources of photons with tunable energy, and the availability of appropriate zone plates (i.e., circular diffraction gratings with radially decreasing linewidths and spacings). There are two main advantages of x-ray microscopy relative to its optical and electron counterparts. First, because of the shorter wavelengths, x rays can provide much higher resolution (about 50 nm and less) than is obtainable with visible light. In addition, with x-ray excitation, the specimens can be studied in their natural state with no need for sample preparation procedures (such as those required in TEM) that may in principle introduce artifacts in the material. In performance, the x-ray microscope falls somewhere between the optical microscope and the electron microscope. (The various configurations of this technique, together with some examples of application, are outlined in Chapter 8.)

Photoelectron spectroscopy employs x rays (or UV photons) for the excitation of the solid and the detection of emitted photoelectrons with characteristic energies that can provide chemical information about the material. It is also referred to as electron spectroscopy for chemical analysis (ESCA), which in principle can be accomplished with either incident x rays (i.e., x-ray photoelectron spectroscopy, XPS) or ultraviolet radiation (i.e., UV photoelectron spectroscopy, UPS). In these cases, the energy spectrum of the photoelectrons, emitted from near the surface of the specimen, permits nondestructive chemical and elemental analysis of the surface. The energies of photoelectrons are much less than 1 keV, and thus the escape depth (and the depth resolution) is within about 50 Å of the surface. Some major applications of ESCA in the examination of solids are in detecting particular elements present at the surface of the material, in determining binding energies, and in the analysis of the band structure of solids (see Chapter 9).

Laser ionization mass spectrometry (LIMS) employs a high-power-density (focused) pulsed laser to vaporize the specimen surface. The species ejected by the laser ionization are analyzed in a mass spectrometer, and a spectrum of the detected ions for each laser pulse is recorded, providing useful information on impurities or surface contaminants in the specimen. Using LIMS, all elements can be detected with a spatial resolution on the order of 1 μm. The elemental detection limits are in the range of parts per million, and the depth profiling can be obtained by using successive vaporization of the same area. The main advantage of this technique is that it avoids charging problems, and thus the microanalysis of a wide range of materials, including insulating specimens, can be performed (see Chapter 10).

In *ellipsometry,* the change in the polarization of the electromagnetic wave striking the interface between two dielectric media is monitored. It is used, for example, to determine the optical constants and the thickness of thin dielectric films. The technique can also be used as a contactless, nondestructive method for monitoring the changes (e.g., caused by ion-implantation-induced damage) in semiconductors. Other important applications of ellipsometry include, for example, real-time monitoring of thin film growth, studies of oxidation of semiconductors, and surface roughness evaluation (see Chapter 11).

Raman spectroscopy is based on measuring the energy shift of the incident photon beam that is scattered off the material. In this technique, the specimen is illuminated with monochromatic light generated (typically) by a laser. Incident photons induce transitions in the material and, consequently, they gain or lose energy. The energy shift during the scattering process is caused by either (1) the photon energy transfer to the lattice (i.e., phonon emission) or (2) the absorption of a phonon by the photon. In the former case, the reduction in photon energy is called the Stokes shift, and in the latter case, when the scattered photon emerges at a higher energy, it is called the anti-Stokes shift. Raman spectra contain information on the vibrational modes in the material. Spatially resolved Raman measurements (with a resolution of about 1 μm) can be obtained by incorporating an optical microscope in the Raman system. In such a case, i.e., the Raman microprobe, the illumination of the sample by a laser is through an optical microscope, which is coupled with the monochromator. The major applications of Raman spectroscopy include, for example, analysis of both the composition and crystal structure. The changes in the shift, the width, and the symmetry of the Stokes line can be used to distinguish between crystalline and amorphous materials. This technique can also be employed in the characterization of various defects and phases, of stresses and induced damage in the material, and of semiconductor processing and structures. One of the important new applications of this technique is time-resolved Raman spectroscopy performed at high temperatures. The latter may be useful, for example, in elucidating temperature-induced effects in solids. (For two recent reviews, see Nakashima and Hangyo, 1989, and Huong, 1991.)

In *modulated optical reflectance imaging,* two low-power laser beams (both focused to the same 1-μm spot on the specimen surface) are employed. One, the modulated argon-ion laser (pump), induces localized periodic temperature variation (up to 10°C) in the specimen, whereas the other, the helium–neon laser (probe), monitors the response of the specimen to the local perturbation. The presence of subsurface defects and damage can hinder the propagation into the sample of pump-laser-induced periodic thermal waves at that location. This affects the local surface temperature and results in a variation of the modulated reflectance which is measured by the probe laser. A spatial resolution of less than 1 μm can be achieved in these measurements. Some of the major applications of this nondestructive technique are in the imaging of subsurface defects in various materials and devices, ion-implantation monitoring, and metallization integrity studies (e.g., see Smith *et al.,* 1987; Hahn *et al.,* 1990).

1.5. Acoustic Wave Excitation

In the last decade, scanning acoustic microscopy has emerged as a valuable technique in the microanalysis of solid-state materials and devices. The technique employs the detection of elastic waves generated inside the solid as a result of the interaction of an acoustic wave with the material. Thus, this method provides a means for the microcharacterization of the mechanical properties of solid-state materials. Acoustic microscopy is especially advantageous in the nondestructive analysis of subsurface properties of those materials that are opaque to optical radiation. Some important applications of this technique are in the spatially resolved examination of such mechanical properties of materials as density and elasticity, and also in the microcharacterization of various defects and different phases in materials. The fundamental principles of scanning acoustic microscopy and some examples of its applications are described in Chapter 12.

1.6. Tunneling of Electrons and Scanning Probe Microscopies

Unlike the other techniques that utilize the effects produced by the irradiation of the material with excitation probes (e.g., electron beam, ion beam, photon beam, and acoustic waves), a different approach is used in a variety of techniques based on the tunneling of electrons from the material. These are "microscopies without lenses."

FIM, (field-ion microscopy) an earlier technique, basically consists of a cooled fine metal tip (i.e., the specimen, or emitter) with a radius of curvature of up to 0.1 μm positioned inside a vacuum chamber which is backfilled with an imaging gas. A positive voltage applied to the emitter results in very high electric field regions (above the protruding surface atoms) which ionize the imaging gas molecules. In this process, the ionization is caused by the tunneling of the electron from the imaging gas atom to an unoccupied surface state. The resulting ions are then accelerated radially onto a fluorescent screen to form an image, consisting of spots that represent the surface atoms. An additional important capability of FIM is the chemical analysis which can be performed in the *atom-probe field-ion microscope* by using a mass spectrometer. The latter allows one to relate the image spots with the chemical nature of the corresponding atoms. The applications of FIM include direct observations of defects, such as vacancies, interstitials, impurity atoms, dislocations, stacking faults, and grain boundaries. Some major limitations include problems with the preparation of a fine tip, high-field-induced specimen fracture, and sampling problems relating to the very small specimen volume being analyzed (see Chapter 13).

In the last decade, the discovery by Binnig *et al.* (1982) of the STM, (scanning tunneling microscopy) which is capable of imaging objects on the atomic scale, catalyzed development of a whole battery of similar techniques, or microscopies without lenses. The basic components of such microscopes are a fine probe (or tip) that is mechanically scanned, using a piezoelectric scanner, in very close proximity to the specimen and a feedback circuit that controls the distance between the tip and the sample surface. An interaction between a probe and a specimen generates a specific signal that can be used to produce an image with the resolution determined by the tip diameter and the tip-to-sample separation. Thus, the common term for this battery of techniques, which offers a powerful means of analyzing surface properties of materials, is *scanning probe microscopy* (or scanning tip microscopy). The basic principle of the operation of the STM is the measurement of the quantum-mechanical electron tunneling current between an ultrasharp tip and the sample. The tip, made of conductive material such as tungsten, can be moved in three dimensions using piezoelectric elements for x, y, and z translators. The tip is positioned in such close proximity (about 1 nm) to the sample surface that at a low operating voltage (on the order of millivolts) the tunneling current (about 1 nA) is detected. The tunneling current, which depends exponentially on the distance between the tip and the sample surface and which is very sensitive to that distance, is kept constant by using a feedback circuit that changes the tip height z by applying the voltage to the z-controlling piezoelectric element. Thus, as the tip is scanned across the specimen surface in the x and y directions, the monitoring of the tunneling current and recording the voltage that controls the tip height allow obtaining an image which reveals the surface topography with atomic resolution.

The general tendency is to apply the basic principles of scanning probe microscopy in the analysis of a wide range of materials properties. In the last several years, a variety of configurations of scanning probe microscopes have been developed (for some recent reviews, see Hansma and Tersoff, 1987; Hansma *et al.*, 1988; Howie, 1989; Wickramasinghe, 1990). Whereas STM requires the surface of the material studied to be electrically conductive, atomic force microscopy (AFM), developed by Binnig *et al.* (1986), can be employed for the examination of insulators as well as conductors. In this method, a tip (e.g., a diamond fragment) is attached to a flexible cantilever, which is deflected because of the interaction force (e.g., interatomic forces) between the tip and the sample surface. The force that is experienced by the tip can be deduced from the deflection of the cantilever, which can be measured using electron tunneling detection or optical detection. The types of optical detection used are interferometry and beam deflection. For example, in the latter case, a diode laser beam is reflected off the cantilever to a position-sensitive photo-detector. A feedback system is used to control the deflection, and thus the interaction force. The topography of the sample surface can be recorded using the constant-force mode of operation, which is analogous to the constant-current mode in STM. In magnetic force microscopy (MFM), a magnetic tip (e.g., iron or nickel fine wire) mounted on a cantilever is used to determine surface magnetic field distributions in the magnetic material. In electrostatic force microscopy (EFM), the charged tip responds to isolated charges on the sample surface resulting from the electrostatic interaction; this allows one to map out the charge distribution on the sample surface.

Scanning probe ballistic-electron-emission microscopy (BEEM) is a non-destructive technique for direct imaging of interfaces (such as metal–semiconductor interfaces) with nanometer-scale spatial resolution (Kaiser and Bell, 1988). BEEM employs three electrodes (the STM emitting tip, a biasing electrode, and the collecting electrode) and the behavior of ballistic electrons traversing the sample surface layer to the interface region. BEEM can provide direct imaging of subsurface interface electronic structure by scanning the STM tip across the sample surface and simultaneously measuring the collector current. Spectroscopic information can also be obtained from the measurements of the collector current as a function of an applied sample-tip voltage.

Some other scanning probe microscopies reported in recent years are *scanning capacitance microscopy, near-field scanning optical microscopy, scanning ion-conductance microscopy,* and *scanning thermal microscopy.* They can provide information on electrical, optical, magnetic, thermal, and mechanical properties of materials with nanometer-scale spatial resolution. The basic principles of the scanning probe microscopies are discussed in Chapter 14.

1.7. Emerging Techniques

1.7.1. Electron Holography

In recent years, with the development of field-emission electron guns which produce beams with a greater coherence length, it became feasible to obtain holographic images on the atomic scale. Off-axis electron holography is a two-step imaging method performed in a field-emission electron microscope equipped with an electron biprism (e.g., see Hanszen, 1982; Tonomura, 1987; Lichte, 1991; and references therein). The first step involves the formation and recording on film of the hologram, and the second step consists of the reconstruction of the image using a laser beam. In the first step, the electron hologram is formed by irradiating the object of interest with a coherent electron beam in such a way that half of the electron beam bypasses the specimen and serves as the reference wave. The electron biprism superposes the image and reference waves to form an interference pattern which is magnified and recorded on film as the electron hologram. In the second step, the holographic image is optically reconstructed by illuminating the electron hologram with a collimated laser beam. In electron holography, not only is the intensity distribution reconstructed in the optical image, but also the phase distribution of the electron wave can be recorded as an interference micrograph. Using holographic techniques, one can also obtain a phase-amplified interference micrograph. This allows phase measurement to a very high precision.

Some recent applications of electron holography include measurement of the electrostatic field distributions associated with reverse-biased $p-n$ junctions (Frabboni *et al.*, 1985; Matteucci, 1991), observation of magnetic fluxons in superconductors (Tonomura, 1990; Matteucci *et al.*, 1991), thickness measurement on an atomic scale (Tonomura, 1990), holography of atomic structures (Lichte, 1991), and confirmation of the occurrence of the Aharonov–Bohm effect (Tonomura, 1990).

A novel method for electron holography, *lensless low-energy electron holography* with atomic resolution, has been developed recently. This technique employs an ultrasharp tip as a coherent point source of electrons with low energies between about 20 and 100 eV and fine mechanical control based on the principles of scanning probe microscopy (Fink *et al.*, 1990, 1991).

1.7.2. Positron Microscopy

Being identical to electrons in all respects but charge, positrons in microscopy offer some distinct advantages over electrons. For example, a primary (incident) electron in the target becomes indistinguishable from other electrons in the material, whereas it should be feasible to deduce the behavior of each positron before it annihilates. Various materials properties, such as the presence of impurities and defects, or spatial variations in composition, or the presence of internal and external fields may all affect the diffusion (and drift) of positrons through the material, and thus these materials properties can be examined. Also, because of their positive charge, positrons can get trapped at negatively charged defects, at impurities, and at surface states, and thus it should be possible to examine these defect structures. Two different types of positron microscopes, the transmission positron microscope (TPM) and a positron reemission microscope (PRM), were developed in recent years (for details and some examples of applications, see, e.g., Hulett *et al.*, 1984; Van House and Rich, 1988a,b; Brandes *et al.*, 1988; Frieze *et al.*, 1990; and references therein).

The first instrument employing positrons in microscopy was a transmission microscope (Van House and Rich, 1988a). This TPM is thought to provide new applications based on expected new contrast mechanisms when using positrons. For example, it was estimated that a strongly atomic-number-dependent fractional difference in the amplitude contrast would be expected between the TPM and TEM; and that comparison of the calibrated contrast differences between their corresponding images could provide information on atomic form factors resulting from the screening differences, and could also demonstrate a sensitive microanalysis method (Van House and Rich, 1988a). Among other predictions, it was also suggested that the comparison of possible differences between electron and positron diffraction contrast could elucidate some issues related to defect formation in materials (for more details on other possible effects and applications, see Van House and Rich, 1988a, and references therein).

The construction and some examples of application of a PRM were also reported recently (Van House and Rich, 1988b; Brandes *et al.,* 1988; Frieze *et al.,* 1990). The basic concept of the PRM concerns spontaneous reemission of low-energy (about 1 eV) positrons from the material as the result of high-energy (up to 10 keV) positron irradiation. Incident positrons diffuse in the material, and those that reach the specimen surface are reemitted from the material having negative positron affinities. The emitted low-energy positrons, which are accelerated, magnified, and focused, form an image. During the diffusion through the specimen, positrons can be trapped by defects and impurities, which then can be observed. The contrast in the PRM image can also be produced by other factors, such as variations in the specimen thickness or in the bulk density or in the work function of the surface, the presence of contaminant thin-film layers, and positron trapping by defects at surfaces and interfaces. Thus, the PRM can provide important surface-related information. Some applications of the PRM in the microcharacterization of defects have been demonstrated recently, and it should also be noted that a spatial resolution of about 0.3 μm has been achieved, and much better resolution is predicted in the future (Van House and Rich, 1988b; Brandes *et al.,* 1988; Frieze *et al.,* 1990).

1.7.3. Neutron Microscopy

One of the important differences in using neutrons (compared with electrons or photons) as probes is that, carrying no electric charge, they are not influenced by the electrons of the atom and thus can penetrate much larger distances into the material. An important requirement for neutron microscopy is the availability of ultracold neutrons (UCN), which can be reflected from suitable mirrors in imaging systems. (For a recent review of the subject, see

Frank, 1989, and references therein.) An additional distinction is made between very slow neutrons (VSN) with velocities below 10 m/s (and a wavelength $\lambda > 400$ Å) and very cold neutrons (VCN) with velocities up to 100 m/s ($\lambda > 40$ Å). The main efforts in recent years were devoted to the construction of the optical systems for focusing neutrons in a neutron microscope. A comparison between optical elements suitable for the focusing of neutrons, such as nuclear and magnetic lenses, zone plates, and mirrors, was reported recently (see Frank, 1989, and references therein). Since the penetration depth of neutrons is large, and since most targets with thicknesses in the range of several microns are expected to be transparent for ultracold neutrons, the amplitude contrast is low. Thus, an important issue is devising methods of phase contrast (Frank, 1989).

It is also possible to use neutron beams for chemical analysis. For example, when neutrons are captured by a specimen nuclide, γ rays are emitted, which can be used for nondestructive chemical analysis. As the result of the absorption of low-energy neutrons, some nuclides may also emit charged particles. From the energy spectrum of these particles, the concentration and the depth distribution of the elements in the material can be determined. For a recent review of the status of the neutron microprobe, see Lindstrom *et al.* (1988).

1.7.4. Scanning Probe Microscopies

As mentioned above, the discovery of STM catalyzed the development of new techniques utilizing a tip for measuring various properties of materials. A variety of configurations of scanning probe microscopies, which provide a wealth of information on a wide variety of materials properties, have been developed. These include AFM, MFM, EFM, scanning capacitance microscopy (SCM), near-field scanning optical microscopy (NSOM), scanning ion-conductance microscopy (SICM), and BEEM. In addition, there have been efforts in combining different techniques in order to expand the capabilities of the analysis. This combination of various techniques would allow various complementary properties of the material to be measured in the same instrument. For example, attempts in combining STM with optical microscopy and with electron microscopy have already been undertaken and such facilities will overcome the lack of a low magnification capability in STM. It would also be desirable to combine different scanning probe microscopes in a multifunctional unit for the complementary analysis of various materials properties with nanometer-scale spatial resolution.

New techniques based on the principles of scanning probe microscopy are also being developed at a rapid pace. Among these are methods for the analysis of the optical properties of materials. For example, the *photon scanning tunneling microscope* (Reddick *et al.,* 1989; Courjon *et al.,* 1989) is based on the detection by a sharpened optical fiber tip of an evanescent tunneling wave which is generated by total internal reflection and which extends above the sample surface. In analogy with the electron scanning tunneling microscope, in this method a probe (an optical fiber tip) is brought close to the specimen surface so as to allow some of the photons to propagate in the fiber. The sharpened fiber tip is mounted on the piezoelectric scanner and the other end of the fiber is coupled with a photomultiplier. The photon current is kept constant by using a feedback circuit that changes the distance between the fiber tip and the sample, and thus the surface topography can be recorded with subwavelength resolution. This method is best suited for studies of dielectric surfaces that are inaccessible with an STM.

In another optical method reported recently, one can measure the light emission caused by the inverse photoelectric effect resulting from electrons injected into a solid using an STM (Coombs *et al.,* 1988). Since the excitation in this case is localized within the dimensions of the tunnel tip, this method can provide spectroscopic information on the sample surface with nearly atomic spatial resolution.

It is also possible to obtain luminescence spectra by using *photon emission spectroscopy and microscopy* (using an STM). In this method, electrons injected (by tunneling) into the bulk conduction band of a semiconductor may recombine radiatively and lead to the emission of characteristic luminescence, which can provide local information on electronic properties of the material with nanometer-scale spatial resolution (Abraham *et al.,* 1990).

There are also promising methods combining the STM technique with optical probes. For example, in *scanning tunneling optical spectroscopy,* the photoenhanced scanning tunneling current measured in semiconductor samples illuminated with monochromatic light provides a means of determining the band gap (and other electronic properties) of the material with nanometer-scale spatial resolution (Qian and Wessels, 1991).

An interesting development may also be expected in *lensless low-energy electron microscopy.* This new technique, which employs a point (i.e., atomic-size) source of low-energy electrons (on the order of 20 eV), is based on the principles of scanning probe microscopy. In other words, a fine tip emitter is mechanically scanned (using a piezoelectric scanner) in very close proximity to the specimen and a feedback circuit regulates the distance between the tip and the sample surface. High-resolution images of surfaces are produced by monitoring the secondary electron count rate (Fink, 1988).

1.7.5. Mechanical Properties Microprobe

The mechanical properties microprobe allows one to evaluate various mechanical properties of materials with a submicron spatial resolution. The major applications of this technique are in the analysis of microstructures and small-scale devices, thin films, and various multiphase structures containing grains of very small size. This method employs microindentation tests for the determination of such mechanical properties of materials as hardness, modulus, stress relaxation, and creep resistance (e.g., see Oliver *et al.*, 1987; Mayo *et al.*, 1990).

Recently, it was demonstrated that the AFM can be configured as a nanoindenter in order to evaluate the nanomechanical properties of materials (Burnham and Colton, 1989).

References

Abraham, D. L., Veider, A., Schonenberger, C., Meier, H. P., Arent, D. J., and Alvarado, S. F. (1990). *Appl. Phys. Lett.* **56**, 1564.
Binnig, G., Rohrer, H., Gerber, C., and Weibel, E. (1982). *Phys. Rev. Lett.* **49**, 57.
Binnig, G., Quate, C. F., and Gerber, C. (1986). *Phys. Rev. Lett.* **56**, 930.
Brandes, G. R., Canter, K. F., and Mills, A. P. (1988). *Phys. Rev. Lett.* **61**, 492.
Burnham, N. A., and Colton, R. J. (1989). *J. Vac. Sci. Technol.* **A7**, 2906.
Cookson, J. A. (1991). *Nucl. Instrum. Methods* **B54**, 433.
Coombs, J. H., Gimzewski, J. K., Reihl, B., Sass, J. K., and Schlittler, R. R. (1988). *J. Microsc. (Oxford)* **152**, 325.
Courjon, D., Sarayeddine, K., and Spajer, M. (1989). *Opt. Commun.* **71**, 23.
Cowley, J. M. (1989). *J. Vac. Sci. Technol.* **A7**, 2823.
Fink, H. W. (1988). *Phys. Scr.* **38**, 260.
Fink, H. W., Stocker, W., and Schmid, H. (1990). *Phys. Rev. Lett.* **65**, 1204.
Fink, H. W., Schmid, H., Kreuzer, H. J., and Wierzbicki, A. (1991). *Phys. Rev. Lett.* **67**, 1543.
Frabboni, S., Matteucci, G., Pozzi, G., and Vanzi, M. (1985). *Phys. Rev. Lett.* **55**, 2196.
Frank, A. I. (1989). *Nucl. Instrum. Methods* **A284**, 161.
Frieze, W. E., Gidley, D. W., Rich, A., and Van House, J. (1990). *Nucl. Instrum. Methods* **A299**, 409.
Goodhew, P. J., and Castle, J. E. (1983). *Inst. Phys. Conf. Ser. No. 68* (EMAG 1983), p. 515.
Grime, G. W., and Watt, F. (1990). *Nucl. Instrum. Methods* **B50**, 197.
Hahn, S., Smith, W. L., Suga, H., Meinecke, R., Kola, R. R., and Rozgonyi, G. A. (1990). *J. Cryst. Growth* **103**, 206.
Hansma, P. K., and Tersoff, J. (1987). *J. Appl. Phys.* **61**, R1.
Hansma, P. K., Elings, V. B., Marti, O., and Bracker, C. E. (1988). *Science* **242**, 209.
Hanszen, K. J. (1982). In *Advances in Electronics and Electron Physics,* Vol. 59, Academic Press, New York, p. 1.
Howie, A. (1989). *J. Microsc. (Oxford)* **155**, 419.
Hulett, L. D., Dale, J. M., and Pendyala, S. (1984). *Mater. Sci. Forum* **2**, 133.
Huong, P. V. (1991). In *Analysis of Microelectronic Materials and Devices* (M. Grasserbauer and H. W. Werner, eds.), Wiley, New York.
Jamieson, D. N., Romano, L. T., Grime, G. W., and Watt, F. (1990). *Mater. Characterization* **25**, 3.
Kaiser, W. J., and Bell, L. D. (1988). *Phys. Rev. Lett.* **60**, 1406.
Lichte, H. (1991). In *Advances in Optical and Electron Microscopy,* Vol. 12 (T. Mulvey and C. J. R. Sheppard, eds.), Academic Press, New York, p. 25.

Lindstrom, R. M., Fleming, R. F., and Rook, H. L. (1988). In *Microbeam Analysis* (D. E. Newbury, ed.), San Francisco Press, San Francisco, p. 407.

Matteucci, G., Missiroli, G., Nichelatti, E., Migliori, A., Vanzi, M., and Pozzi, G. (1991). *J. Appl. Phys.* **69**, 1835.

Mayo, M. J., Siegel, R. W., Narayanasamy, A., and Nix, W. D. (1990). *J. Mater. Res.* **5**, 1073.

Michette, A. G. (1988). *Rep. Prog. Phys.* **51**, 1525.

Nakashima, S., and Hangyo, M. (1989). *IEEE J. Quantum Electron.* **25**, 965.

Oliver, W. C., McHargue, C. J., and Zinkle, S. J. (1987). *Thin Solid Films* **153**, 185.

Qian, L. Q., and Wessels, B. W. (1991). *Appl. Phys. Lett.* **58**, 1295.

Reddick, R. C., Warmack, R. J., and Ferrell, T. L. (1989). *Phys. Rev.* **B39**, 767.

Smith, W. L., Rosencwaig, A., Willenborg, D., Opsal, J., and Taylor, M. (1987). *Nucl. Instrum. Methods* **B21**, 537.

Tonomura, A. (1987). *Rev. Mod. Phys.* **59**, 639.

Tonomura, A. (1990). *J. Vac. Sci. Technol.* **A8**, 155.

Van House, J., and Rich, A. (1988a). *Phys. Rev. Lett.* **60**, 169.

Van House, J., and Rich, A. (1988b). *Phys. Rev. Lett.* **61**, 488.

Wickramasinghe, H. K. (1990). *J. Vac. Sci. Technol.* **A8**, 363.

Part II

Electron Beam Techniques

2

Scanning Electron Microscopy

B. G. Yacobi and D. B. Holt

2.1. Introduction to Basic Principles

Since the appearance of the first commercial scanning electron microscope (SEM) three decades ago, major advances were achieved in resolution, new electron probe sources, the development of additional modes, and computerization (including built-in image store capabilities in the current generation of instruments). These advances were accompanied by an increasing number of applications in a wide variety of fields, and the SEM became an indispensable tool in various fields of research providing a wealth of information on physical, structural, and compositional properties of a wide range of materials.

The major advantages of SEM microcharacterization of materials are (1) the presence of several modes that can provide important complementary information on various properties of the material and (2) its ability to accommodate (examine) macroscopic specimens with no special sample preparation steps in most cases.

The basic components of the SEM are illustrated in Fig. 2.1. The fundamental principle of the SEM is analogous to the coupling of two cathode-ray tubes (CRTs). Thus, the electron-optical column (including the electron gun, the electromagnetic lenses, and the specimen chamber) plays the role of

B. G. Yacobi • EMTECH, Toronto, Ontario, Canada. *D. B. Holt* • Department of Materials, Imperial College, London SW7 2BP, England.

Microanalysis of Solids, edited by B. G. Yacobi *et al.* Plenum Press, New York, 1994.

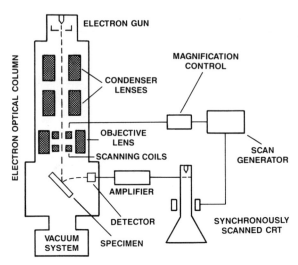

Figure 2.1. Schematic diagram showing the basic components of a scanning electron microscope (SEM).

one CRT containing a specimen; and the signal detection, display, and operating unit plays the role of the other CRT. The electron-optical column produces a finely focused electron beam that can be scanned in a raster fashion over the specimen surface. Various forms of physical energy generated by the specimen as a result of electron beam–solid interactions are converted into electrical signals that can be amplified, processed, and displayed or stored in some fashion.

As an electron source, either a *thermionic electron gun* or a *field emission gun* is used. The former may contain a tungsten or a lanthanum hexaboride (LaB_6) cathode. This heated cathode is maintained at a high negative potential relative to the Wehnelt cylinder, and the anode is maintained at earth potential. The field emission electron gun consists of a pointed cathode, i.e., a single-crystal tungsten tip, and two anodes. In this case, a voltage applied between the cathode and the first anode regulates the field strength at the tungsten tip and causes the emission of electrons that are accelerated by the second anode voltage. The advantages of the field emission electron source are (1) the higher brightness that allows a smaller probe size for a given beam current and (2) a smaller energy spread. The brightness of the field emission source is two orders of magnitude higher than that of the LaB_6 thermionic source and three orders of magnitudes higher than that of the tungsten thermionic source. The electron probe size (down to 20 Å) of the field emission source is smaller than those of the thermionic sources. Also the service life (in excess of 1000 h) of the field emission source should be compared with values of several hundred hours and about 50 h for the LaB_6 and for the tungsten thermionic sources, respectively. However, a higher operating vacuum is required for the field emission source.

Typically, in SEM electron-optical columns two condenser and one objective electromagnetic lenses are employed to form a fine electron beam and focus it onto the specimen surface. Scanning coils, positioned in the bore of the objective lens, deflect the electron beam so that its spot scans line by line a square raster over the specimen surface. Usually, the rasters in the SEM may contain from 100 to more than 1000 horizontal lines (the latter value is used for obtaining a micrograph).

The electron probe diameter decreases with increasing beam voltage and decreasing beam current (e.g., Goldstein *et al.*, 1981). Because of the electron-optical column limitations of the SEM, the beam current (I_b) is related to the minimum diameter (d_m) of electron probe as:

$$I_b \cong C\beta C_s^{-2/3} d_m^{8/3} \tag{2.1}$$

where C is a constant, β is the gun brightness, and C_s is the spherical aberration coefficient for the final lens. This relationship shows that under similar conditions, a brighter electron source yields more current. (As mentioned above, in order of decreasing brightness, the electron sources used are the field emission gun, the LaB_6 emitter, and the tungsten filament.) For the same beam voltage and beam current, the electron probe diameter obtained using a tungsten filament is about twice as large as the diameter for a LaB_6 gun. For example, for a 10-kV electron beam, the probe diameter of the LaB_6 filament is about 60 Å for an electron beam current (I_b) of 10^{-11} A, increasing to about 350 Å for $I_b = 10^{-9}$ A and to about 5000 Å for $I_b = 10^{-6}$ A. For a 30-kV electron beam, the probe diameter of the LaB_6 filament is about 40 Å for $I_b = 10^{-11}$ A, increasing to about 250 Å for $I_b = 10^{-9}$ A and to about 3000 Å for $I_b = 10^{-6}$ A.

As mentioned above, several different signals, which are produced as a result of the dissipation of the electron beam energy in the material into other forms of energy, can be detected and converted into an electrical signal, which is subsequently amplified and fed to the grid of the synchronously scanned display CRT. The amplified signal modulates the brightness of the CRT and produces an image of the specimen. The raster scan of the electron spot over the specimen surface results in a one-to-one correspondence between picture points on the display CRT screen and points on the specimen. Thus, the variations in the strength of the particular signal being detected result in variations in brightness on the CRT screen, i.e., contrast on the micrographs.

A certain minimum signal-to-noise ratio is required for each mode of the SEM. This depends on particular detector sensitivities and electronic system noise levels, which determine the minimum signal level required to get pixel brightnesses above the noise level (i.e., reliable data for processing). A certain electron beam power is required in order to generate that signal level in any particular specimen.

The *image magnification* is the ratio of the size of the square area scanned on the display CRT screen to the size of the area scanned on the specimen. The magnification can be varied by varying the currents in the SEM scanning coils. Note that the magnification is increased when the currents in the scanning coils are reduced. In the modern generation of SEMs, the magnification can be continuously varied from 10× to 300,000×. In many cases, the SEM is equipped with two display CRTs, or a split-screen CRT, so that signals providing complementary information can be displayed simultaneously.

In addition to its excellent spatial resolution for secondary electron imaging, the SEM also has a large *depth of field* (relative to an optical microscope). The depth of field is the thickness on the specimen surface for which the magnified image is in focus. The large depth of field results in a three-dimensional appearance of the SEM images (e.g., Fig. 2.2). The depth of field, D, for secondary electron imaging is determined by the variation of the electron-beam radius, r, above and below the optimum focus. This variation in r is caused by the electron-beam divergence. The semiangle of divergence of the electron beam, α, is very small (i.e., 10^{-2} to 10^{-3} rad). To a first approximation, $r = \alpha D/2$. The smallest distance that can be resolved by eye on the CRT, d (or the pixel size), is approximately 0.1 mm (the corresponding resolution on the specimen is d/M, where M is the magnification). The resolution on the CRT will not degrade if $2r \leq d/M$. Thus, the depth of field is

$$D = \frac{d}{\alpha M} \qquad (2.2)$$

10 μm
⊢——⊣

Figure 2.2. Secondary electron image of a patterned GaAs on InP substrate.

For example, for $M = 1000\times$ and $\alpha = 10^{-2}$ rad, $D = 10$ μm; for $M = 100\times$ and $\alpha = 10^{-3}$ rad, $D = 1$ mm. For a given electron probe size and magnification, the depth of focus can be varied by selecting (1) the final aperture size and (2) working distance, which affect the divergence, α, of the electron beam (α increases with the former and decreases with the latter).

In recent years, major advances in environmental scanning electron microscopy (ESEM) have also been made (e.g., Danilatos, 1988, 1990, and references therein). In the ESEM, the characterization of materials in a gaseous environment can be performed. The necessary modifications of the conventional (vacuum) SEM for ESEM include incorporation of differential pumping and use of the gaseous detector device (Danilatos, 1990). The high-vacuum electron-optics column is separated from the high-pressure specimen chamber by pressure-limiting apertures. Thus, the examination of the specimen at high pressures can be realized. The ESEM provides a powerful means for the microcharacterization of a wide range of materials, such as wet specimens and plastics, which are incompatible with the constraints of a vacuum.

In addition to the developments in the hardware, great efforts in recent years were also directed toward the quantitation of SEM analysis. These efforts, as well as numerous examples of SEM applications in the microcharacterization of solid-state materials, are outlined in detail in the list of books on the SEM provided in the bibliography section at the end of this chapter.

Also of importance are developments in computer-aided image processing and storage. This is especially important in the case of images with inferior signal-to-noise ratio (e.g., those obtained using low beam voltages on the order of 1 kV). In general, computer-aided image processing systems can (1) store the image, (2) reduce the noise level, and (3) convert SEM slow scans into TV rates for convenient viewing. Thus, the use of frame stores with signal averaging facilities greatly improves the signal-to-noise ratio and makes it possible to use lower electron-beam powers. This also implies that with a frame store, the images can be examined and processed without further electron bombardment of the sample, so that possible beam damage and contamination do not continue.

In this chapter, the basic principles and modes of the SEM are described. First, the important subject of the interaction of electrons with the solid will be outlined. The following sections will review the fundamentals of the SEM modes that are currently used in the analysis of solid-state materials and devices.

2.2. Interaction of Electrons with Solids

Electron-beam–solid interactions, electron beam energy dissipation, and the generation of electron–hole pairs in the solid are important for understanding the effects produced as a result of electron irradiation and in interpreting electron microscopy observations. (For detailed descriptions of the modeling of the interaction of the electron beam with the specimen, see Newbury *et al.,* 1986, and Newbury, 1989).

Two types of scattering mechanisms, i.e., elastic and inelastic, have to be considered. The elastic scattering of the primary (incident) electrons by the nuclei of the atoms, which are partially screened by the bound electrons, can be analyzed by using the Rutherford model. The elastic scattering of the incident electrons by atoms gives rise to high-energy backscattered electrons and (in the SEM) to atomic number contrast and channeling effects in the emissive mode (see Section 2.3). Inelastic scattering can be described by the Bethe expression for the mean rate of energy loss per unit of distance traveled in the solid. Inelastic interaction processes result in a variety of signals in electron probe instruments. These are the emission of secondary electrons, Auger electrons, and characteristic x rays, the generation of electron–hole pairs, cathodoluminescence, and thermal effects. As a result of the interaction between the primary electrons and the solid, the incident electron undergoes a successive series of elastic and inelastic scattering events in the material. Because of these scattering events within the material, the original trajectories of the incident electrons are randomized. The range of the electron penetration is a function of the electron-beam energy E_b:

$$R_e = (k/\rho)E_b^\alpha \qquad (2.3)$$

where k and α depend on the atomic number of the material and on E_b, and ρ is the density of the material. One can estimate the so-called generation (or excitation) volume in the material. According to Everhart and Hoff (1971),

$$R_e = (0.0398/\rho)E_b^{1.75} \qquad (\mu m) \qquad (2.4)$$

where ρ is in g/cm^3 and E_b is in keV. This expression was found for the electron beam energy range between 5 and 25 keV and atomic numbers $10 < Z < 15$. A general expression for the range was derived by Kanaya and Okayama (1972) and it agrees well with experimental results in materials covering a wide range of atomic numbers:

$$R_e = (0.0276A/\rho Z^{0.889})E_b^{1.67} \qquad (\mu m) \qquad (2.5)$$

where A is the atomic weight in g/mol, Z is the atomic number, ρ is in g/cm^3, and E_b is in keV. Figure 2.3 presents a comparison of the electron generation range calculated according to these two models for several materials. Because of its wider applicability and good agreement with experimental data,

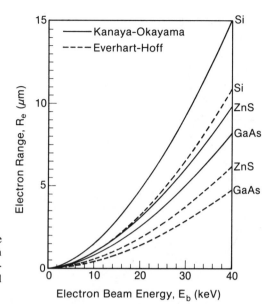

Figure 2.3. The electron beam range R_e in various materials as a function of electron beam energy E_b, calculated from the Kanaya–Okayama and Everhart–Hoff models.

it appears that the Kanaya–Okayama model is generally more appropriate for quantitative SEM analysis. It should be noted at this juncture that, in principle, the diffusion of minority carriers can increase the generation volume. This could affect the spatial resolution in cathodoluminescence (CL) and charge-collection (CC) measurements. However, the observed spatial resolution in both CL and CC is in fact very close to R_e. This is basically the result of the rapid fall in the diffused carrier density with distance of the form $(1/r)\exp(-r/L)$, where L is the minority carrier diffusion length. Calculations show that the minority carrier density is highest (and most relevant) inside a volume that is commensurate with the electron generation range (Donolato, 1978). Thus, the spatial resolution in measurements that involve the minority carrier diffusion should not depend substantially on the diffusion length L. It should also be noted that the dissipated energy density in the generation volume is concentrated up toward the incident beam impact point.

It is also important to emphasize the difference in the excitation volumes of thin specimens and bulk samples (Fig. 2.4). But even in thick (bulk) samples, signals actually originate from different effective depths in the material. Thus, secondary electrons originate from material within about 100 Å of the specimen surface, backscattered electrons are emitted from within about the upper one-half of the excitation volume, and the signals in CL and CC originate from within the whole excitation volume.

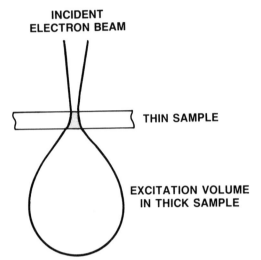

INCIDENT
ELECTRON BEAM

THIN SAMPLE

EXCITATION VOLUME
IN THICK SAMPLE

Figure 2.4. Schematic illustration of the volumes analyzed in a thin specimen and a bulk sample.

The generation factor, i.e., the number of electron–hole pairs generated per incident beam electron, is given by

$$G = E_b(1 - \gamma)/E_i \qquad (2.6)$$

where E_i is the ionization energy (i.e., the energy required for the formation of an electron–hole pair) and γ represents the fractional electron beam energy loss relating to all of the backscattered and emitted electrons. The ionization energy E_i is related to the band gap of the material ($E_i = 2.8E_g + M$, where $0 < M < 1$ eV) and is independent of the electron beam energy (Klein, 1968). The local generation rate of carriers is $g(r, z) = \langle g \rangle GI_b/e$, where $\langle g \rangle$ is the normalized distribution of the ionization energy in the generation volume, I_b is the electron beam current, and e is the electronic charge. The local generation rate has been determined experimentally for silicon by Everhart and Hoff (1971), who proposed a universal depth–dose function $g(z)$, which can be expressed as a polynomial

$$g(z) = 0.60 + 6.21z - 12.40z^2 + 5.69z^3 \qquad (2.7)$$

(Note that in general, this function is written in terms of the normalized depth $y = z/R_e$.) This expression, which is shown for different electron beam energies in Fig. 2.5, represents the number of electron–hole pairs generated per electron of energy E per unit depth and per unit time.

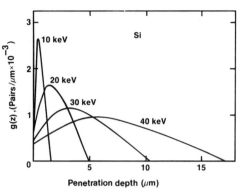

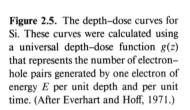

Figure 2.5. The depth–dose curves for Si. These curves were calculated using a universal depth–dose function $g(z)$ that represents the number of electron–hole pairs generated by one electron of energy E per unit depth and per unit time. (After Everhart and Hoff, 1971.)

An important method of calculating the electron beam energy dissipation in the solid is that of Monte Carlo trajectory simulation, which can be used, for example, in x-ray analysis computations, and also in CL and CC microanalysis. In Monte Carlo electron trajectory calculations, each electron can undergo elastic and inelastic scattering, and it can be backscattered out of the target. The path of an electron is calculated in a stepwise manner (i.e., each electron travels a small distance in a straight line between random scattering events). At each step, the type of scattering event (i.e., elastic or inelastic) and the appropriate scattering angle are chosen by using random numbers. The angular scattering events are the result of elastic interaction, and the energy loss is continuous. The Monte Carlo method can also compute the backscatter coefficient for any particular case. It should be emphasized that a few trajectories calculated by the Monte Carlo method do not represent the electron beam–solid interaction volume. In order to achieve statistical significance in Monte Carlo computations, about 1000 trajectories have to be calculated. These calculations also help to visualize the interaction volume and energy distribution. This is best seen for smaller numbers of calculated trajectories. An example is presented in Fig. 2.6, which illustrates Monte Carlo calculations of 100 electron trajectories of 10, 20, and 30 keV in GaAs. Similar representations of the interaction volume as a function of such parameters as tilt, thickness, and atomic number of the specimen can be obtained. An important result of Monte Carlo calculations is that the boundaries of the generation volume are not well defined. Both Monte Carlo calculations and experiments indicate that the density of electron trajectories varies significantly and that the energy dissipation is not uniform (see Fig. 2.7). To summarize, single values of parameters for the generation volume obtained from Eq. (2.4) or (2.5) are only approximations. In fact, contours of equal energy dissipation, determined experimentally and calculated by the Monte Carlo method, in-

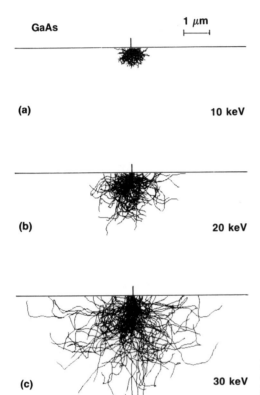

GaAs

1 μm

(a)   10 keV

(b)   20 keV

(c)   30 keV

Figure 2.6. Trajectories of 100 electrons of (a) 10 keV, (b) 20 keV, and (c) 30 keV calculated by the Monte Carlo method for GaAs.

dicate that a substantial fraction of energy dissipation occurs in a small volume near the electron beam impact point (see Fig. 2.7).

The *spatial resolution* of the various SEM modes is determined mainly by the electron-probe size, the size of the generation volume which is related to the electron-beam penetration range in the material, and the minority carrier diffusion. In practice, spatial resolution may be degraded by a poor signal-to-noise ratio, mechanical vibrations, and electromagnetic interference. For bulk specimens, the spatial resolution of the secondary electron mode is the highest (i.e., about 30 to 50 Å), since the low-energy secondary electrons can only escape from material within about 100 Å of the specimen surface (for such depths, the beam spreading is not substantial, as seen in Fig. 2.4), and since the electron-probe size can be made much smaller than the size of the generation volume. In the "bulk" modes, such as CL and CC, signals are typically obtained from the entire excitation volume by using electron beams of 15–30 kV. Consequently, the size of the excitation volume, and thus the

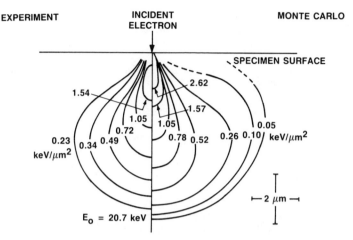

Figure 2.7. Energy dissipation profiles for a low-atomic-number solid as a function of depth, as obtained experimentally by the etching technique and as calculated by Monte Carlo simulation. (After Shimizu *et al.*, 1975.)

resolution, is on the order of 1 μm. The spatial resolution in these cases may also be affected by the minority carrier diffusion. However, as mentioned above, the minority-carrier density is high, and thus most relevant, inside a volume of about the same magnitude as that corresponding to the generation volume (Donolato, 1978). Thus, in practice, for the CL and CC modes, the spatial resolution is determined basically by the size of the generation volume, i.e., it is on the order of 1 μm. In the x-ray mode, similarly to the CL and CC modes, the signal is detected from the whole energy dissipation volume, and, thus, the spatial resolution is also on the order of 1 μm.

As mentioned above, different forms of energy produced by the electron bombardment of the material can be used as signals for the various SEM modes of operation that will be outlined in the following sections. Thus, (1) emitted electrons produce signals for the emissive mode, (2) in thin samples, transmitted electrons form the signals for scanning transmission electron microscopy (STEM), (3) x rays are the basis of the x-ray mode (i.e., electron probe microanalysis, EPMA), (4) ultraviolet, visible, and near-infrared photons are the basis of the CL mode, and (5) CC currents or voltages are the signals of the CC (or conductive) mode. Two other SEM techniques that have emerged recently are the electron acoustic mode and scanning deep-level transient spectroscopy (SDLTS). Table 2.1 summarizes briefly applications and spatial resolutions of the various SEM modes.

The issue of possible electron beam-induced damage should be addressed at this juncture. The size of the electron beam dissipation volume increases

Table 2.1. Information Obtainable and Spatial Resolutions
of the Various SEM Modes

SEM mode	Information obtained	Spatial resolution
Emissive		
Secondary electrons	Specimen topography; magnetic contrast	~5 nm
	Voltage contrast	~0.5 μm
Backscattered electrons	Atomic number contrast; specimen topography; magnetic contrast	~0.01–1 μm
	Crystal structure	~0.2–1 μm
Transmitted electrons (in STEM)	Crystal structure; defects; elemental analysis	~0.5 nm
X-ray mode	Elemental analysis of materials	~1 μm
Cathodoluminescence (CL)	Electronic properties of luminescent materials; luminescent and nonluminescent (dark) defects	~0.1–1 μm
Charge-collection (CC)	Electrical properties of semiconductor devices; electrically active defects	~0.1–1 μm
Electron acoustic mode	Subsurface defects	~0.1–1 μm
SDLTS	Deep levels of semiconductors	~1 μm

with the beam voltage V_b. Thus, for a given electron beam current I_b, an increase in V_b results in a reduction in the electron beam energy dissipated per unit volume, since $R_e \propto E_b^{1.75}$ but the beam total energy $\propto E_b$. In contrast, an increase in I_b at a given V_b results in a corresponding increase in the electron beam energy dissipated in that volume. Therefore, a general rule for the *minimization of electron beam damage* can be formulated as *maximizing V_b and minimizing I_b* so that a usable signal of interest is obtained, and also *using the fastest line and frame speeds* that provide sufficient interline scan separation at a given magnification.

One should also emphasize the inverse relation of spatial and signal resolution. When the signal is unacceptably low, the beam power must be increased. This can be realized by increasing the electron beam current or/and the beam energy. The former will result in an increase of the probe size, which makes the spatial resolution worse. The increase of the beam energy results in an increase in the size of the excitation volume and thus in the degradation of resolution for bulk modes.

2.3. The Emissive Mode

Three types of emitted electrons can be selected for detection by appropriate detectors. These are the *secondary, backscattered,* and *Auger electrons.*

As mentioned in Chapter 1, Auger electrons emitted from about the top 10 Å of the solid have energies characteristic of the elements in the material, and thus they can provide a powerful surface analytical technique in ultrahigh-vacuum electron probe instruments (for details on Auger electron spectroscopy and scanning Auger electron microscopy, see Chapter 4).

A simplified diagram of the emission spectrum of the secondary and backscattered electrons is shown in Fig. 2.8. The angular spread of both secondary and backscattered electrons emitted from surfaces is a cosine distribution. In general, the total number of emitted electrons depends on a variety of factors, such as the excitation conditions, the topography and crystallographic orientation, and the electronic band structure of a particular material. The total electron yield can be written as ($\delta + \eta$), where δ is the secondary electron coefficient (i.e., the average number of secondary electrons generated by each primary electron) and η is the backscattered electron coefficient (i.e., the fraction of primary electrons that are backscattered). If I_b is the incident electron beam current and I_s is the specimen current (i.e., the absorbed electron current that flows from the primary electron beam to earth via the specimen), one can write:

$$I_b = \delta I_b + \eta I_b + I_s = I_b(\delta + \eta) + I_s \qquad (2.8)$$

Note that if ($\delta + \eta$) = 1, then $I_s = 0$, i.e., no current flows to earth via the specimen. In this case, the primary electron beam is not injecting (or extracting) charge into (or from) the specimen. If ($\delta + \eta$) < 1, the incident electron beam is injecting charge into the specimen, and in the case of electrical nonconductors, the specimen will charge negatively. A typical dependence of total electron yield ($\delta + \eta$) on the primary electron beam energy is shown in Fig. 2.9. At the crossover points (e.g., at the electron beam energy E_2) the total electron yield is unity. Thus, this condition will allow the formation of an image from an insulating specimen. This is also important in analyses of electron beam-sensitive materials. Effective use of low-voltage SEM (LVSEM)

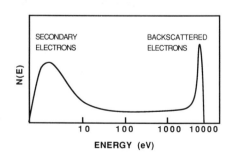

Figure 2.8. Energy spectrum of electrons emitted from the solid. (The incident beam energy is 10 keV.)

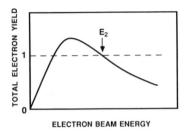

Figure 2.9. Dependence of total electron yield ($\delta + \eta$) on the primary electron beam energy. At the crossover points (e.g., at the electron beam energy E_2) the total electron yield is unity. For many materials E_2 is in the range between about 1 and 3 keV.

became possible with the realization of high spatial resolution and electron source brightness at low energies in the current generation of instruments. For many materials E_2 is between about 1 and 3 keV.

2.3.1. Secondary Electrons

Secondary electrons, which provide the most widely used SEM technique, are those emitted with energies less than 50 eV and they produce topographic contrast with much better resolution and depth of field than do optical microscopes. As mentioned above, having relatively low energies, secondary electrons can only escape from the material from within about 100 Å of the specimen surface. Since for such depths the beam spreading is not significant, and since the electron-probe size can be made much smaller than the size of the excitation volume, the spatial resolution of the secondary electron mode is the highest (i.e., on the order of 50 Å) for bulk specimens.

The most common method of detecting secondary electrons is by using the Everhart–Thornley detector consisting of a control grid, a scintillator, a light guide, and a photomultiplier tube (Fig. 2.10). A potential of about +200

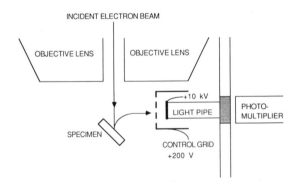

Figure 2.10. Schematic diagram of the Everhart–Thornley detector.

V on the grid attracts secondary electrons that are then accelerated to the aluminum coating on the scintillator which is held at a positive potential of about 10 kV. By acquiring high energies, these electrons produce a large number of photons in the scintillator. These photons are guided to the photocathode of the photomultiplier where they cause photoelectric emission of electrons. The latter are multiplied by secondary electron emission along a chain of dynodes.

The secondary electron coefficient δ depends strongly on the electron beam energy. At lower beam energies, the secondary electrons are generated closer to the specimen surface, and thus they have a high escape probability. At higher electron beam energies, however, although the number of secondary electrons increases, they are now excited deeper in the specimen, and thus their escape probability diminishes significantly. The secondary electron coefficient also depends on the specimen tilt. The secondary electron yield increases with increasing specimen tilt angle. This is because of the fact that when the sample is tilted to some angle relative to the primary electron beam, a greater number of the secondary electrons are produced within the escape region. Thus, more secondary electrons are produced from tilted regions of the specimen, and this provides an important mechanism for the surface topography imaging.

The secondary electrons can also provide important information on local electric and magnetic fields that are present in the material and that affect the energy and direction of emission of these low-energy electrons.

The emission of the secondary electrons is dependent on the electrical potential present at the electron-beam impact point. Thus, it is possible to image (and measure) variations in electrostatic potential in different regions of a sample. This is referred to as *voltage contrast* and it can be used in the analysis of electronic devices (e.g., Davidson, 1989). Voltage contrast with high-frequency stroboscopy is especially useful in the analysis of integrated circuits (IC). In such measurements, IC devices are mounted on a special holder and vacuum feedthroughs are used for external control of the IC operation. Thus, using stroboscopic voltage contrast in the SEM, a device can be observed operating at its normal frequency, and the faulty regions in the working device can be located.

The secondary electron emission can also provide important information on magnetic fields that are present above the surface of magnetic materials and that affect the emission of the secondary electrons. In this case, secondary electrons exiting the sample surface and passing through the external magnetic field are deflected by the Lorentz force. The deflection produces an asymmetry in the angular distribution of the secondary electrons. Thus, the number of secondary electrons collected from certain regions will be different as a result of the trajectory modification by the external magnetic field. This magnetic contrast, which is referred to as *type I magnetic contrast*, is thus a contrast mechanism produced by trajectory effects and not by the variations in the secondary electron emission (e.g., Newbury *et al.*, 1986).

Recent advances in the development of spin polarization detectors for secondary electrons in *spin-polarized electron microscopy* made it possible to image magnetic field distributions at the surface of ferromagnetic materials with a spatial resolution on the order of 50 nm and less (e.g., Pierce, 1988). This method offers a powerful means of analysis of the shape of magnetic domains of a ferromagnet with high spatial resolution. In SEM with polarization analysis (SEMPA), magnetization images can be obtained by measuring the magnitude and direction of the spin polarization of the secondary electrons emitted from a ferromagnet caused by the primary (unpolarized) beam.

2.3.2. Backscattered Electrons

Backscattered electrons are emitted with energies close to that of the incident (primary) electron beam. Typically, 10 to 30% of the primary electrons become backscattered electrons, which emerge from within about the upper one-half of the excitation volume; and in general, the spatial resolution is on the order of the diameter of the excitation volume. Backscattered electrons can provide information on the specimen topography, as well as atomic number contrast since regions of higher atomic number Z backscatter more primary electrons. In fact, the backscatter coefficient η increases monotonically with Z. Areas that appear brighter are regions with a larger average atomic number, and thus, this type of contrast can provide qualitative information on compositional uniformity.

Backscattered electrons, in principle, can be detected with the Everhart–Thornley detector by turning off the attractive grid bias (or placing a negative bias on the grid). In such a case, the secondary electrons are rejected and backscattered electrons (moving toward the detector) are collected. Because of the small solid angle of collection, however, the collection efficiency is small. Having high energies, backscattered electrons are emitted from the specimen surface in straight lines. Thus, they can be efficiently collected by positioning a detector in their path (usually mounted under the pole piece of the electron-optical column). One detector configuration in such a case employs the Everhart–Thornley detector with the scintillator placed concentrically around the primary beam. A standard alternative is a solid-state detector, such as a Si diode (Fig. 2.11). In order to intercept a large fraction of the backscattered electrons, these solid-state detectors are made sufficiently large in size. But such large detectors also have a large capacitance, and that limits observations to slow scan imaging. For a nonflat sample, in order to discriminate between the topographic information and the compositional information,

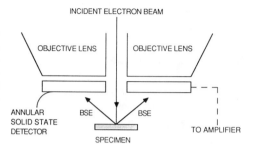

Figure 2.11. Schematic diagram of a solid-state (a semiconductor) electron detector (a Si diode). High-energy backscattered electrons generate in a semiconductor electron–hole pairs that are separated by a *p–n* (or a Schottky) junction; a current produced in an external circuit is amplified to produce a signal for a video display.

the detector is divided into two halves (or four quadrants for more flexibility). So, when the signals from the two halves of the detector are added, the topographic components are largely canceled out, and thus, the compositional information can be derived. When the detector signals are subtracted, the atomic number variations are canceled out, and the topographic information can be obtained.

Under appropriate operating conditions, backscattered electrons can also provide crystallographic information because of electron channeling. The latter is based on the fact that the backscattered electron yield varies as the angle of incidence of the scanned electron beam passes through the Bragg angle to crystal lattice planes. Thus, selective channeling of primary electrons between the crystal planes at specific incidence angles can produce SEM pictures consisting of series of bands and fine lines. Such SEM pictures, which are called electron channeling patterns (ECPs), depend on the crystal structure and orientation of the material (e.g., Joy *et al.,* 1982; Joy, 1989). These patterns can be used for the determination of crystal orientation and lattice parameter, and their sharpness is a measure of local crystal perfection. ECPs can be obtained from selected small areas of the specimen by rocking the incident electron beam at the site of interest. It is also possible to obtain crystallographic information by using Kossel patterns (Dingley and Steeds, 1974) and electron backscattering patterns (Dingley, 1981). Kossel patterns are produced by the x rays that are generated in the specimen as a result of the electron beam irradiation. These x rays can be diffracted by the crystal lattice planes and consequently detected (on a photographic plate above the sample) as a series of cones. The information in x-ray Kossel patterns can be used for the determination of lattice parameters and crystal orientation. In the electron backscattering pattern (EBSP) method, the sample is inclined to the electron beam at an angle of about 20° or less from glancing incidence. A stationary electron beam irradiates the area of interest, which results in the angular variation of intensity of the backscattered electrons because of the strong reflection of the electrons incident at the Bragg angle (the EBSP is a high-angle Kikuchi pattern).

The EBSP can be observed on a suitably positioned phosphor screen (which is parallel to the specimen) coupled with the appropriate TV camera or recorded directly on film. The EBSP provides a powerful means for deriving crystallographic information with a high spatial resolution (about 0.2 μm).

As mentioned above, backscattered electrons are emitted from within about the upper one-half of the excitation volume. One can, however, reduce the information depth (and hence, improve the resolution) by employing the *energy filtering* of low-loss electrons. In this method, a high-pass filter allows transmission only of electrons with very low energy losses, so that an information depth comparable to that of the secondary electrons can be obtained.

In the cases of magnetic materials with no external fringing fields, the type of magnetic contrast produced with the secondary electrons (described above) will not be applicable. Such materials, however, can be studied using *type II magnetic contrast*. In this case, the contrast mechanism is based on the Lorentz deflection of the primary electron beam entering the tilted sample. This deflection resulting from the internal magnetic field (depending on the primary beam direction and the magnetization direction) causes the primary electrons to move either closer to or farther from the surface, which results in varying the backscattering coefficient. In other words, depending on the domain magnetization, the number of backscattered electrons will vary, providing type II magnetic contrast. Note that this type of contrast is very weak (e.g., Newbury *et al.*, 1986).

2.4. Transmitted Electrons

As mentioned in Chapter 1, in thin specimens on the order of 1000 Å or less, transmitted electrons can be analyzed. The basic information in these electron beams is related to the crystal structure, composition, and defects in the material. Typical maximum electron beam energies in SEMs are 30 to 40 keV, which can only penetrate very thin specimens. Since the SEM is best suited for the analysis of macroscopic (bulk) samples, most of the work employing transmitted electrons is performed in TEMs and STEMs where typical electron beam energies are 100 keV or higher. The energies of the transmitted electrons depend basically on the thickness and the structure of the specimen. Electron energy losses are characteristic of the elements present in the material, and thus electron energy-loss spectroscopy (EELS) can be used for chemical analysis. For the latter to be useful, however, electron energies greater than about 100 keV have to be employed, so it will not be applicable to SEMs (since the maximum electron beam energy is usually 30 to 40 keV).

Dedicated STEMs equipped with a field emission gun have also been developed and provide a resolution of 5 Å in the STEM mode. The images and diffraction patterns obtained in the STEM are equivalent to those of conventional TEMs. However, the electrons are detected as a serial signal as the incident beam scans over the specimen. The possibility of using energy-selective electron detectors (electron spectrometers) and of processing the signal for display creates many new possibilities. One of these is EELS in which characteristic energy losses are used to identify the elements present. Another is the recently developed "Z-contrast imaging" (see below).

Since STEMs are ultrahigh-vacuum instruments, they can also employ surface analytical techniques such as Auger electron spectroscopy and microscopy. Since very thin specimens are used for the transmitted electrons, electron beam spreading is negligible, so for the CL and EBIC modes in the STEM, the spatial resolution can be greatly improved. This is, however, limited by the inverse relation of spatial-to-signal (and spectral) resolution. This is related to the fact that since in STEMs the excitation volume is small the intensity of the signal is also low, and it is often necessary to increase the electron beam diameter (in order to increase the beam current) and excite a larger volume until a detectable signal is obtained.

Pennycook and co-workers (Pennycook, 1989; Pennycook and Jesson, 1991) use for transmitted electrons in STEM an annular detector with a large central hole to detect only the electrons that have been (Rutherford) scattered through large angles. Rutherford scattering occurs only close to the nucleus of atoms and depends essentially on Z, the atomic number. Hence, atomic resolution is obtained, with the great advantage over atomic-resolution TEM that the contrast is independent of the exact focus and thickness of the specimen. (The STEM Z-contrast images are incoherent so interference effects do not occur. The reversals of image contrast with thickness, focus, etc. in atomic-resolution TEM require skill, experience, and persistence to obtain reliably-interpretable atomic-resolution images.)

Most recently, Franchi et al. (1990) and Ogura et al. (1990) demonstrated that it is possible to obtain high-resolution backscattered electron images of bulk specimens using a field-emission-gun (FEG) SEM with an in-lens electron detector. (In-lens detection of backscattered electrons uses a detector above the objective lens. The detected electrons follow curved paths up through the lens bore.) Again, it is the Rutherford-scattered electrons that are detected to obtain Z-contrast images. Some degradation of resolution results from beam (lateral) spreading in the massive specimen but less than might be expected. These workers were able to demonstrate the resolution of quantum wells (QWs) down to 2 nm in thickness and with composition differences between semiconductor alloy QWs and confining layers differing in average atomic number by less than one.

2.5. The X-ray Mode

The basis of the x-ray mode (also called electron probe microanalysis, EPMA) is Moseley's law that relates the energy, E, of the characteristic x-ray emission line to the atomic number, Z, i.e.,

$$E = C_1(Z - C_2)^2 \qquad (2.9)$$

where C_1 and C_2 are constants for a particular line type (e.g., the K_α lines). For the analysis of characteristic x rays, two methods can be used. These are energy-dispersive spectrometry (EDS) and wavelength-dispersive spectrometry (WDS). In EDS, which employs a solid-state detector, all x rays incident simultaneously on the detector are sorted by differences in their energy. In WDS, a crystal spectrometer is used to preselect a particular x-ray wavelength for processing, and thus for the analysis of various elements in the sample this method is sequential. (The basic principle of the crystal or wavelength spectrometer is Bragg's law of x-ray diffraction, i.e., $n\lambda = 2d \sin\theta$.)

EDS offers the higher collection efficiency and lower spectral resolution. Other advantages of EDS include relative simplicity of mechanical design with no moving parts, and rapid acquisition of the whole spectrum. One of the disadvantages of EDS is the need for cryogenic cooling. WDS provides the better spectral resolution but lower collection efficiency. Additional advantages of WDS include its better sensitivity to trace elements and light elements. A disadvantage of WDS is that only one element can be analyzed at a time, and, with crystal spectrometers, there is also the need for assuring mechanical alignment between measurements. Thus, while the EDS system can be used more efficiently, since it can detect a wide range of elements simultaneously, the WDS method should be used for more precise quantitative microanalysis of the specimen. In both the EDS and WDS methods, the relative arrival rates of characteristic x-ray photons from different elements are measures of their concentrations in the excitation volume.

For the determination of actual composition, a set of corrections is applied. The corrections are for the atomic number Z that affects the efficiency of characteristic x-ray emission, for the absorption A that reduces the count rate for element Y if element X absorbs its characteristic x-ray photons, and for fluorescence F that increases the count rate for element Y if it absorbs x-ray photons and reemits them at its own characteristic energy. In modern instruments, these so-called ZAF corrections are performed by microcomputers.

In addition to quantitative local analysis, micrographic displays of the elemental distributions (x-ray maps) can also be obtained. In favorable cases (i.e., for heavy elements and the absence of other elements that emit x-ray photons at energies close to those of the elements of interest), the x-ray mode can detect down to about 10^{18} atoms/cm^3 with a spatial resolution of about 1 μm.

2.6. Cathodoluminescence Mode

In CL analysis, electron beam bombardment of a solid causes the emission of photons in the ultraviolet, visible, and near-infrared ranges because of the recombination of electron–hole pairs generated by the incident energetic electrons. The signal provides a means for CL microscopy and spectroscopic analysis of luminescent materials. In CL microscopy, luminescent images or maps of regions of interest are displayed on the CRT, whereas in CL spectroscopy, luminescence spectra from selected areas of the sample are obtained. The latter is analogous to a *point analysis* in the x-ray microanalysis mode. However, unlike x-ray emission, CL does not identify the presence of specific atoms. The lines of characteristic x rays, which are emitted because of electronic transitions between sharp inner-core levels, are narrow and are unaffected by the environment of the atom in the lattice. In contrast, the CL signal is generated by detecting photons that are emitted as a result of electronic transitions between the conduction band, and/or levels resulting from impurities and defects lying in the fundamental band gap, and the valence band. These transition energies and intensities are affected by a variety of defects, by the surface of the material, and by external perturbations, such as temperature, stress, and electric field. Thus, no universal law can be applied in order to interpret and to quantify lines in the CL spectrum. Despite this limitation, the continuing developments of the CL technique are motivated by its attractive features: (1) CL is a contactless method that provides information on electronic properties of luminescent materials with a spatial resolution of less than 1 μm; (2) the detection limit for impurity concentrations can be as low as 10^{14} atoms/cm^3, which is several orders of magnitude better than that of the x-ray microanalysis mode in the SEM; (3) CL is especially useful in the microanalysis of optical properties of wide-band-gap materials, such as diamond; (4) depth-resolved information can be obtained by varying the range of electron penetration that depends on the electron-beam energy; (5) CL-SEM is a powerful tool in the microcharacterization of optoelectronic materials and devices, since in these cases it is the luminescence properties that are of practical importance. (For recent reviews on CL, see Yacobi and Holt, 1986, 1990). It should be emphasized that, being a contactless method with no requirements for device fabrication steps, CL is of great value in providing an assessment of a starting material; thus, it may help to elucidate problems related to processing steps that cause device failure.

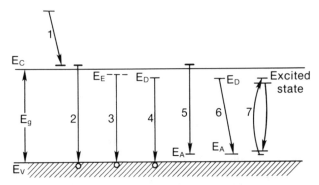

Figure 2.12. Schematic diagram of luminescence transitions between the conduction band (E_C), the valence band (E_V), and exciton (E_E), donor (E_D), and acceptor (E_A) levels in the luminescent material. Process 1: an electron excited above the conduction band edge dribbles down, resulting in phonon-assisted photon emission, or, more likely, the emission of phonons only. Process 2: intrinsic luminescence resulting from direct recombination between an electron in the conduction band and a hole in the valence band. Process 3: the exciton decay observable at low temperatures. Processes 4, 5, and 6: transitions that start and/or finish on localized states of impurities (e.g., donors and acceptors) in the gap produce extrinsic luminescence, and these account for most of the processes in many luminescent materials. Process 7: the excitation and radiative deexcitation of an impurity with incomplete inner shells, such as a rare earth ion or a transition metal.

Luminescence spectra of inorganic solids can be divided between *intrinsic* and *extrinsic*. Intrinsic luminescence, which appears at elevated temperatures as a near-Gaussian-shaped band of energies with its peak at a photon energy $h\nu_p \cong E_g$, is the result of recombination of electrons and holes across the fundamental energy gap E_g. Extrinsic luminescence, on the other hand, depends on the presence of impurities and defects. In the analysis of optical properties of inorganic solids it is also important to distinguish between *direct-gap* materials (e.g., GaAs) and *indirect-gap* materials (e.g., Si). This distinction is based on whether the valence band and conduction band extrema do or do not occur at the same value of the wave vector **k** in the energy band, $E(\mathbf{k})$, diagram of the particular solid. In direct-gap materials, no phonon participation is required during the electronic transitions. In indirect-gap materials, in order to conserve momentum during the electronic transitions, the participation of an extra particle (i.e., a phonon) is required; the probability of such a process is significantly lower as compared with direct transitions. Thus, fundamental emission in indirect-gap materials is relatively weak compared with that due to impurities or defects.

Recombination centers with energy levels in the gap of a luminescent solid are radiative or nonradiative depending on whether recombination leads to the emission of a photon or not. A simplified schematic diagram of transitions that lead to luminescence in materials containing impurities is shown in Fig. 2.12. It should be noted that lattice defects, such as dislocations, vacancies, and their complexes with impurity atoms, may also introduce localized levels in the band gap, and their presence may lead to changes in the recombination rates and mechanisms of excess carriers in luminescence processes. Recombination of electron–hole pairs may also occur via nonradiative processes, such as multiple phonon emission, the Auger effect, and recombination related to surface states and defects. It should also be noted that nonradiative surface recombination is a loss mechanism of great importance for some materials (e.g., GaAs). This effect, however, can be minimized by increasing the electron-beam energy in order to produce a greater electron penetration range.

The description of the formation of the CL signal involves the analysis of the generation, diffusion, and recombination of minority carriers (e.g., Yacobi and Holt, 1990). The diffusion of the stationary excess minority carriers for continuous irradiation can be treated in terms of the differential equation of continuity. When competitive radiative and nonradiative centers are both present, the observable lifetime is given by $1/\tau = 1/\tau_r + 1/\tau_n$, where τ_r and τ_n are the radiative and nonradiative recombination lifetimes, respectively. The radiative recombination efficiency is defined as $\eta = \tau/\tau_r = [1 + (\tau_r/\tau_n)]^{-1}$. The CL intensity can be derived from the overall recombination rate $\Delta n(r)/\tau$ by noting that only a fraction $\Delta n(r)\eta/\tau$ recombines radiatively and assuming a linear dependence of the CL intensity L_{CL} on the stationary excess carrier density Δn:

$$L_{CL}(r) = \int_V f[\Delta n(r)/\tau_r]\, d^3r \qquad (2.10)$$

where f is a function containing correction parameters of the CL detection system and factors that account for optical absorption and internal reflection losses (e.g., Yacobi and Holt, 1990). The excess carrier density $\Delta n(r)$ can be obtained from the solution of the differential equation of continuity for the diffusion of the excess minority carriers, and for the simplified case of a point source or a sphere of uniform generation, one can obtain the value of f and write:

$$L_{CL} = f\eta G I_b/e \qquad (2.11)$$

where I_b is the electron beam current and e is the electronic charge. This equation indicates that the rate of CL emission is proportional to η, and from the definition of the latter we conclude that in the observed CL intensity one cannot distinguish between radiative and nonradiative processes in a quantitative manner. One should also note that η depends on such factors as temperature, the presence of defects, and the particular dopants and their concentrations.

Early analytical models of the CL in semiconductors provided a description of the recombination process as a function of excitation conditions (Wittry and Kyser, 1967). One important result of the analysis of the dependence of the CL intensity on the electron-beam energy indicates the existence at the surface of a *dead layer,* where radiative recombination is absent because of the presence of a space-charge depletion region caused by the pinning of the Fermi level by surface states (Wittry and Kyser, 1967).

In recent years, CL microscopy and spectroscopy techniques have been extensively employed in the microcharacterization of electronic properties of a wide range of luminescent materials, such as semiconductors, minerals, phosphors, ceramics, and biological-medical samples. Some important applications of CL microcharacterization techniques include (for a general review and details, see Yacobi and Holt, 1990): (1) uniformity analysis of luminescent materials (e.g., obtaining distributions of various defects and impurities) and recombination studies in the vicinity of defects; (2) deriving information on the electronic band structure related to the fundamental band gap; (3) measurements of the dopant concentration, and of the minority carrier diffusion length and lifetime (minority carrier lifetime can be obtained from time-resolved CL measurements employing a beam blanking system and a fast detector); (4) microcharacterization of semiconductor devices (e.g., degradation of optoelectronic devices); (5) microcharacterization of stress distributions in epitaxial layers; (6) depth-resolved studies of defects in ion-implanted samples and of interface states in heterojunctions (CL depth profiling can be performed by varying the range of electron penetration that depends on the electron-beam energy; the excitation depth can be varied from about 10 nm to several micrometers for electron-beam energies in the range between about 1 and 30 keV).

In CL microscopy images of regions of interest can be displayed on the video CRT, whereas in CL spectroscopy an energy-resolved spectrum corresponding to a selected area of the sample can be obtained. CL detector designs differ in the combination of components used (e.g., Holt, 1981; Trigg, 1985). Although most of these are designed as SEM attachments, several CL collection systems were developed in dedicated STEMs and TEMs (Petroff *et al.,* 1978; Pennycook *et al.,* 1977; Steeds, 1989). In these cases, since thin specimens are used, electron beam spreading is relatively small and thus improved spatial resolution can be obtained. However, since the excitation volume is smaller, CL intensities are also lower. A schematic diagram of basic components used in a typical CL detection system for the visible range is shown in Fig. 2.13. The signal from the photomultiplier can be used to produce

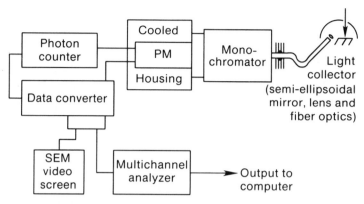

Figure 2.13. Schematic diagram of a typical early CL detection system for the visible range.

both micrographs and spectra. When the grating of the monochromator is bypassed, photons of all wavelengths falling on the photomultiplier produce the panchromatic (integral) CL signal. In the dispersive mode, for a constant monochromator setting and a scanning electron beam condition, monochromatic micrographs can be obtained; and when the monochromator is stepped through the wavelength range of interest and the electron beam is stationary or scans a small area, CL spectra can be derived. The evolution of light collectors of increasing efficiency (around 90% collection) is presented in Fig. 2.14. It should be mentioned that, in addition to the mirror-based systems, a relatively simple and inexpensive but powerful CL system utilizing an optical fiber light collection system was also developed (Hoenk and Vahala, 1989). The proper choice of a detector is important in CL measurements. In the visible range, photomultipliers are the most efficient detectors. For luminescence in the infrared range, solid-state detectors, as well as Fourier transform spectrometry (FTS) can be used. For detailed quantitative analysis, the calibration of the CL detection system for its spectral response characteristics is important.

It should be emphasized that for detailed CL microcharacterization of defects and impurities in materials, it is essential to employ dispersive systems with a high efficiency of light collection (and transmission and detection) and capable of sample cooling, preferably to liquid helium temperatures. Sample cooling results in an increase in CL intensity, sharpening of the CL spectrum into lines corresponding to transitions between well-defined energy levels that allow more reliable interpretation of CL spectra (see Fig. 2.15), and reductions in the rate of electron bombardment damage in electron-beam-sensitive materials.

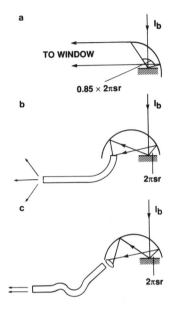

Figure 2.14. The evolution of light collectors of increasing efficiency. (After Steyn *et al.*, 1976.) (a) Parabolic Mirror with Electron Beam Hole and Window. (b) Semiellipsoidal Mirror and Solid Light Guide. (c) Semiellipsoidal Mirror with Lens and Fiber Optic Bundle Light Guide.

The quantification of CL analysis is difficult, because the interpretation of CL cannot be unified under a simple law and because of the lack of information on the competing nonradiative processes present in the material. In addition, the influence of defects, of the surface, and of external perturbations, such as temperature, electric field, and stress, have to be taken into account in quantitative CL analysis. Correlations between dopant concentrations and such band-shape parameters as the peak energy and the half-width of the CL emission can be used as means for the quantitative analysis of the carrier concentration. The development of quantitative CL analysis is a challenging issue. A recent effort at a more systematic quantification of CL (Warwick, 1987), which is based on accounting for the effects of the excess carrier concentration, absorption, and surface recombination, can be used for further developments of quantitative CL.

Some specific examples of the applications of CL techniques were recently reviewed (Yacobi and Holt, 1990). An example of uniformity characterization, as well as of the analysis of the electrically active defects, is shown in Fig. 2.16. These CL micrographs demonstrate two different forms of dislocation contrast, *dark dot* and *dot and halo* contrast, for GaAs crystals doped with Te to concentrations of 10^{17} cm^{-3} (Fig. 2.16a) and 10^{18} cm^{-3} (Fig. 2.16b). The latter case shows variations in the doping concentration around dislocations. This example also demonstrates that CL microscopy is a valuable tool for the determination of dislocation distributions and densities in lu-

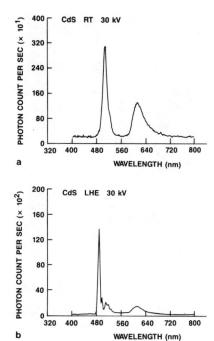

Figure 2.15. CL spectra from a CdS crystal (a) at room temperature and (b) at liquid helium temperature. The near-band-gap emission in (b) has a peak count rate about 4.5 times that of the fundamental band in (a) and is resolved into a series of narrow lines, known as the *edge emission.* (After Holt, 1981.)

minescent materials. Reliable measurements of dislocation densities up to about 10^6 cm^{-2} can be made with the CL image. Among various applications of the CL mode, the power of high-resolution CL in the analysis of optical properties of complex artificial structures has been demonstrated recently in the case of quantum well wires and boxes based on III–V compounds (e.g., Bimberg *et al.,* 1985; Petroff, 1987; Lebens *et al.,* 1990; Warwick, 1991) and in the quantitative analysis of spatial variations of stress in mismatched heterostructures (Yacobi *et al.,* 1989).

2.7. The Charge-Collection Mode

In the CC mode (also referred to as the conductive mode), three basic types of signal can be identified (Fig. 2.17):

a. The specimen current, or the absorbed electron current, that flows from the primary electron beam to earth via the specimen (Fig. 2.17a). In this case, the electron beam current $I_b = I_e + I_s$, where I_e is the total emissive current and I_s is the specimen current.

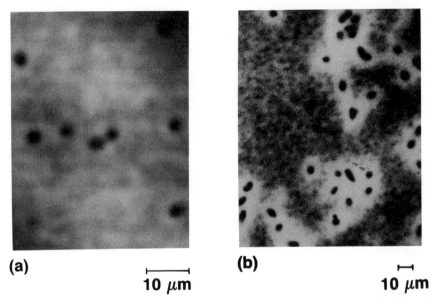

(a)   **(b)**

10 μm   10 μm

Figure 2.16. CL micrographs of Te-doped GaAs. (a) *Dark dot* dislocation contrast in GaAs doped with a Te concentration of 10^{17} cm^{-3}. (b) *Dot and halo* dislocation contrast in GaAs doped with a Te concentration of 10^{18} cm^{-3}.

b. A current flowing in a closed loop between two contacts in the absence of an external voltage source (Fig. 2.17b). In this case, CC current will flow if electron beam irradiation generates a potential difference between the contacts, i.e., an *electron-voltaic effect*. Such a signal is often referred to as electron beam-induced current (EBIC), which requires the presence of an electrical junction that can be a *p–n* junction or a Schottky barrier. In these measurements, electron–hole pairs generated in the depletion region, or within minority carrier diffusion range of it, are separated by the built-in electric field and the CC current is measured in the external circuit.

c. A CC current is obtained between two contacts with the application of external biasing of the specimen. This case does not require the presence of an electron-voltaic effect, and it is often referred to as β-conductivity (Fig. 2.17c).

The CC arrangements for the analysis of *p–n* junctions and Schottky barriers are the perpendicular (edge-on) and planar geometries which are illustrated in Fig. 2.18.

The EBIC technique is routinely used in examinations of *p–n* junction and Schottky barrier characteristics and also in the analysis of various defects, such as dislocations. The EBIC contrast in a SEM image is caused by variations

Figure 2.17. Schematic illustration of the three basic types of charge-collection (or conductive) mode signal. (a) The specimen (or the absorbed electron) current flowing from the incident electron beam to earth via the specimen. (b) The signal is a current flowing in a closed loop between two contacts in the absence of any external voltage source. In this case, the charge-collection current will flow only if the specimen exhibits electron-voltaic effects to generate potential differences between the contacts and drive currents around the external circuit. (c) Under external bias, charge-collection currents will flow. Variations (contrast) will be obtained even in the absence of electron-voltaic effects, as the beam scans, if the conductivity is non-uniform. This is referred to as β-conductivity.

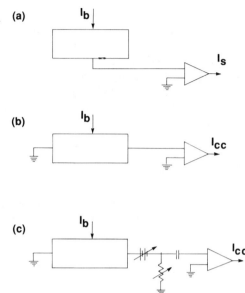

in CC efficiency that may arise from recombination at defects. In EBIC measurements, regions with a high carrier recombination efficiency will appear darker and regions with low carrier recombination will appear bright, providing a means for direct imaging of electrically active defects in semiconductor devices. In general, two types of signal can be used: the EBIC and the electron

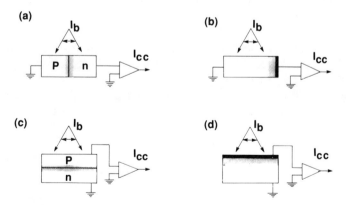

Figure 2.18. Schematic illustration of charge-collection geometries. Panels (a) and (b) show perpendicular p–n junction and Schottky barrier geometries, respectively. Panels (c) and (d) show planar p–n junction and Schottky barrier geometries, respectively.

beam-induced voltage (EBIV). In the EBIC case, the short-circuit current of a device is measured at zero bias, and the signal, amplified with a low-input-impedance amplifier, modulates the intensity of the SEM video display. In the EBIV method, the open-circuit voltage is measured across the diode by a voltmeter with high input impedance. One can describe these methods analytically by noting that the electron beam irradiation of a device, such as a p–n junction or a Schottky barrier, produces a CC current

$$I_{cc} = \eta_{cc} G I_b \qquad (2.12)$$

where G is the generation factor, i.e., the number of electron–hole pairs produced per incident electron, and η_{cc} is the CC efficiency, i.e., the fraction of the pairs that is separated by the built-in field. This CC current flows in the direction opposite to the forward-biased barrier diode current, I_d, and the externally observed EBIC signal current is

$$I = I_{cc} - I_d = I_{cc} - I_0[\exp(eV/nkT) - 1] \qquad (2.13)$$

where I_0 is the saturation current and n is the ideality factor of the barrier. The short-circuit current, I_{sc}, is obtained by setting $V = 0$, i.e., $I_{sc} = I_{cc}$. By setting $I = 0$, the open-circuit voltage, i.e., EBIV signal, is measured:

$$V_{oc} = (nkT/e) \ln(1 + I_{cc}/I_0) \qquad (2.14)$$

Some major applications of the EBIC and EBIV techniques include (1) microcharacterization of the electrical junction characteristics of active areas of semiconductor devices, (2) microcharacterization of the concentration and distribution of electrically active defects and detecting subsurface defects and damage, (3) measuring the minority carrier diffusion length and lifetime, the surface recombination velocity, and the width and depth of depletion zones, (4) measuring the Schottky barrier height, (5) quality control and failure analysis of electronic devices. (For details, see Schick, 1981; Leamy, 1982; Holt and Lesniak, 1985; Holt, 1989).

The minority carrier diffusion length, L, in EBIC measurements is derived from the dependence of I_{cc} on the distance x of the SEM electron beam away from the junction, as depicted in the geometries of Fig. 2.18a,b. The simple expression that describes the measurement is

$$I_{cc}(x) = I_0 \exp(-x/L) \qquad (2.15)$$

in which surface recombination is neglected. However, it should be emphasized that, even with negligible surface recombination, difficulties with the analysis can arise. A considerable effort in recent years was directed at deriving accurate values of L (e.g., Wu and Wittry, 1978; Oelgart et al., 1981; Kittler et al., 1985; Puhlmann et al., 1991). Davidson et al. (1982) showed that the results obtained in devices could be injection level dependent. A method of analysis using the method of moments was proposed by Donolato (1983) and employed by Cavalcoli et al. (1991). A method for dealing with specimens such as silicon wafers in which the diffusion length varies with depth was developed by Donolato and Kittler (1988). A treatment including the effect of surface recombination was also developed by Donolato (1982) and successfully employed by Hungerford (see Holt, 1989, pp. 276–277).

Using EBIC measurements, one can also determine the doping concentration for a p–n junction or a Schottky barrier. This can be accomplished by measuring the depletion width as a function of the reverse bias, and from the well-known relationship between depletion width, reverse bias, and doping concentration, the latter can be determined (Schick, 1981). In such a measurement, the spatial resolution of the depletion width measurement as a function of the reverse bias can be improved by differentiating EBIC line scans (Schick, 1981).

As an example, for the quality control of semiconductor devices, one can select a planar geometry for the CC arrangement (as in Fig. 2.18c,d). During the scanning of a junction (with constant electron-beam energy), contrast in the signal will originate if the electrical junction characteristics, such as the depletion region width, vary. Any nonuniformities in the barriers in complex devices, including those related to electrically active defects, can be seen.

CL and EBIC modes are routinely used for the characterization of electrically active defects in semiconductors and semiconductor devices. It has been demonstrated that simultaneous CL and EBIC contrast studies can provide a powerful means for characterizing the orientation, shape, and depth of dislocations in semiconductors (Jakubowicz, 1986; Jakubowicz et al., 1987). In this method, experimental results are compared with theoretical curves derived for the ratio of the CL and EBIC contrast (for the case of a pointlike source and a pointlike defect):

$$C_{CL}/C_{EBIC} = (1 - e^{-H(\alpha-1/L)})/(1 - e^{-h(\alpha-1/L)}) \qquad (2.16)$$

where C_{CL} and C_{EBIC} are the CL and EBIC contrast, respectively, H is the depth of the defect, h is the effective penetration depth of the electron beam (the depth of an assumed point source of CL), α is the absorption coefficient, and L is the minority carrier diffusion length. By obtaining the best fit to the experimental line scans for different inclination angles of an extended defect in the material, the geometric properties of the defect can be derived. Equation (2.16) indicates that $\sigma = C_{CL}/C_{EBIC}$ depends mainly on the geometry, and thus one can separate the contribution of the defect configuration to the contrast from that caused by local changes in recombination properties. Applying this analysis to dislocations, one can see that if during the electron beam scanning along a defect σ remains constant and C_{CL} and C_{EBIC} vary, this is caused by local variation of the recombination properties. Variations in σ, on the other hand, would imply changes in the depth of a dislocation segment. Figure 2.19 presents EBIC and CL images of the same area of Sn-doped GaAs. The bright and dark bands are dopant striations, and the dark irregular regions are associated with nonradiative dislocations. Although the EBIC and CL images appear similar, differences are clearly revealed in the contrast profiles, measured along the line O–O′ (shown in Fig. 2.19), presented in Fig. 2.20,

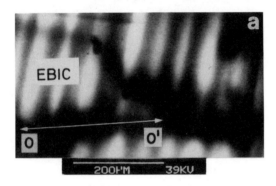

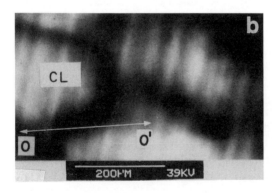

Figure 2.19. EBIC and CL images of the same area of a Sn-doped GaAs specimen. (After Jakubowicz et al., 1987.)

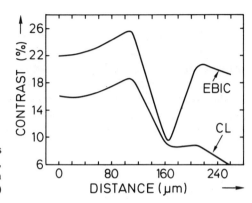

Figure 2.20. EBIC and CL contrasts as functions of the electron beam position, measured along the line O–O′ shown in Fig. 2.19. (After Jakubowicz *et al.,* 1987.)

confirming qualitatively the behavior expected from the model (Jakubowicz *et al.,* 1987). Quantitative comparison between experimental CL and EBIC results and calculations, performed on a stacking fault in a GaAs epitaxial layer, showed a good agreement for the ratio of CL and EBIC contrast profiles (Jakubowicz *et al.,* 1987). These results demonstrate that simultaneous CL and EBIC measurements can be used to obtain information about the geometry and recombination properties of dislocations.

2.8. The Electron Acoustic Mode

The (thermal wave) electron acoustic mode employs a chopped electron beam to produce intermittent heating (Brandis and Rosencwaig, 1980; Cargill, 1980; Rosencwaig, 1982). As a result of the periodic heating, spherical heat waves spread from the heated volume and are damped out. The periodic thermal expansion and contraction produces propagating acoustic waves with the chopping frequency (in the approximate range from 0.1 to 5.0 MHz). These waves can be detected by a piezoelectric transducer in contact with the specimen. In the SEM, this constitutes scanning electron acoustic microscopy (SEAM). The spatial resolution of this mode is on the order of 1 μm. (For recent reviews and examples of applications, see Balk, 1988, 1989.)

One of the important applications of this mode is in the microcharacterization of subsurface defects and inhomogeneities in bulk materials.

2.9. Scanning Deep-Level Transient Spectroscopy

SDLTS, developed by Petroff and Lang (1977), is a derivative of the DLTS technique that is based on the capture and thermal release of carriers at traps. A voltage bias pulse is employed to fill the traps in the DLTS technique, whereas an electron-beam injection pulse is used in the SDLTS method. The SDLTS technique allows one to determine the energy levels and the spatial distribution of deep states in semiconductors. This method is an important addition to the SEM techniques, since it complements CL spectroscopy for the assessment of nonradiative centers (as mentioned above, the lack of information on competing nonradiative processes constitutes one of the fundamental difficulties in quantitative luminescence analysis). With the development of capacitance meters of higher sensitivity, SDLTS became more widely applicable (e.g., Breitenstein and Heydenreich, 1989; Heydenreich and Breitenstein, 1986).

The SDLTS technique fills the deep levels using an electron beam pulsed *on* for several microseconds and *off* for about 10 μs. In the *on* state, the beam-induced electron–hole pair density will change the trap occupation level. After each excitation pulse, the deep levels are in a nonequilibrium state. In the *off* state, thermal equilibrium is restored by the thermal emission of captured carriers and, if the levels are in the space-charge region of a p–n junction or a Schottky barrier, the relaxation process will produce a measurable current transient or transient in the capacitance of the system. In principle, two different types of signal, current transient or capacitance transient, can be used in SDLTS measurements. The rate of decay depends on the energy of the deep level and the temperature. By measuring the time constant of the transients as a function of the excitation pulse repetition rate (i.e., rate window) at various temperatures, the energy levels and the concentrations of deep traps in the material can be determined. The major distinction between the two techniques is that capacitance transient SDLTS can distinguish between electron and hole traps from the sign of the signal which is independent of the rate window, whereas with current transients, the sign of the signal which depends on the rate window is the same for electrons and holes.

A schematic illustration of the basic components of an SDLTS system is shown in Fig. 2.21. The spatial distribution of the traps can be obtained by selecting a given energy level corresponding to a defect center in the material by setting an appropriate temperature and rate window and scanning the pulsed electron beam over the specimen surface. Conversely, a DLTS spectrum can be obtained by keeping the pulsed electron beam stationary and varying the temperature, e.g., between -150 and $300°C$.

An example of an SDLTS image is presented in Fig. 2.22, which shows both the EBIC image (Fig. 2.22a) and the SDLTS map (Fig. 2.22b) of an n-GaAs (Au) Schottky diode containing a 400-meV hole trap level (the so-called A-level) (Breitenstein and Heydenreich, 1989). This center is considered to be associated with a Ga_{As} antisite defect, i.e., a Ga atom on an As lattice site. Many dark dots in the EBIC image (Fig. 2.22a) are most probably related to

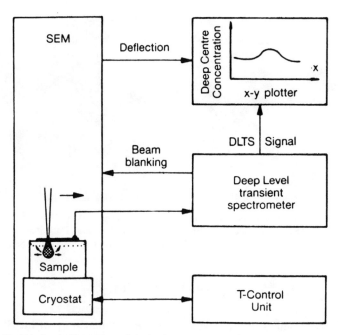

Figure 2.21. Schematic illustration of the basic components of an SDLTS system. (After Breitenstein and Heydenreich, 1989.)

dislocations. In the EBIC image, there are also several larger dots, e.g., one denoted A. As observed in the SDLTS image (Fig. 2.22b), this type of defect is accompanied by an increase of SDLTS signal in the vicinity. (Note that the bright contrast in the SDLTS image corresponds to a higher concentration of the A-level in this case). However, in the extended dark region labeled C, the SDLTS image indicates a much lower A-level concentration in that vicinity. These observations were explained as follows (Breitenstein and Heydenreich, 1989). The material they studied was grown under Ga-rich conditions that favor Ga_{As} antisite defect formation. Thus, if the defects in the vicinity of region C are local Ga precipitates or act as sites for gettering Ga atoms, the excess of Ga around these defects is reduced, so Ga_{As} antisite defect formation is not favored anymore, leading to a much lower A-level concentration and substantially reduced SDLTS signal.

The main difficulty with SDLTS is its low sensitivity, which is inevitable, since, in order to obtain the desired high spatial resolution on the order of 1 μm, one must actually excite only a very small area defined by the electron probe size. Thus, a typical SDLTS signal is much smaller than the standard DLTS signal obtained from a larger area (on the order of 1 mm).

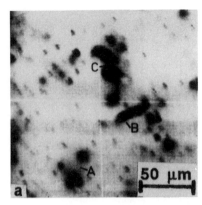 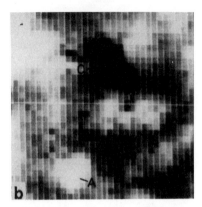

Figure 2.22. (a) EBIC image and (b) scanning deep-level transient spectroscopy (SDLTS) image of an n-GaAs (Au) Schottky diode containing a 400-meV hole trap level. Dark areas in the SDLTS image indicate a low trap concentration. (After Breitenstein and Heydenreich, 1989.)

2.10. Conclusions

Major advances in the developments in the hardware of the SEM (i.e., new electron probe sources, the development of additional modes, and computerization) were accompanied by an expansion in applications in diverse fields of research. These advances have established the SEM as a versatile tool consisting of a family of several microcharacterization techniques that provide a wealth of information on physical, structural, and compositional properties of a wide range of materials. The presence of various techniques (modes) in one instrument is of great convenience since it can provide important complementary information on different properties of the solid. The basic classification of these modes is summarized in Table 2.1.

Great efforts are directed toward the quantitation of SEM analysis, and also at further developments of the various SEM modes. These efforts are supported by developments in the computer-aided image processing and storage.

Bibliography: Books on Scanning Electron Microscopy

Ehrenberg, W., and Gibbons, D. J. (1981). *Electron Bombardment Induced Conductivity,* Academic Press, New York.

Goldstein, J. I., and Yakowitz, H. (eds.) (1975). *Practical Scanning Electron Microscopy,* Plenum Press, New York.

Goldstein, J. I., Newbury, D. E., Echlin, P., Joy, D. C., Fiori, C., and Lifshin, E. (1981). *Scanning Electron Microscopy and X-ray Microanalysis,* Plenum Press, New York.

Heinrich, K. F. J. (1981). *Electron Beam X-ray Microanalysis,* Van Nostrand, Princeton, N.J.

Holt, D. B., and Joy, D. C. (1989). *SEM Microcharacterization of Semiconductors,* Academic Press, New York.

Holt, D. B., Muir, M. D., Grant, P. R., and Boswarva, I. M. (eds.) (1974). *Quantitative Scanning Electron Microscopy,* Academic Press, New York.

Joy, D. C., Romig, A. D., and Goldstein, J. I. (eds.) (1986). *Principles of Analytical Electron Microscopy,* Plenum Press, New York.

Newbury, D. E., Joy, D. C., Echlin, P., Fiori, C. E., and Goldstein, J. I. (1986). *Advanced Scanning Electron Microscopy and X-ray Microanalysis,* Plenum Press, New York.

Oatley, C. W. (1972). *Scanning Electron Microscopy. I. The Instrument,* Cambridge University Press, Oxford.

Reimer, L. (1985). *Scanning Electron Microscopy,* Springer-Verlag, Berlin.

Thornton, P. R. (1968). *Scanning Electron Microscopy—Application to Materials and Device Science,* Chapman & Hall, London.

Wells, O. C. (1974). *Scanning Electron Microscopy,* McGraw–Hill, New York.

Yacobi, B. G., and Holt, D. B. (1990). *Cathodoluminescence Microscopy of Inorganic Solids,* Plenum Press, New York.

References

Balk, L. J. (1988). *Adv. Electron. Electron Phys.* **71,** 1.

Balk, L. J. (1989). In *SEM Microcharacterization of Semiconductors* (D. B. Holt and D. C. Joy, eds.), Academic Press, New York.

Bimberg, D., Christen, J., Steckenborn, A., Weimann, G., and Schlapp, W. (1985). *J. Lumin.* **30,** 562.

Brandis, E., and Rosencwaig, A. (1980). *Appl. Phys. Lett.* **37,** 98.

Breitenstein, O., and Heydenreich, J. (1989). In *SEM Microcharacterization of Semiconductors* (D. B. Holt and D. C. Joy, eds.), Academic Press, New York.

Cargill, G. S. (1980). *Nature* **286,** 691.

Cavalcoli, D., Cavallini, A., and Castaldini, A. (1991). *J. Appl. Phys.* **70,** 2163.

Danilatos, G. D. (1988). *Adv. Electron. Electron Phys.* **71,** 109.

Danilatos, G. D. (1990). *Scanning* **12,** 23.

Davidson, S. M. (1989). In *SEM Microcharacterization of Semiconductors* (D. B. Holt and D. C. Joy, eds.), Academic Press, New York.

Davidson, S. M., Innes, R. M., and Lindsay, S. M. (1982). *Solid State Electron.* **25,** 261.

Dingley, D. J. (1981). In *Scanning Electron Microscopy* 1981/IV (O. Johari, ed.), SEM Inc., Chicago, p. 273.

Dingley, D. J., and Steeds, J. W. (1974). In *Quantitative Scanning Electron Microscopy* (D. B. Holt, M. D. Muir, P. R. Grant, and I. M. Boswarva, eds.), Academic Press, New York.

Donolato, C. (1978). *Optik* **52,** 19.

Donolato, C. (1982). *Solid State Electron.* **25,** Stuttgart 1077.

Donolato, C. (1983). *Appl. Phys. Lett.* **43,** 120.

Donolato, C., and Kittler, M. (1988). *J. Appl. Phys.* **63,** 1569.

Everhart, T. E., and Hoff, P. H. (1971). *J. Appl. Phys.* **42,** 5837.

Franchi, S., Merli, P. G., Migliori, A., Ogura, K., and Ono, A. (1990). In *Proc. XIIth Int. Congr. Electron Microscopy,* San Francisco Press, San Francisco, p. 380.

Goldstein, J. I., Newbury, D. E., Echlin, P., Joy, D. C., Fiori, C., and Lifshin, E. (1981). *Scanning Electron Microscopy and X-ray Microanalysis,* Plenum Press, New York.

Heydenreich, J., and Breitenstein, O. (1986). *J. Microsc.* **141,** 129.

Hoenk, M. E., and Vahala, K. J. (1989). *Rev. Sci. Instrum.* **60**, 226.

Holt, D. B. (1981). In *Microscopy of Semiconducting Materials,* IOP, Bristol, p. 165.

Holt, D. B. (1989). In *SEM Microcharacterization of Semiconductors* (D. B. Holt and D. C. Joy, eds.), Academic Press, New York.

Holt, D. B., and Lesniak, M. (1985). In *Scanning Electron Microscopy* 1985/I (O. Johari, ed.), SEM Inc., Chicago, p. 67.

Jakubowicz, A. (1986). *J. Appl. Phys.* **59**, 2205.

Jakubowicz, A., Bode, M., and Habermeier, H. U. (1987). In *Microscopy of Semiconducting Materials,* IOP, Bristol, p. 763.

Joy, D. C. (1989). In *SEM Microcharacterization of Semiconductors* (D. B. Holt and D. C. Joy, eds.), Academic Press, New York.

Joy, D. C., Newbury, D. E., and Davidson, D. L. (1982). *J. Appl. Phys.* **53**, R81.

Kanaya, K., and Okayama, S. (1972). *J. Phys. D.: Appl. Phys.* **5**, 43.

Kittler, M., Seifert, W., Schroder, K.-W., and Susi, E. (1985). *Cryst. Res. Technol.* **20**, 1435.

Klein, C. A. (1968). *J. Appl. Phys.* **39**, 2029.

Leamy, H. J. (1982). *J. Appl. Phys.* **53**, R51.

Lebens, J. A., Tsai, C. S., Vahala, K. J., and Kuech, T. F. (1990). *Appl. Phys. Lett.* **56**, 2642.

Newbury, D. E. (1989). In *SEM Microcharacterization of Semiconductors* (D. B. Holt and D. C. Joy, eds.), Academic Press, New York.

Newbury, D. E., Joy, D. C., Echlin, P., Fiori, C. E., and Goldstein, J. I. (1986). *Advanced Scanning Electron Microscopy and X-ray Microanalysis,* Plenum Press, New York.

Oelgart, G., Fiddicke, J., and Reulke, R. (1981). *Phys. Status Solidi A* **66**, 283.

Ogura, K., Ono, A., Franchi, S., Merli, P. G., and Migliori, A. (1990). In *Proc. XIIth Int. Congr. Electron Microscopy,* San Francisco Press, San Francisco, p. 404.

Pennycook, S. J. (1989). *Ultramicroscopy* **30**, 58.

Pennycook, S. J., and Jesson, D. E. (1991). *Ultramicroscopy* **37**, 14.

Pennycook, S. J., Craven, A. J., and Brown, L. M. (1977). In *Developments in Electron Microscopy and Analysis,* Inst. Phys. Conf. Ser. No. 36, London, p. 69.

Petroff, P. M. (1987). In *Microscopy of Semiconducting Materials,* Inst. Phys. Conf. Ser. No. 87, Bristol, p. 187.

Petroff, P. M., and Lang, D. V. (1977). *Appl. Phys. Lett.* **31**, 60.

Petroff, P. M., Lang, D. V., Strudel, J. L., and Logan, R. A. (1978). In *Scanning Electron Microscopy* 1978/I (O. Johari, ed.), SEM Inc., Chicago, p. 325.

Pierce, D. T. (1988). *Phys. Scr.* **38**, 291.

Puhlmann, N., Oelgart, G., Gottschalch, V., and Nemitz, R. (1991). *Semicond. Sci. Technol.* **6**, 181.

Rosencwaig, A. (1982). *Science* **218**, 223.

Schick, J. D. (1981). In *Scanning Electron Microscopy* 1981/I (O. Johari, ed.), SEM Inc., Chicago, p. 295.

Shimizu, R., Ikuta, T., Everhart, T. E., and DeVore, W. J. (1975). *J. Appl. Phys.* **46**, 1581.

Steeds, J. W. (1989). *Rev. Phys. Appl.* **24**, C6-65.

Steyn, J. B., Giles, P., and Holt, D. B. (1976). *J. Microsc. (Oxford)* **107**, 107.

Trigg, A. D. (1985). In *Scanning Electron Microscopy* 1985/III (O. Johari, ed.), SEM Inc., Chicago, p. 1011.

Warwick, C. A. (1987). *Scanning Microsc.* **1**, 51.

Warwick, C. A. (1991). *J. Phys. IV* (Suppl. J. Phys. III) **1**, C6-117.

Wittry, D. B., and Kyser, D. F. (1967). *J. Appl. Phys.* **38**, 375.

Wu, C. J., and Wittry, D. B. (1978). *J. Appl. Phys.* **49**, 2827.

Yacobi, B. G., and Holt, D. B. (1986). *J. Appl. Phys.* **59,** R1.
Yacobi, B. G., and Holt, D. B. (1990). *Cathodoluminescence Microscopy of Inorganic Solids,* Plenum Press, New York.
Yacobi, B. G., Zemon, S., Jagannath, C., and Sheldon, P. (1989). *J. Cryst. Growth* **95,** 240.

3

Transmission Electron Microscopy

A. J. Garratt-Reed

3.1. Introduction

The ultimate resolution of any imaging technique is limited to about the wavelength of the radiation which carries the information. For light, the wavelength is on the order of 500 nm. However, modern science demands information from samples at far higher resolution—in the limit, it would be ideal to image individual atomic positions in crystals, requiring a resolution of about 0.1 nm.

All moving objects have associated with them a wavelength, given by the deBroglie relationship $\lambda = h/p$, where λ is the wavelength, h is Planck's constant, and p is the momentum of the particle. This equation is valid for all particles, including those traveling at relativistic speeds. For nonrelativistic particles, $p = mv$, where m and v are respectively the mass and velocity of the particle. Since the energy E is related to the mass and velocity by $E = 0.5$ $(mv)^2$, and since also $E = eV$, where e is the charge on the electron and V is the accelerating voltage, it can be seen that, in this restricted case, $p = (2meV)^{0.5}$. Hence, λ decreases as the electron energy increases. For 100-kV electrons (for which relativistic effects are almost negligible), λ is 3.7 pm, decreasing to 1.97 pm for 300-kV electrons, at which energy relativistic corrections must be made. Such electrons, therefore, would be capable of con-

A. J. Garratt-Reed • Center for Materials Science and Engineering, Massachusetts Institute of Technology, Cambridge, Massachusetts 02139.

Microanalysis of Solids, edited by B. G. Yacobi *et al.* Plenum Press, New York, 1994.

veying information at very high resolution in a suitably designed instrument. Since they can travel significant distances through matter (typically on the order of tens of nanometers) they are capable of probing the structure of materials. Not surprisingly, therefore, electron microscopy has become a vital tool for scientists in many disciplines.

It is the purpose of this chapter to explore, in an introductory manner, the present and projected capabilities of transmission and scanning transmission electron microscopes, particularly (but not exclusively) as they are applied in the study of materials. If the reader appreciates that substantial books have been written covering in detail small parts of the field summarized here, it will not be necessary to explain or apologize for the limited depth to which we are able to delve into the topics in this chapter, or for the omission of any historical perspective. Since little of what we discuss here is beyond what would be regarded as background information, we have also chosen to omit references to sources. A number of works are listed in the bibliography (Section 3.8) to which the interested reader may turn for more information and extensive references.

In what follows, we shall first describe briefly the fundamentals of the various forms of electron microscope, then we shall look at how the electrons interact with the sample and convey information to the image. Following this we shall discuss in more detail the instrumental and experimental considerations, and after presenting a few examples of applications of electron microscopy, we shall end by briefly looking at what developments might be expected in the future.

3.2. Fundamentals

All imaging systems have three fundamental parts, namely a source of radiation, a focusing system, and a display and/or recording system. The conventional transmission electron microscope is a close analogue of the slide projector or photographic enlarger: a source of electrons (or light) emits radiation which is "condensed" onto the sample (slide) by a compound condenser lens; the spatially modulated electron (light) intensity transmitted by the sample is focused onto a viewing screen (or recording film) by a compound objective/projector lens system. The sample must, of necessity, be thin enough so that a suitably large fraction of the incident electrons pass through; we shall discuss this requirement again later.

An analogue of the camera, where a light source floods a scene with illumination, some of which is reflected and focused onto a film by a lens, is not practical for electron images because the scattered electrons have too large a spread of energy and could not effectively be focused to a high-quality image. However, the analogue of the flying-spot scanner, in which a beam of radiation is focused into a fine spot and rastered over the sample, with the output signal being displayed on a cathode-ray tube (CRT) scanned in synchronism, is implemented in the scanning electron microscope (which is not part of the subject of this chapter) and the scanning transmission electron microscope, and is of the utmost importance, as we shall see, in modern science.

As we have implied above, the microscope that is the analogue of the slide projector is known as the conventional transmission electron microscope (CTEM) and the probe-forming kind is known as the scanning transmission electron microscope (STEM). The CTEM is "conventional" only in the sense that it has been manufactured commercially (of course, in steadily improving quality) for over 50 years, whereas the first commercial STEMs were marketed about 35 years later. Now, many instruments are designed to operate in either mode, as the investigation in hand warrants; these are given the somewhat illogical designation "TEM/STEM," and are said to work either in "TEM mode" (when flooding the sample with electrons and imaging with lenses) or in "STEM" mode. There is a principle, known as reciprocity, which states that the physics of image formation is identical in either case, and there should be no difference in the image details observable in the different modes. In what follows, when we refer to formation of an image, we shall (unless the context makes clear otherwise) mean either the focusing of electrons onto the screen of the CTEM, or the focusing of the scanning probe of electrons on the sample in the STEM and the electronic formation of the image on the CRT. (There frequently are differences in the images actually obtained in TEM and STEM modes of TEM/STEM instruments. These differences are a result of very different electron-optical parameters in the implementation of the different modes by the instrument manufacturer in response to commercial needs to compromise in the design, not to any fundamental physical difference between the modes of operation.)

We shall return to consider in far greater detail the design and operation of instruments, after we have discussed how electrons interact with matter, and what type of information we might hope to acquire from different types of samples.

3.3. Interaction of Electrons with Matter, and Contrast

The interaction of the electrons with a thin sample results in transmitted electrons (some having hardly interacted with the sample, some having been scattered elastically, and some having been scattered inelastically), x rays, Auger electrons, and light. All of these output signals convey information about the sample, and have been used in commercially available instruments as the basis for image formation or microanalysis of one type or another. However, systems detecting the Auger electrons or light are beyond the scope of this chapter, and we shall limit ourselves to considering only the various transmitted electrons and the x rays. In the case of the transmitted electrons, it is necessary for some mechanism of the interaction of the electrons with the sample to change the uniform incident illumination in such a way that the information carried by the transmitted electrons can be processed by the imaging system so that the desired details about the sample may be deduced. The final output from the microscope is an image conveying the information in the brightness variations from point to point; since variations in brightness are called contrast, the mechanisms which generate these variations are termed contrast mechanisms.

3.3.1. Mass Thickness Contrast

The simplest mode of image formation occurs when some areas of the sample absorb more electrons than other parts. The number of electrons emerging from any point in the exit face of the sample then mirrors the density of the sample to the electrons, projected in the direction of the electron beam, at that point. Straightforward imaging of these electrons then results in a map of the electron density of the sample. Very few samples exactly match this rather simple model; however, many biological samples, which typically have been treated ("stained") with a heavy element (frequently uranium) so that certain parts of the sample preferentially absorb the heavy element, and others reject it, closely approximate it. This is because the electrons are scattered much more strongly by the heavy element than by the light atoms which comprise the vast majority of organic material. If the electrons are scattered by a sufficiently large angle, they can be prevented (by use of a suitable aperture later in the microscope) from reaching the image. It is as if they are absorbed by the sample, and the image is thus dark in areas rich in the stain. The scattering power of a sample is approximately proportional to the product of the square of the average atomic weight and the thickness; hence, this form of contrast is known as mass thickness contrast.

As has been implied, mass thickness contrast is principally a tool of the biologist. There are, however, a few specialized applications (which will be discussed later) in which this mode of contrast formation is extremely valuable as a high-resolution mode of image formation.

3.3.2. Diffraction Contrast

A second method of image formation is by diffraction contrast. Electrons employed for transmission microscopy, we recall, have a wavelength on the order of a few picometers (3.7 pm for 100-kV electrons). Since a typical interplanar spacing for densely packed planes in most solids is about 0.2 nm, the angle for electron diffraction (half the angle between the incident beam and the diffracted beam) is on the order of 1°. (We recall that Bragg's law states that the condition for strong diffraction is given by the equation

$$n\lambda = 2d \sin(\theta) \tag{3.1}$$

where n is an integer, λ is the wavelength of the radiation, θ is the Bragg angle, and d is the periodic spacing of the diffraction grating.)

Consider a parallel beam of electrons passing through a lens parallel to its optic axis. By the definitions of lens action, the electrons will be focused into a spot at the back focal point of the lens. Beams traveling in other directions will be focused in the back focal plane of the lens at a distance from the axis equal to $f \tan(\alpha)$, where f is the focal length of the lens and α is the angle between the optic axis and the beam. If the beam is a diffracted beam, as discussed above, and the lens has a focal length typical of many in current use, about 2 mm, this distance will be on the order of 50 μm. Hence, an aperture of 50-μm diameter centered around the image of the direct beam in the back focal plane will prevent the diffracted beams from passing and contributing to the image. This aperture is termed the *objective aperture* (by analogy with optical microscopy, the first part of the image-forming system is called the *objective lens*). Consider now a sample placed in the beam in front of the lens. If the material of the sample is crystalline, then it is possible that in some areas electrons are diffracted from the incident beam into other directions. For example, near a dislocation, the lattice planes may be sufficiently distorted that locally, strong diffraction occurs, although the bulk of the sample is not in a diffracting orientation. These diffracted electrons are taken from the direct beam, which is therefore locally reduced in intensity. Since only the directly transmitted electrons contribute to the image, the areas where diffraction was occurring appear dark. Since holes in the sample appear bright in the image (all incident electrons pass onto the viewing screen or film), this method is termed *bright-field* microscopy.

A variation is to tilt the incident beam in such a way that a diffracted beam of interest passes through the lens on the optic axis, and hence through the objective aperture. This technique is called *dark-field* microscopy, because holes in the sample are dark in the image. Now the image is formed by those electrons that were diffracted (together with a small number of electrons that were incoherently scattered and happen to pass through the objective aperture), and hence the diffracting regions appear bright, on an otherwise dark background. The contrast in dark-field images can be extremely high, enhancing the visibility of weak features. It is possible to identify ordered regions in an otherwise disordered matrix by forming an image with a diffracted beam originating in the ordered phase. An example would be the identification of γ' regions in a superalloy.

Diffraction contrast images strain (or, more strictly, displacement) in a crystal lattice. As such, it is not a very high-resolution imaging tool (because the strain is not localized), although a modified form of the dark-field imaging technique described above, known as "weak-beam dark-field" imaging, can limit the imaged regions to those with the highest strains. However, it is possible to determine the direction of the strain (and hence, for example, the Burgers vector of a dislocation), and diffraction contrast remains a powerful tool for the materials scientist.

3.3.3. Phase Contrast

Phase contrast is the most difficult contrast mechanism for a novice to imagine, but it is also the most powerful mechanism for providing images with ultrahigh resolution. While mass thickness contrast can be explained by invoking a classical model of electrons, and simple models of diffraction serve at least to introduce diffraction contrast, phase contrast depends fundamentally on the wave-mechanical nature of electrons. The incident illumination is considered to be a series of waves which, since the direction of propagation is defined as the normal to the wave front, are plane waves if the illumination is parallel. If a small region of the wave front is, for example, delayed more than the rest of the front as it passes through a sample, then the front will no longer be planar, and it will be as if near the distortion, some energy travels in directions away from the incident beam. The term *phase contrast* given to this mode of image formation arises because the delay introduced by the sample is in fact a change in the phase of the electron wave front.

Phase contrast may be illustrated by a simple optical experiment, the results of which are shown in Fig. 3.1. The sample is a glass coverslip, onto which droplets of hydrofluoric acid have been sprinkled, thus creating slight dimples. The illumination source is a focused laboratory lamp with a pinhole and lens acting as a parallelizer, thus creating the plane wave front. A camera was used as the image-forming system for convenience of recording the images on film, but for a demonstration one or two simple lenses can be used and the result will be essentially the same. Since light travels more slowly in glass than in air, and since the sample thickness varies from place to place, the phase (but not the intensity) of the transmitted light also depends on the part of the sample through which it travels. It can be seen that in the "underfocused" condition, the thinned regions (where the acid droplets landed) appear brighter than the surrounding area, the converse being true in the overfocused condition. At the geometric focus the etched regions become almost invisible. In

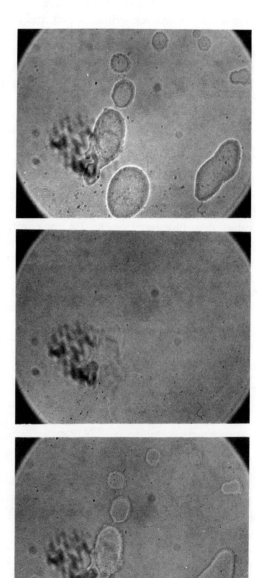

Figure 3.1. Illustration of phase contrast. Photographs of slight indentations made by sprinkling hydrofluoric acid on a glass coverslip, and then rapidly washing in water. Full details of the experiment are given in Section 3.3.3. Top: underfocus; center: in focus; bottom: overfocus. While the dust particles are seen sharply in the in-focus image, the indentations are almost invisible. The contrast in the under- and overfocus images of the indentations is approximately complementary. The other rather diffuse features that are essentially unchanged in the three images are caused by dust on lenses in the system.

this demonstration, the "phase objects," as the depressions may, quite accurately, be called, are well separated, and as the image goes out of focus, the various regions of light and dark do not interact. Phase-contrast microscopy of typical materials samples is perhaps more realistically modeled by photographing a glass diffraction grating, where a complex interaction occurs, illustrated schematically in Fig. 3.2. (For simplicity in visualization, in the figure we imagine the lens to have a fixed focal length, and we move the viewing screen. In the electron microscope, of course, the screen is fixed and the focal length of the lens is adjusted; the results are identical.) It can be seen that at focus there is no detail observed in the image. At small deviations from focus, dark and bright bands, with a spacing equal to the spacing of the grating, are observed. The bright bands in the underfocused image become dark in the overfocused image, and vice versa. It is left to the interested reader to show (by extending the ray paths in the illustration) that at larger defocus values than illustrated in Fig. 3.2, the contrast can vanish and then reverse several times. In the circumstances of the illustration (i.e., parallel illumination and no lens aberrations) there is no apparent limit to the number of such reversals that can be observed. In practice the beam divergence and spherical aberration (see Section 3.4.2.1, paragraph c) limit the number of observable contrast reversals to two or three in most cases.

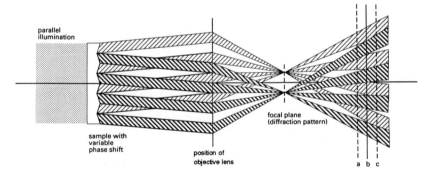

Figure 3.2. Ray diagram illustrating a simple form of phase contrast. The sample is imagined to be a faceted plate. Parallel illumination incident from the left is split into two series of beams, represented by the dark or light cross-hatching. (In an effort to simplify the rather complex drawing, the heavily cross-hatched beams are drawn as if they are in front of the other set. The reader must understand that in fact the beams overlap.) The two series are seen to be focused at separate points in the back focal plane of the lens. On a viewing screen at position a, it can be seen that some areas are illuminated by one beam, others by two overlapping beams, and still others receive no illumination, resulting in a pattern of dark and light fringes. As the screen is moved to position b, all areas receive equal illumination so no detail is visible, while at position c, the fringes become visible again, but the areas that were previously light are now dark, and vice versa.

The electron analogue of the optical refractive index is the potential, which is a function of position within the unit cell. If a crystal is oriented with a low-index direction parallel to the electron beam, the rows of atoms are positions of lower potential, and act as a two-dimensional diffraction grating. By suitable choice of defocus, details of the potential in the unit cell (and, hence, the structure of the crystal) may be inferred. However, since the sample thickness (and, hence, the potential) and the defocus are in general not precisely known, the correlation usually is not direct, and it is necessary in almost all cases to perform computer calculations to "match" an electron image with a computed image before this result is obtained.

The advantage of phase-contrast imaging is that extremely high-resolution detail is transmitted from the sample to the recording medium (viewing screen, CRT, film, or whatever). The limit is set by imperfections in the lenses and other instabilities in the microscope (together termed *aberrations*), and is given approximately by the expression

$$R = 0.6\lambda^{0.75}C_{s}^{0.25} \tag{3.2}$$

where R is the minimum spacing that can be resolved and C_s is the spherical aberration coefficient of the objective lens (typically roughly equal to the focal length). Currently the resolution is about 0.15 nm in the best commercial instruments.

3.3.4. High-Angle Annular Dark-Field Imaging

A recent development in high-resolution microscopy has been the practical demonstration of ultrahigh-resolution imaging using only those electrons that have been scattered through large angles. The full theory is beyond the scope of this chapter, but a qualitative description is appropriate.

Normal phase-contrast imaging requires that an aperture be placed after the sample to limit the angular range of the electrons that contribute to the image. It follows that the momentum of the electrons that pass through the aperture is quite well defined. The Heisenberg uncertainty principle states that the product of the uncertainty in the momentum and the uncertainty in the position equals Planck's constant. Hence, if the momentum is precisely known, the position cannot be measured precisely. If, on the other hand, electrons scattered through large angles are detected, and provided we do not attempt to measure exactly *where* on the detector the electron was found, then the uncertainty in the momentum becomes large, so the uncertainty in the position is reduced. (The rigorous theories do not rely on this argument; however, it is precise enough for our purposes, and serves to illustrate that it is vital to treat the electrons as waves and not as particles.)

Figure 3.3 illustrates the experimental arrangement. As a practical matter, it is only possible to generate high-angle images in a scanning instrument, and it is only worthwhile doing so in the highest-resolution microscopes, i.e., those fitted with field-emission electron guns (see Section 3.4.1.2). It is necessary to eliminate from the image the diffracted electrons (which are coherent); this is simply accomplished by making the detector in the form of an annulus, whose inner diameter is large enough so that diffracted and other low-angle electrons do not hit it. Of course, where there is a hole in the sample no electrons are scattered, so the corresponding part of the image is dark. Hence, this form of imaging is known as high-angle annular dark-field (HAADF) imaging.

The expected improvement in resolution is by the square root of 2, in a properly optimized microscope, and recent experiments in 100-kV instruments have confirmed that this improvement is realized. Instruments operating at

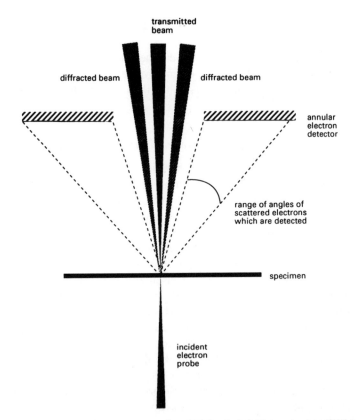

Figure 3.3. Schematic arrangement of an annular dark-field detector in a STEM.

300 kV have not yet been equipped to perform this type of imaging, but theoretically a point-to-point resolution of slightly better than 0.1 nm should be achievable in them simply by applying current design practices. If this turns out to be practical, it may be expected that in the future STEMs will become the instruments of choice for high-resolution imaging, as they are already for microanalysis.

The HAADF image has a further advantage, in that it is critically sensitive to the mass of the scattering atom. Hence, regions in the sample of differing atomic number are quickly detected, and, in some special cases, even semi-quantitative composition analyses are possible by measuring the changes in brightness of the image.

Although HAADF imaging has only recently been employed by materials scientists, and incorporated in commercial microscopes, the technique has been used for many years by biologists, and the first electron microscope images of individual atoms were also made by HAADF imaging, the instruments used for these experiments having been constructed by the researchers.

3.3.5. Ionization and X-ray Generation

One way in which energetic electrons interact with matter is by ionizing atoms of the sample. As the ions decay, they can radiate excess energy in the form of x rays, whose energy is characteristic of the sample. A spectrometer capable of analyzing the energy (or, equivalently, the wavelength) of the x rays can therefore provide the basis for a chemical analysis of the volume of sample illuminated by the electron beam. In ideal cases, this volume may be as small as a few cubic nanometers.

The incident electrons which generate these ionizations continue on their way having lost an amount of energy equal to the ionization energy of the target atom plus whatever kinetic energy the ejected electron may be given. Spectroscopy of the energy distribution of the transmitted electrons can there-fore also serve as a microanalytical tool. Indeed, in some cases, such "electron energy-loss spectroscopy" (EELS) can be the most sensitive tool available, and is demonstrably capable of single-atom detection.

Microanalysis using these methods is discussed in Section 3.5.

3.4. Construction Details

3.4.1. Electron Gun

There are two types of electron source of practical use for electron mi-croscopy, namely, thermionic sources and field-emission sources. Before about

1970 all electron microscopes used thermionic sources. Recently, however, the field-emission source has become more common, although still specialized in its applications.

A criterion for describing the performance of an electron source is its *brightness*. The brightness of any source of radiation is the amount of that radiation emitted per unit area of the emitter into a unit solid angle of space. It is an elementary theorem of optics (and here is left for the reader to prove) that the brightness of the beam of radiation at any subsequent point can never exceed that of the source. We shall defer discussion of the influence of the gun brightness on the capabilities of electron microscopes until later sections; for now, we will merely note that these influences exist.

A further criterion against which an electron source may be judged is its coherence. The coherence of an illumination system determines how closely the electron beam approaches the ideal assumed in Section 3.3.3 above that the wave front is parallel. Put in a rather oversimplified way, the smaller the source and the smaller the diameter of the beam-limiting aperture in the illumination system, the better the coherence is. Both strategies, unfortunately, reduce the flux of electrons on the sample, making it difficult to see to focus the image, and increasing the demands on the image stability by increasing the time required for image recording. A small, high-brightness source is clearly preferable to an extended, low-brightness one.

3.4.1.1. Thermionic Sources

Figure 3.4 illustrates the essentials of a thermionic electron gun. A heated emitter (classically, a hairpin of tungsten wire) is positioned behind a small aperture in an electrode known as a Wehnelt cylinder, often abbreviated simply to Wehnelt. The entire assembly (which is often called the "gun") is connected to the negative terminal of the high-voltage power supply, the positive terminal of which is grounded. The Wehnelt is at a negative potential, termed the "bias" (typically, several hundred volts) with respect to the filament. In front of the Wehnelt is the anode, at ground potential, with an aperture through which the beam of electrons passes. In higher-voltage systems (operating at electron energies above 100 keV) an electron accelerator is employed between the gun and the anode to ease the engineering difficulties associated with such high voltages, but the principle remains the same. The filament is heated, usually simply by passage of a direct current through it, until a sufficient number of electrons are thermally excited from the conduction band over the work-function barrier and into the vacuum. If the filament position is correctly adjusted, enough of the electric field caused by the high voltage "protrudes" through the aperture in the Wehnelt to accelerate the electrons away from the filament. The area of the filament over which this condition

Wehnelt Cylinder

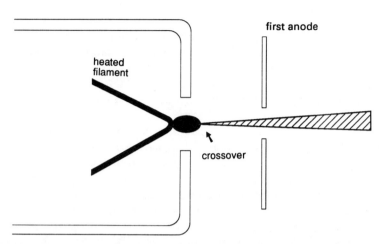

Figure 3.4. Schematic drawing of the essentials of a thermionic electron gun. The electrons leave the filament and are focused by the electric field to a small crossover just outside the gun; this crossover acts as the electron source for the subsequent optics.

exists (and which, therefore, contributes to the electron beam) is governed by the spacing between the Wehnelt and the filament, and by the bias voltage. Assuming that the geometry has been set up correctly, the number of electrons emitted per unit area per second, which in turn determines the brightness of the source (for a given accelerating voltage), is determined by the temperature of the filament. Unfortunately, as the filament temperature is increased, so the evaporation rate of the material of the filament increases, shortening the lifetime (in the extreme, of course, the filament melts). Interestingly, the ratio of the number of electrons emitted per evaporated atom is—very roughly— constant for a particular filament material. Thus, a twofold increase in brightness corresponds to a halving of the filament lifetime. At a given temperature, the number of electrons emitted is determined by the work function of the surface. However, if a material with low work function also has a high evaporation rate, constraints of acceptable filament lifetime might not allow the filament to operate at as high a brightness as another filament made of a material with higher work function but also of more refractory nature. This discussion has so far ignored the effect of the imperfect vacuum on the operation of the gun. Residual gas in the gun area can affect the performance in a number of ways, all of which either reduce the brightness or the lifetime of the filament. Electron microscopes built in the 1950s typically operated with a gun vacuum around 10^{-4} mbar. Tungsten hairpins, being rugged and relatively insensitive to the vacuum, were invariably used for the electron emitter, but nevertheless provided a lifetime of only a few tens of hours, governed entirely by effects of the residual gas. In subsequent designs, the vacuum gradually improved, and in modern designs is better than 10^{-6} mbar, at which pressure tungsten filaments last, under operating conditions, a few hundred hours, the lifetime being governed by the evaporation. The brightness of a gun with a tungsten filament operating at 100 kV is on the order of 5 $\times 10^5$ A/cm^2 per sr, and a typical effective source size is a few tens of microns.

With the improved vacuum it became possible to consider other materials as the electron emitter in thermionic guns, the most useful (and now ubiquitous) being lanthanum hexaboride. There have been a number of difficulties to overcome in designing satisfactory LaB$_6$ electron sources, with the result that they have acquired a reputation for being unreliable, or at least unforgiving in operation, so until recently such electron guns were employed only when their principal attribute, higher brightness than the tungsten hairpin, was essential. Continuing research and development has addressed the problem of obtaining truly satisfactory performance from LaB$_6$ electron guns, and they are now in fact a reliable and indispensable part of many modern high-performance electron microscopes. It has been suggested that by operating in a vacuum of $<10^{-7}$ Mbar the temperature (and hence the brightness) could be reduced, leading to a prolongation of the lifetime to many thousands of hours, but still with a brightness perhaps 20 times that of the tungsten gun.

3.4.1.2. Field-Emission Sources

Of the various electron sources, the field-emission gun has the highest brightness and smallest size (a cold field-emitter brightness has been measured as 10^9 A/cm^2 per sr, and the source size is 5–10 nm), and is the gun of choice for the highest-resolution analytical electron microscopes, among others. The field-emission source operates on principles quite different from those for the thermionic gun. Figure 3.5 shows the energy levels of the electron at various parts of each system. In the thermionic gun, the potential gradient away from the filament (i.e., the electric field) is very small. Although the Fermi level (the highest occupied level at a temperature of absolute zero) inside the filament is below the energy of free space by an amount equal to the work function, heat excites electrons into higher energy states within the conduction band, with the result that the most energetic are able to leave the surface and be accelerated gently toward the anode.

In the cold field-emission gun, in contrast, an extremely intense electric field (on the order of several volts per nanometer) is applied to the cathode. As a result, the vacuum level falls below the Fermi level within a very short distance (a few tenths of a nanometer) of the surface of the tip. The electrons

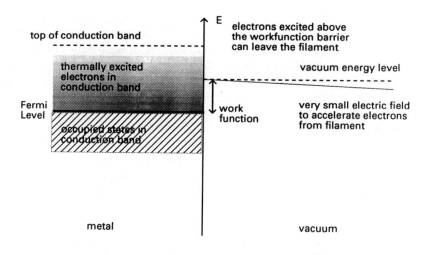

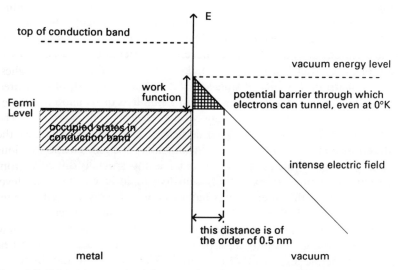

Figure 3.5. Schematic energy-level diagrams of (above) a thermionic electron source and (below) a field-emission source. The former requires thermal excitation of the electrons in the conduction band, the latter would work satisfactorily at absolute zero.

still see a potential barrier whose height equals the work function, but now the width is so small that the probability of electrons tunneling from the Fermi level into free space becomes large, resulting in intense electron emission, easily controlled by varying the field. Besides having high brightness, the field-emission source also gives a very low energy spread to the electrons (although the explanation of this assertion will be omitted here).

There are, of course, trade-offs—the benefits of the field-emission gun are not obtained without cost (literally!). In the thermionic gun, the surface of the cathode, being very hot, is thoroughly degassed, and any residual gas atom or molecule which happens to arrive at the filament is almost instantly reevaporated. In the cold field-emitter, in contrast, the tip is at room temperature, and several monolayers of adsorbed gas are stable. While the cathode can be degassed by heating it briefly, as soon as it cools down again, a new layer of adsorbed gas will start to accumulate, the rough rule of thumb being that at 10^{-6} mbar a monolayer accumulates in 1 s. Since gas on the surface of the emitter increases the work function (and hence reduces the emission drastically), the pressure must be low enough so that a useful time passes between the degassing events. For example, at 10^{-10} mbar, 10% of a monolayer is built up in about 20 min, and at 10^{-11} mbar the same amount of gas is adsorbed in about 3 h. Experience with such microscopes indicates that guns operating in this pressure range do indeed have useful operating times corresponding roughly to these estimates. Hence, very expensive ultrahigh-vacuum technology is essential in the construction of microscopes which have field-emission guns, and even in these instruments, long-term stability of the emission current is impossible to achieve. The extreme electric field at the surface of the emitter is obtained by making the cathode in the form of an extremely sharp tip (usually of single-crystal tungsten of [310] orientation), with a radius on the order of a few tens to a hundred or so nanometers—it will be recalled that a sharp point concentrates electric fields. The tip is then placed a few millimeters from an anode held at a relative potential of 3–5 kV, the emission being controlled by adjusting this potential. The actual manufacture of the tips, while certainly not trivial, is not especially difficult. However, such sharp tips are exquisitely sensitive to local electric discharges ("arcs"), with even the smallest arcs having enough energy to melt the end of the tip into a little ball with a radius far too large to allow the required field to be achieved. Hence, the high-voltage design must be far more conservative than in a microscope employing a thermionic electron source, which can withstand moderate arcs or other discharges. This constraint, too, increases the complexity and cost of the instrument. Again, because of the extremely sharp tip, while the brightness is very high, the actual area of the tip which emits electrons is small, so the total current which a field-emission gun can generate is limited (another way of describing this is to say that the gun emits electrons only into a very small cone). Thus, although the field-emission source can generate a very small, highly coherent probe with high brightness, for high-current applications the thermionic (LaB_6) sources are still preferred.

There are variants of the field-emission source, including heated sources (to minimize the buildup of contamination on the surface of the tip), and sources with specially treated tips. There is at present no general consensus as to which variant, if any, is the best choice for an electron microscope.

3.4.2. Lenses

3.4.2.1. Aberrations

Before discussing lenses specifically, we will first mention the various imperfections that can degrade an electron lens.

Astigmatism. An astigmatic lens does not have cylindrical symmetry. A point source of radiation is not imaged at a point, but at two mutually perpendicular lines at different distances from the lens. Astigmatism is accurately correctable, and should not be a determining factor in the performance of a modern electron microscope, but the adjustment of the correction is difficult, requiring skill and experience, so in practice astigmatic images are obtained far more often than most microscopists will admit. In principle, only one stigmator is required in an imaging system; however, in CTEMs there are two "imaging systems" (the illumination system and the image-forming system) and hence at least two stigmators, and in field-emission analytical electron microscopes, the source itself can have significant astigmatism which is conveniently corrected by a dedicated stigmator independent of that used to correct the astigmatism of the probe-forming (objective) lens.

Chromatic aberration. Chromatic aberration is the manifestation of the change in focal length of a lens as the wavelength of the incident radiation changes. It is significant in the electron microscope for several reasons. First, there is an energy distribution in the emission from any electron gun, as a result of both the physics of operation of the gun and the small instabilities or noise in the high-voltage power supply; it is not desirable that these electrons be focused at different places in the microscope. Again, all practical power supplies for electron lenses, be they electrostatic or magnetic, have some degree of instability; a small change in excitation of the lens produces the same effect on the electron optics as a small change in the electron energy. Further, in CTEMs the electrons are focused after they have passed through the sample, during which passage they have a significant (often essentially unity) probability of losing energy in inelastic collisions, with a concomitant change in wave-

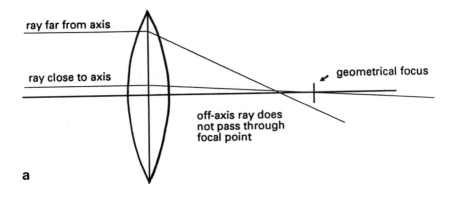

ray far from axis

ray close to axis

geometrical focus

off-axis ray does
not pass through
focal point

a

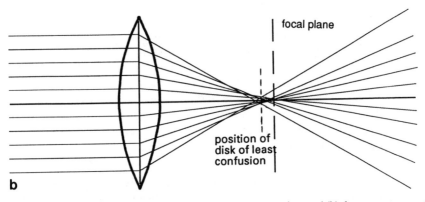

focal plane

position of
disk of least
confusion

b

Figure 3.6. (a) Illustrating the phenomenon of spherical aberration, and (b) the consequences of spherical aberration on the formation of an electron probe.

length. Chromatic aberration in the image-forming system will result in an image with details blurred by the inelastically scattered electrons being focused at indeterminate positions.

The chromatic aberration of a lens is determined by the design, and is specified by a distance, designated C_C, the chromatic aberration coefficient, of the same order as the focal length of the lens. In most cases, for study of thin samples, in well-designed modern microscopes, the chromatic aberration is not the determining factor in the ultimate performance of the instrument.

Spherical aberration. Consider Fig. 3.6a in which is illustrated schematically a lens suffering spherical aberration through which two electrons, both traveling parallel to the axis, are passing. The incident electron close to the optic axis emerges from the lens and crosses the axis at the focal point (by definition!). The incident electron far from the axis emerges from the lens and crosses the axis at some distance from the focal point. (Electron lenses are always converging, and in all known designs, the lens action is stronger the farther the ray is from the axis.) While, for glass lenses used to focus light, it is possible to grind the lens to minimize the spherical aberration, or to design a correction system, no equivalent scheme has yet been devised for electron lenses, and the electron microscopist is obliged to accept spherical aberration in the instrument. Figure 3.6b illustrates the consequence of spherical aberration when attempting to form a probe of electrons.

The spherical aberration is described by a coefficient, designated C_s, with the dimension of length, whose magnitude is very similar to the focal length of the lens. The effect of spherical aberration is most easily visualized when considering the formation of an electron probe; however, the resolution limit is exactly the same if the lens is used for image formation. As can be seen from Figure 3.6b, one can define a "disk of least confusion" which is the smallest region illuminated (assuming a point source of radiation and no other aberrations). The diameter d of the disk is given by

$$d = 0.25C_s\alpha^3 \qquad (3.3)$$

where α is the half-angle of convergence of the cone of radiation. Hence, there is incentive to minimize the convergence angle of the probe (which is equivalent to minimizing the diameter of an objective aperture placed before the lens) in order to minimize the effect of the spherical aberration. However, diffraction theory states that the disk of least confusion resulting from the wave nature of the electrons is given by

$$d = 0.61\lambda/\alpha \qquad (3.4)$$

Hence, to maximize the diffraction-limited resolution, we need a large angle of convergence. The optimum, given when the two expressions are equal, can (depending on how exactly the expressions are combined) be shown to result in the expression given in Eq. (3.2), which predicts the best resolution obtainable for a given electron wavelength and lens. It can be seen that the resolution increases faster with reduction of the wavelength than it does with improving the spherical aberration. Since the wavelength reduces with increasing energy, there is thus considerable incentive to operate high-resolution electron microscopes at as high an operating voltage as possible. Unfortunately, as the voltage increases, so does the weight and complexity, leading to practical difficulties in operation. Hence, at present all successful commercial high-resolution instruments operate in the range 200–400 kV, and are termed *intermediate-voltage* microscopes, in contrast to the earlier generations of instruments operating near 100 kV, and the specialized high-voltage machines which operate near or above 1 MV. Similar considerations lead to the conclusion that the optimum voltage for analytical microscopes is also, with currently available technology, in the same range, although few dedicated intermediate-voltage AEMs (i.e., instruments fitted with field-emission electron guns) have yet been installed.

3.4.2.2. Electron Lenses

Electrons can be focused either by electric fields or by magnetic fields. All electron microscopes contain an example of an electrostatic lens in the electron gun; however, for critical purposes, magnetic lens designs perform better than electrostatic lenses, and are universally employed in high-performance microscopes. The essential feature of a magnetic lens is a pair of pole pieces with accurately parallel faces, each containing a hole, coaxial with the optical axis, through which the electron beam passes. A solenoid, through which a current is passed, and a yoke to complete the magnetic path (the details of which will not be expanded on here) are the other parts of a practical lens. The designer can vary the field (limited by the saturation flux of the magnetic materials), the shape of the pole pieces, their gap, and the diameter of the holes (or "bores"), within constraints set by the intended application, to modify the properties of the lens.

3.4.2.3. The Objective Lens

It can be shown that, in most cases, and with proper design, the aberrations of the objective lens dominate all others in the microscope. We shall, therefore, only discuss this lens. As has already been discussed, the performance of a lens in imaging a sample onto a screen, and in forming a probe of electrons onto the sample are identical; hence, we do not differentiate between probe-forming and imaging lenses.

As was shown above, the spherical aberration constant of the lens must be minimized for best resolution. The spherical aberration constant is on the same order as the focal length, so the lens must also have a short focal length (although, in other ways, this is a nuisance). The best performance is obtained with the sample inside the gap between the polepieces, and with the gap minimized. However, there must be room to insert (and, generally, to tilt) the sample, either from the side of the lens, or through the top bore, so (since the proportions of the ideal lens are relatively fixed) in either case there is a limit to how small the gap may be made. Other considerations may apply, too; for example, if the microscope is to be equipped with an x-ray detector there must be room to allow a clear view of the sample. It has for many years been the practice of microscope manufacturers to offer a choice of different pole-piece designs so that users may select the appropriate compromise between resolution and other factors in microscope performance.

The perfection of the pole pieces is critical in determining the performance of the instrument. The greatest care must be taken, from ensuring the homogeneity of the iron or alloy used for the blank, through manufacture, to day-to-day use, to protect the pole piece from damage. The dimensions, and especially the concentricity, must be maintained to within fractions of a micron, and a single scratch on the face (caused, for example, by a sample holder hitting the pole piece) can irretrievably ruin the imaging capability of the lens. Microscope managers are, therefore, usually reluctant to change pole pieces in a microscope which must be used for high-resolution work.

3.4.3. Detectors

Since the earliest days of electron microscopy, conventional transmission images have been recorded by exposing conventional film directly to the electrons. Science has been, and continues to be, well served by this method. Limitations in most high-resolution images arise from such details as the statistical nature of the limited number of electrons contributing to an image feature; in this case, the film can faithfully record these fluctuations and hence does not significantly degrade the information available. The film does have a limited dynamic range which is easily saturated in high-contrast images or in diffraction patterns. This limitation is simply overcome by taking multiple exposures for different times, to record the different features.

Recent developments in electronic detection of images, using, for example, charge-coupled devices, give promise of detectors which are capable both of detecting low-signal images with the same efficiency as film, and of achieving a far greater dynamic range for recording diffraction patterns, at the same time providing directly an electronic output suitable for transmitting the image to a computer for storage and processing. Since much modern microscopy involves computer manipulation of images, and since the cost of photographic materials is substantial, the advantages are obvious. On the other hand, present-day detectors are limited to recording images with, typically, 1024×1024 pixels. In contrast, photographic emulsions can record at least 100 pixels/mm, or, over a standard piece of film 8 cm $\times$ 10 cm, an image of about $8000 \times 10,000$ pixels. Hence, there is no prospect that photography will be eliminated from electron microscopy laboratories in the near future.

Detectors for STEM do not have to resolve the spatial positions of the electrons striking them. Simple designs consisting of a scintillator, some sort of arrangement for transmitting light out of the microscope column, and a photomultiplier work well.

3.4.4. Cleanliness

Cleanliness is a function of sample preparation, microscope use and maintenance, and microscope design. For convenience, we will discuss all of these features in this section of the text.

Large hydrocarbon molecules are easily cracked by electron bombardment; the end result is essentially pure carbon and lighter, gaseous fractions which are pumped away. Should there be other elements present (silicon and oxygen, for example) insulating materials are readily synthesized in the microscope, coating all surfaces which are subject to electron bombardment, including the gun components, the anode, and the various apertures, as well as the sample.

At the very least, the presence of a carbon layer on the sample degrades the image or microanalysis. The presence of even tiny amounts of insulating compounds can be devastating to the performance of the instrument. The insulating layers allow charges to build up unpredictably in the column, thwarting any attempt to obtain a stable, well-stigmated image. Clearly, the higher the resolution sought in the image, the more essential scrupulous cleanliness becomes.

All modern microscopes are designed with vacuum systems which are adequately clean for their intended applications. Nevertheless, the microscope manager will be critically interested in maintaining the cleanliness of the instrument. In the first place, silicone oil will not be used in the microscope's pumps. It is impossible to prevent some degree of backstreaming from diffusion pumps, and once silicon is present in the column it is almost impossible to remove, even in fully bakeable systems. Hence, a pure hydrocarbon oil, such as polyphenyl ether (Santovac 5, for example), is always used. (Some specially built or modified instruments eschew the use of diffusion pumps altogether, relying on turbo-molecular pumps and/or ion pumps; it goes without saying that these systems are very expensive.) Checking for leaks (especially in the gun area, where oxygen from a leak can cause oxidation of a LaB_6 emitter and deposition of lanthanum oxide on the Wehnelt) will also be a priority. Even so, some contamination will build up, and routines for cleaning the instrument will be established.

No amount of care by the microscope manager can prevent the contamination of the column by a dirty or unsuitable sample (many samples, e.g., many polymers, can emit vapors which can condense on nearby surfaces and hence degrade the instrument's performance). When there are only one or two users on an instrument, there is usually sufficient incentive for those users to be careful in their specimen preparation and microscope operation. However, it is more difficult to control a microscope which must serve a wider community, including novice users and those with new, unknown samples. The highest-quality high-resolution results can only be obtained reproducibly when access to the top-performance instrument is limited to highly qualified, motivated operators, whose samples are known to be stable, and whose experimental technique has already been refined by practice on other microscopes.

3.5. Microanalysis

A major application of electron microscopy in materials science is microanalysis. The most widely used technique is x-ray analysis, which will be the subject of Section 3.5.1. EELS has a number of unique capabilities, and is discussed in Section 3.5.2, both of these methods yielding chemical information. Microdiffraction, which gives crystallographic information about the sample, is briefly covered in Section 3.5.3.

A transmission microscope designed for microanalysis is termed an analytical electron microscope (AEM). Most instruments in this class are TEM/STEMs, and are capable of forming conventional images of the sample as well as performing microanalysis. Although surprisingly few compromises have to be made in the design of such instruments, commercial pressures have dictated that most of them are not fully optimized for microanalysis. Dedicated STEMs, on the other hand, are designed from the start for uncompromised microanalysis, and in most demanding applications provide the highest-quality data. The principles of operation are, however, identical for either type of instrument, and in the following discussion, we will not differentiate between them.

3.5.1. Energy-Dispersive X-ray Spectroscopy

Commonly abbreviated as "EDXS" or "EDX analysis," spectroscopy of the x rays emitted by a sample under electron bombardment provides an extremely high-resolution method (albeit one with only moderately good sensitivity) for performing chemical analysis. In this chapter we will confine our discussion to analysis of thin samples, but we will mention that microanalysis in the SEM and its near-cousin the electron microprobe not only is practicable and very valuable, but also was developed before thin-sample analysis.

If an isolated atom of element A is in a region of space containing a current density of J amperes per square centimeter of electrons, then it will undergo N ionization events per second, N being given by

$$N = Q_A J/e \qquad (3.5)$$

where e is the charge on the electron and Q_A is termed the ionization cross section. Q_A has the dimension of area and is a function of the energy of the incident electrons, and varies from element to element, but is not, at least to the first order, dependant on the chemical state of the atom. It is not difficult to show from this that the number N_x of x rays generated per second in a foil of density ρ and thickness t, containing a weight fraction C_A of the element A, is given by

$$N_x = (i/e)(N_0/A_A)\rho t \omega Q_A C_A \qquad (3.6)$$

where i is the electron beam current, N_0 is Avogadro's number, ω is the fluorescence yield (i.e., given an ionized atom of element A, ω is the probability that it will decay and emit an x ray), and A_A is the atomic weight of element A.

As the electrons propagate through the foil, they acquire transverse momentum, as a result of collisions with the atoms of the sample, and the beam broadens. Hence, even if the beam diameter was very small at the entrance surface of the foil, the volume of material sampled by the beam can be large. As a result, we might think that we need to minimize the sample thickness to keep the spatial resolution as high as possible. Unfortunately, as is shown by Eq. (3.6), reducing the sample thickness reduces the x-ray count rate. It is not possible to compensate by increasing the beam current, because, as can be easily shown, the probe current i is related to the probe diameter d by

$$i = \pi^2 d^{8/3} B/(4C_s)^{2/3} \qquad (3.7)$$

where C_s is the spherical aberration constant of the probe-forming system (normally dominated by the final lens), and B is the brightness of the electron source, from which it can be seen that as the required probe current increases, the probe diameter must increase, in turn degrading the spatial resolution.

The x-ray count rate is significant because the precision of the final analysis is determined by the total number of x rays detected. (This conclusion follows from a consideration of the statistics of the x-ray spectrum—a subject that will not be pursued here.) One way that the x-ray count rate may be increased is to employ an electron gun with the highest possible brightness. It is for this reason that dedicated AEMs are built with field-emission electron sources. The other way to maximize the count rate is to optimize the x-ray detector. It will be recalled that Eq. (3.6) gave the number of counts generated per second from a given element in a sample. These x rays are emitted isotropically. In practice, an x-ray detector intercepts only a quite small proportion of the emitted total. The energy-dispersive detector detects a larger fraction than does the wavelength-dispersive detector (the type fitted to electron microprobes), and also has some engineering and economic advantages, so is universally the system of choice for AEMs. Even so, the latest, most sensitive detectors have an area of 30 mm^2 and are placed around 10 mm from the sample, so collect only about 2.5% of the x rays that are generated.

Increasing the energy of the electrons reduces the beam broadening and increases the brightness of the electron gun, so despite a modest reduction in the ionization cross sections at higher beam voltages, the current thought is that it will be advantageous to use beam voltages of about 300 kV. Microscopes operating at these voltages have been available for a number of years, but all have been designed primarily as high-resolution imaging instruments, taking advantage of the short wavelength of the higher-energy electrons. At the time of writing, however, several of the major electron microscope manufacturers are offering analytical microscopes fitted with field-emission guns operating at 300 kV. The first of these has not been in routine use, so it is too early to say whether the promise will be fulfilled, or if perhaps difficulties, such as radiation damage to the samples, will limit the application of these instruments to specialized projects only.

X-ray microanalysis may be employed to analyze regions smaller than 2-nm diameter in foils about 20 nm thick with a sensitivity generally of well under 1 wt%, or to detect segregation at boundaries where only a few atoms in a thousand are of the segregant species. In boundaries with larger amounts of segregation (above about 2×10^{13} atoms of segregant per square centimeter of boundary) a useful estimate of the actual number of atoms on the boundary may be obtained. An alternate way of presenting more qualitative data is in the form of an elemental map, generated by rastering the beam across the sample under digital control, pausing for a predetermined time at each pixel, and recording the numbers of x rays generated by each element of interest. The results are then displayed as a two-dimensional plot, termed an *x-ray map*, for each element, the intensity of each pixel depending on the number of x rays recorded for that element at that point. Examples of each of these techniques will be presented later.

3.5.2. Electron Energy-Loss Spectroscopy

Consider an (imaginary) atom with a single electron which can exist in only one bound state, and with an ionization energy E. Imagine also that this atom is rigidly fixed so that no translational energy may be given to it. If the atom is subject to electron bombardment, then the only way that the incident electrons could lose energy would be by ionizing the bound electron. By definition, the electron causing the ionization must lose at least an amount of energy equal to E, but could lose any amount more up to its own total energy, by giving kinetic energy (which is not quantized) to the ejected electron. Hence, the probability of the incident electron losing an amount of energy less than E is zero, while the probability of losing E or more (up to the incident energy) is finite. If the energy E is characteristic of the target atom, then by measuring the energy spectrum of the transmitted electrons in an electron microscope, it becomes possible to identify which elements are present. This description is, of course, a simplification, but it illustrates the principle of analysis by EELS.

The attraction of EELS is that potentially vastly more signal is available than in an x-ray system. Most of the electrons that ionize an atom can be collected and measured, but only a very small fraction result in the detection of an x ray. Older electron energy analyzers detect the spectrum serially; i.e., the spectrometer is tuned to a particular energy loss, the signal is measured and recorded, then the spectrometer is readjusted and a new measurement made, and so on. Much signal is therefore lost and the time required for an analysis is prolonged, so EELS by this method is only useful for a few limited applications (where, for example, peak overlaps made EDX analysis difficult, or for analysis of very light elements, including lithium). Within the last few years, however, parallel detection systems have been developed and marketed, greatly expanding the potential applications of EELS.

Another reason that EELS is rapidly becoming more popular as a general-purpose microanalytical tool is the move toward higher operating voltages for electron microscopes. High-quality EELS spectra can only be obtained from samples whose thickness is no more than about the electron mean free path for inelastic collisions. This quantity ranges from a few to a few tens of nanometers for 100-kV electrons (depending on the sample composition), and is a linear function of the operating voltage. Thus, it is much easier to make good samples for EELS measurements at intermediate voltages.

It is known that EELS has a better minimum detectable mass than does EDXS (single atom sensitivity—the ultimate goal—has been demonstrated). It has also been suggested that, at least for the lighter elements in the periodic table, EELS is sensitive to about an order of magnitude lower mass fraction than is EDXS. Unfortunately, these two analyses cannot be performed in the same instrument (the single atom analysis requires the brightness of the field emitter, and the low mass-fraction analysis requires the high current of the lanthanum hexaboride emitter). Also, the quantification of EELS data remains, after much work, rather problematic. Simple analysis of the most common elements is usually quite satisfactory, but analysis of complex systems, or those containing the lightest or heaviest elements has not yet become routine. Thus, EELS will not, at least in the near future, supplant EDXS on the general-purpose analytical electron microscope. Fortunately, no significant compromises have to be made in fitting microscopes with both detectors, so the analyst will be in the happy position of being able to choose the technique best suited to the problem under investigation.

EELS does have potential as a microanalytical tool on microscopes fitted with objective lens pole pieces which are too small to allow access for an x-ray detector, for example, STEMs designed for the highest-resolution HAADF images. A further potential application for EELS is in the analysis of radioactive samples, which have been a perennial difficulty for the microscopist, because the flux of radiation from them enters the x-ray detector and, by a variety of mechanisms, interferes with the detection of the electron-generated x rays. The difficulty is completely circumvented by using EELS analysis, for the only signal reaching the detector is the desired electron signal, there being no line-of-sight path between the sample and the detector, and the adequate radiation shielding required for the operator's safety also preventing stray radiation from reaching the EELS detector.

3.5.3. Electron Diffraction

By revealing details of the crystallography of a sample, electron diffraction adds another dimension to the range of tools available to the microscopist.

We recall that diffraction occurs when electrons traveling in specific directions interact with the periodic sample and, by interference effects, results in the emission of beams of electrons with unchanged energy, but traveling in different directions, the change in direction (strictly, the change in momentum) being determined by the crystallography of the sample.

The traditional way to obtain diffraction information from an electron microscope has been to form a near-parallel beam of electrons, defined by the so-called "selected area diffraction aperture" on the sample. By the nature of lens action, the various diffracted beams are focused as spots in the back focal plane of the objective lens. If the projector lens system is adjusted to image this plane onto the viewing screen or film, the diffraction pattern can be observed and recorded. This works well for samples with relatively large flat areas available for examination in a holder with smooth tilting capabilities; unfortunately, by the nature of microscopy, most samples do not meet the first of these criteria. If the selected area aperture is reduced in size to define a smaller area for study, the intensity of the pattern is reduced until it becomes impossible to record it. The electron intensity can be restored by increasing the convergence of the electron beam, at the expense of increasing the size of the diffraction spots (a point in the diffraction pattern represents a direction in space; therefore, a convergent beam in real space is represented by a finite disk in the diffraction pattern). In the limit the area selected for diffraction study can be defined by the electron probe rather than by the aperture. This mode of operation is known as "convergent-beam diffraction" (CBED) or "microdiffraction" (there are subtle differences in these terms which need not concern us now).

It turns out that the convergence of the resulting probe, far from being a hindrance, actually makes possible the derivation of far more information about the sample. A description of the theory is sufficiently complex to place it beyond the terms of reference of this chapter. However, while a conventional selected area diffraction pattern results from diffraction occurring perpendicular to the incident electron beam (i.e., it provides information only on the periodicity of the sample in a plane normal to the beam), and is limited in resolution by the residual convergence of the electron beam, a convergent beam pattern contains other information (in various fine lines and spots) about the three-dimensional crystallography. In some cases, for example, it is possible to identify uniquely the crystal structure from a single diffraction pattern, while in others a change in lattice parameter of a few parts per thousand may be detected from one area to another in the sample. It is often far easier to determine what phase is present from a qualitative chemical microanalysis (to determine the elements present and their approximate ratios) and a microdiffraction pattern, than it would be by attempting to determine the chemistry alone with sufficient precision to define the phase. Hence, CBED is a powerful tool for the microscopist.

3.6. Examples

Since, in a chapter of this nature, it is clearly not possible to give extensive examples of applications of electron microscopy, we shall only very briefly discuss the types of problems that may be addressed with the technique. We shall not even attempt to do justice to the ingenuity of the many practitioners of microscopy as they have adapted the techniques to derive seemingly obtuse pieces of information, which have enabled them to solve an astonishing array of problems in science, engineering, and medicine.

As has been pointed out, because electrons travel through the sample, transmission electron microscopy studies the bulk of a sample (albeit a very small bulk) as opposed to, for example, the surface. Since contrast can arise from strain in the lattice, changes in the mass thickness, changes in the projected potential, or changes in the chemistry, anything which results in any of these effects, can, in principle at least, be studied. In addition, electron diffraction allows the derivation of structural information.

Materials scientists and engineers use TEM extensively; for example, to identify precipitates and other second phases present in samples; to investigate the structural defects present in samples (dislocations, for example); to study the structure and chemistry of grain and phase boundaries; and to derive the atomic structure of complex crystals. As an illustration of one such application, Fig. 3.7a shows a portion of a micrograph of a superconducting $YbBa_2Cu_3O_{7-x}$ sample, with a defect. In Fig. 3.7b is illustrated a computer-generated simulation of an image of the material, with an extra copper oxide plane in one-half of the region. The image and the simulation are sufficiently similar for it to be concluded that the simulation represents the actual defect. This example is reproduced by courtesy of R. Kontra.

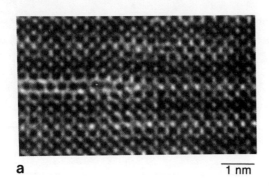

a   1 nm

Transition
↓

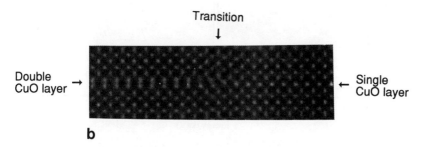

Double
CuO layer →

← Single
CuO layer

b

Figure 3.7. (a) A micrograph and (b) a simulation of an extra copper oxide plane terminating in a sample of $YbBa_2Cu_3O_{7-x}$. (Courtesy of R. Kontra.)

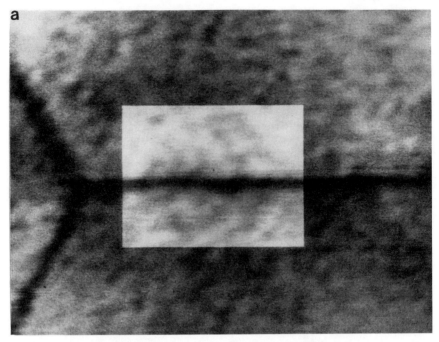

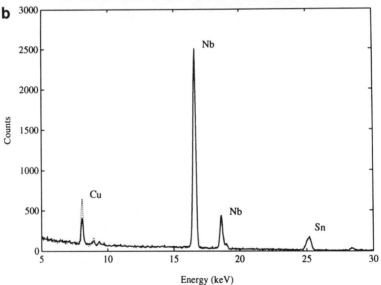

Figure 3.8. (a) A micrograph of a grain boundary in a Nb$_3$Sn sample doped with Cu. The highlighted region (dimensions 12.5 × 17.5 nm) was analyzed to generate the dotted spectrum in the x-ray plot shown in (b). The solid line is the average of the spectra recorded 12.5 nm on either side on the boundary. (From unpublished work of Durval Rodrigues, Jr, and A. J. Garratt-Reed.)

An example of grain boundary analysis is given in Fig. 3.8. The sample is Nb_3Sn with a small copper addition. Figure 3.8a shows the boundary, and Fig. 3.8b the x-ray spectra recorded from the 12.5×17.5-nm region highlighted in Fig. 3.8a (dotted line) and the average from two similar regions, offset approximately 12.5 nm either side of the boundary (solid line). There is clearly excess copper on the boundary. Without going into the calculations, it is possible to derive from these data that the excess corresponds approximately to 1.7×10^{15} atoms of copper per square centimeter of boundary. This example is from the work of Durval Rodrigues, Jr., of Centro de Materiais Refratários, Lorena, Brazil, in collaboration with the present author.

Analytical electron microscopy, as well as high-resolution microscopy, is widely employed to study finely divided matter, such as catalysts dispersed on a substrate, or dust (such as fly ash or other air-borne particulates). As an example, Fig. 3.9 shows a finely divided industrial residue. The particular interest is the location of iron and molybdenum. The x-ray maps clearly show that the iron is strongly associated with sulfur, and the association of molybdenum is less clear. Further investigation leads to the conclusion that essentially all of the iron is present as sulfides, but that the molybdenum is quite uniformly distributed among the phases present.

In the biological sciences, TEM has been an invaluable tool, allowing, amongst many other things, details of the so-called ultrastructure of cells to be elucidated, viruses to be visualized, DNA to be weighed, and immunologically active sites to be located. In medicine, the TEM is close to becoming a routine tool in pathology.

The typical sample for TEM must be a few tens to a very few hundreds of nanometers thick. Techniques for making such samples have not been discussed in the present chapter. This omission does not signify that specimen preparation is a trivial matter; indeed, quite the reverse. A prerequisite for quality electron microscopy is a specimen prepared in such a way as to preserve the details of interest and without the introduction of artifacts. Each new study will require careful evaluation of existing preparation methods, and perhaps the derivation of modifications to them, to ensure that valid information will be obtained in the microscope. It has been the author's experience that few, if any, specimen preparation problems cannot be solved in some way, assuming, of course, that the investigation is appropriate for electron microscopy in the first place; careful discussion with colleagues, reading, and thought, followed by some experimentation, will usually find a method capable of making useful samples.

3.7. Final Comments and Future Trends

Electron microscopy has proved over many years to be applicable to a wide range of scientific investigation. This chapter cannot begin to do justice

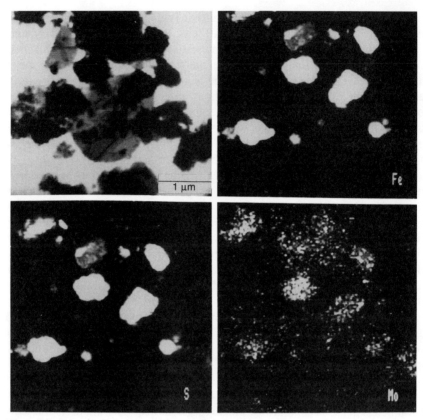

Figure 3.9. (Above left) Image, and (proceeding clockwise) x-ray maps of Fe, Mo, and S from a sample of industrial dust. The Fe, but not the Mo, is seen to be strongly related to S. Suitable locations for detailed probe analysis are easily selected from this montage.

to the inventiveness of the practitioners or to the nuances of technique available. Nor have we had the space to explore the vital area of specimen preparation. It is most certainly true that high-quality microscopy is critically dependent on provision of a high-quality specimen, but necessity here has been the mother of invention, and the author can remember no occasion in his research or in any known to him, when, if an electron microscopic examination of a sample was essential to obtain the desired information, it was not finally possible to produce a useful thin sample.

Over the last fifty-odd years the progress in electron microscopy has been continuous. We have now, perhaps, reached the point where we can see the physical limitations in electron optics, practical limitations in the design of mechanical and electrical stability in our instruments, and limitations in the stability of our samples under the energetic electron bombardment beginning to crowd in on each other. Undoubtedly there is still room for continued improvement. It would be wonderful if a way could be found to correct spherical aberration in the objective lens (there is no known physical reason why this should not be possible). The application of computer control and analysis to electron microscopes will progress rapidly. Digitally implemented manual control has been in widespread use for many years, and digital image analysis is another important and widely used part of electron microscopy that we have completely omitted from our discussion, but the operator has always provided the skill. Techniques of automatic image recognition, microscope optimization, and control already exist. It is, for example, at least 10 years since the author heard a colleague musing reflectively that the computer could stigmate his microscope better than he could himself. Surely it is only a matter of time before these methods become integrated routinely into ordinary commercial instruments.

Whatever the details, though, it would seem to be a safe prediction that TEM will remain a vital tool for scientists of many disciplines for years to come.

3.8. Further Reading

The author can no more hope to do justice to the literature on electron microscopy than he could the subject itself in the space of this chapter. The field continues to move forward, in instrumental capability, theoretical understanding, and areas of application. The proceedings of the various electron microscopy society meetings (for example, the Electron Microscopy Society of America, published annually by San Francisco Press, or, before 1983, by Claitors Publishing, New Orleans, La., the biennial meetings of the Electron Microscopy and Analysis Group of the Institute of Physics, published by IOP Publishing, Bristol, UK, or the Microbeam Analysis Society, which has also sponsored an approximately triennial series of meetings on analytical electron microscopy, all published by San Francisco Press) are good places to start looking for recent developments. Journals such as *Ultramicroscopy, Journal of Microscopy, Microscopy Research and Technique* (formerly the *Journal of Electron Microscopy Technique*), and *Philosophical Magazine A* are popular places for the publication of full papers on electron microscopy techniques. The *Bulletin of the Electron Microscopy Society of America* frequently carries articles describing new or improved techniques in microscopy. Most of the major electron microscope manufacturers publish house magazines which highlight developments and applications of their own products. The majority of papers reporting results of work performed on electron microscopes, even if some new or modified technique is involved (for example, in specimen preparation), are published in the literature of the subject matter of the investigation.

The reader looking for basic information presented in greater depth than in this chapter will find a bewildering array of textbooks available. The following list is, in common with the rest of this chapter, biased heavily toward materials science, and is greatly abbreviated. The reader will notice that the titles of some of these volumes are similar to the headings of some subsections of this chapter.

Experimental High Resolution Electron Microscopy (2nd ed.), John C. H. Spence, Oxford University Press, 1988. As well as the subject matter implied by the title, this book presents a useful discussion of the design and operation of modern electron microscopes.

Electron Energy-Loss Spectroscopy in the Electron Microscope, Raymond F. Egerton, Plenum Press, 1986. Written by one of the principal practitioners of the technique, there is no better exposition on this topic available.

Principles of Analytical Electron Microscopy, edited by David C. Joy, Alton D. Romig, Jr., and Joseph I. Goldstein, Plenum Press, 1986. A useful text, but it tends to understate the potential of field-emission instruments. An earlier but still more widely available volume, *Introduction to Analytical Electron Microscopy,* edited by John J. Hren, Joseph I. Goldstein, and David C. Joy, Plenum Press, 1979, is significantly out-of-date, and is not recommended.

Quantitative Electron Microscopy, edited by John N. Chapman and Alan J. Craven, SUSSP Publications, Department of Physics, University of Edinburgh, 1984. Written by the faculty of a NATO Advanced Study Institute, this is possibly the best all-around book on the subject, but is very difficult to find.

Electron Microscopy of Thin Crystals (2nd ed.), P. B. Hirsch, A. Howie, R. B. Nicholson, D. W. Pashley, and M. J. Whelan, Kreiger Publishing Company, 1977. A very old text (the first edition was published in 1963), it has the major advantage of having been written by the workers principally responsible for the development of the field they describe. Although it is not current in terms of the performance of instrumentation, there is still no better source for the basic theory of the interaction of electrons with periodic samples.

4

Auger Electron Spectroscopy

L. L. Kazmerski

4.1. Introduction and Background

Auger electron spectroscopy (AES) is one of the more useful, most utilized, and generally recognized surface-sensitive analytical techniques for the investigation of the chemical and compositional properties of materials. Major literature covering this surface analysis method is included in the Bibliography. Several reasons account for the familiarity with and widespread application of this technique. (1) All elements, except hydrogen and helium, can be detected; with relatively rapid acquisition time and reliable quantification of results. (2) The interpretation of data is not difficult because of the uniqueness of the spectral features and the large base of scientific literature available. Handbooks are available that facilitate the identification of elemental species. In most applications, the spectra from individual elemental components do not interfere with one another, providing for unambiguous identification of the composition. (3) One-, two-, and three-dimensional depth analysis (with approximately 5–50 Å depth resolution) can be performed using simultaneous sputter (ion) etching and other methods. These depth analysis modes provide for the identification of impurities at intramaterial or device regions and interfaces. (4) The modularity and expandability of the technique enable its combination with other analytical methods (e.g., x-ray photoelectron spectroscopy and secondary ion mass spectrometry) for complementary analyses

L. L. Kazmerski • National Renewable Energy Laboratory, Golden, Colorado 80401.

Microanalysis of Solids, edited by B. G. Yacobi *et al.* Plenum Press, New York, 1994.

of the same areas; to use it within materials growth systems (such as molecular beam epitaxy) for *in situ* determination of surface properties without exposure to external environments; and to expand the inherent AES capabilities, providing other analyses (including electron energy-loss spectroscopy, electron beam induced current analysis, Auger voltage contrast) that extend the microanalytical features to include electrical, metallurgical, and electronic determinations. (5) Commercial instrumentation has evolved extremely rapidly, and can now provide computer-controlled data acquisition with a minimum expectation of operator-induced analysis error. The evolution of the instrumentation has been a major factor in establishing AES as a routine and user-friendly analytical tool.

AES is an analytical technique with firm foundations and numerous applications in the materials-related industries, ranging from metallurgy to integrated electronics. This chapter includes applications of AES to several electronic materials investigations, and the examples are illustrative of the capabilities of this analysis technique to provide information on required chemical and compositional properties. The principles, experimental procedures, applications, strengths, and limitations are covered to provide an *introduction* to this spectroscopy, and the reader is referred to the Bibliography for more in-depth treatments.

4.2. Fundamental Principles

Several basic physical aspects of AES provide both advantages and limitations for actual materials and component analysis. Some of the inherent properties are summarized in Table 4.1. A fundamental understanding of the process must be gained before a useful and correct application is made to the solution of a specific problem, and the major principles underlying AES are discussed in this section.

4.2.1. The Auger Process

AES involves the energy analysis of characteristic electrons, called Auger electrons (Auger, 1925), that are emitted from a solid via a radiationless process as a result of the ionization of the atomic core levels by an incident, energetic electron (or other ionizing) beam. Three fundamental mechanisms are involved in AES:

- The ionization of the core level of the sample atoms under investigation by the incident electron beam
- The radiationless Auger transition

- The detection of the Auger electrons that escape from the sample into the vacuum

The Auger process can be easily visualized by considering the energy-level diagram of Fig. 4.1. The incident, high-energy (2–10 keV) electron beam ionizes a core level, creating a vacancy in the inner level (e.g., the K level in Fig. 4.1). This vacancy is filled immediately by another electron from a higher energy level, such as one of the L levels. One of two processes can now occur. The energy associated with the transition $(L \rightarrow K)$ in Fig. 4.1 can be released in the form of radiation. If the energy difference $E_K - E_L$ is large enough, characteristic x rays are emitted, providing the basis for x-ray fluorescence (Muller and Kiel, 1972). However, another event is probable. The energy can be transferred to another electron in the same level (or in a level close to it), and this electron can gain enough energy to escape from the atom if the difference $E_K - E_L$ is sufficiently large. The escaping electron, produced by the radiationless energy transfer, is call an Auger electron. Its energy, for the example of Fig. 4.1, is approximately equal to $(E_K - E_L) - (E - q\phi)$, where ϕ is the work function of the analyzer material [e.g., 3–4 eV for typical analyzers) (Joshi et al., 1975)].

The process is usually written in terms of the three levels involved, ABC. With this notation, A is the level in which the vacancy is created by the incident electron beam; B, the level from which the electron which fills the vacancy originates; and C, the level from which the Auger electron is emitted. Therefore, the process in Fig. 4.1 is the $KL_1L_{2,3}$ transition. The Auger electron energies are characteristic of the sample material, and independent of the incident electron energy. This kinetic energy may be estimated for the transition ABC as

$$E_{ABC} = [E_A(Z) - E_B(Z)] - [E_C(Z + \zeta) - q\phi] \qquad (4.1)$$

where Z is the atomic number of the atom involved. The term ζ is included since the energy of the final double ionized state is somewhat larger than the sum of energies for individual ionization of the same levels. The value for ζ has been determined experimentally, and lies between 0.5 and 0.75 (Carlson, 1975). More exact expressions for the Auger kinetic energy have been considered in the literature (Bergstrom and Nordlind, 1965), but Eq. (4.1) provides sufficient understanding of the fundamental physics. It can be seen that the Auger electron energies are characteristic of the sample material. Although incident electron beams are usually used in AES, any ionizing source can give rise to Auger electrons. Auger transitions are found in XPS spectra and can be usefully generated with ion beams (Hiraki et al., 1979).

The Auger electrons emitted from a sample can be monitored by placing a suitable detector near the sample surface. The total distribution function,

Table 4.1. Comparative Strengths and Limitations of Selected
Surface Analysis Methods

Auger electron microscopy (AES)
Strengths
- Elemental sensitivity and range (Li and above)
- Surface sensitivity (approximately 3–60 Å)
- Rapid analysis (seconds to minutes)
- Minimum spectral interferences among elements for most material systems
- Semiquantitative, with concentration proportional to signal intensity (simple analytical exercise)
- Sensitivity factors of elements have relatively small range (within one order of magnitude of one another)
- Excellent spatial resolution
- High spatial mapping capabilities
- Two- and three-dimensional analysis (depth profiling, volume mapping)
- Operator-friendly
- Very good reproducibility of analysis
- Metal, semiconductor analysis; some insulator
- Literature-rich background and data base
- Excellent commercial equipment and assistance; user-friendly software and handbooks

Limitations
- Charging (especially with insulators) for some surfaces
- Beam damage from energetic electron beam
- Chemical state information influenced/ controlled by beam artifacts
- Requires excellent, controlled environment (UHV)
- Not sensitive to trace or low-level concentrations (less than 0.1%)
- H and He excluded from analysis
- Spectral interferences for complex systems and low concentrations
- Beam spreading in sample limits ultimate spatial resolutions
- Slow mapping due to high background signals

Ion scattering spectroscopy (ISS)
Strengths
- Excellent surface sensitivity
- Nondestructive analysis
- Surface structure information on single crystals

Limitations
- Small user base; expertise required
- Subsurface analysis (depth profiles) extremely slow
- Small commercial equipment base and support
- Poor spatial resolution

X-ray photoelectron spectroscopy (XPS)
Strengths
- Direct generation of analyzed (information containing) species (photoelectron)
- Excellent, reproducible and predictable chemical information from chemical shifts
- Minimal sample charging
- Minimal beam damage from very penetrating input beam
- Very good quantitative analysis
- Rapid analysis (1–10 min typical)
- Very good reproducibility of analysis
- Sensitivity factors of elements have relatively small range (within one order of magnitude of one another)
- Literature-rich background and data base
- Excellent commercial equipment and assistance; user-friendly software and handbooks

Limitations
- Poor spatial and depth resolution
- Difficult depth profiling due to beam sizes and noise generation
- Interpretation relatively more difficult with some technique artifacts (shape-up lines, plasmon losses, etc.)

Table 4.1. (*Continued*)

Secondary ion mass spectrometry (SIMS)	• Wide range of commercial equipment with wide range of cost
Strengths	
• Direct detection of hydrogen	**Limitations**
• Isotope detection	• Destructive analysis method
• High sensitivity to low concentrations	• Difficult quantification, especially complex systems
• Chemical information from molecular fragments	• Detection sensitivity factors cover wide range (six orders of magnitude)
• Rapid collection of mass data	• Mass interferences make identification difficult
• Very good depth resolution	
• Inherent depth profiling	• High mass resolution requires expensive equipment
• Analysis of fragile materials	
• Charging problems minimal during rapid sputtering	• Complex experimental apparatus; less user-friendly

$N(E) \cdot E$ or $N(E)$ versus E, provides characteristic peaks for the various Auger transitions predicted by Eq. (4.1). This is represented in Fig. 4.2 for superconducting YBaCuO (Nelson *et al.,* 1988). The transitions are more evident in the differentiated distribution function, $dN(E)/dE$ versus E, also shown in this figure. Either of these representations provides identification "fingerprints" of the composition of the sample being investigated.

First principles calculations of the Auger energies are very difficult since the complex analytical treatments must include spin–orbit coupling, transition probabilities, relaxation energies, Coulomb and exchange electron interactions, electron configurations, ionization states, spectroscopic and other quantum-

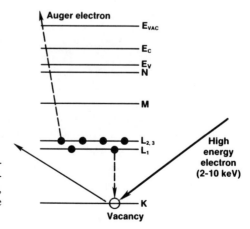

Figure 4.1. Representation of the fundamental process involved in the generation of Auger electrons by an incident, high-energy electron bombardment. The process shown is the $KL_1L_{2,3}$ transition.

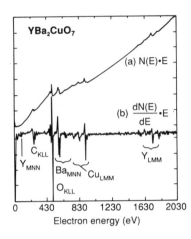

Figure 4.2. AES spectra for $YBa_2Cu_3O_7$: (a) Direct or $N(E)$ versus E spectrum obtained by pulse-counting technique; (b) differentiated or $dN(E)/dE$ versus E spectrum obtained by lock-in amplifier detection.

mechanical terms (Shirley, 1973). The dominant Auger transitions involve electrons in neighboring orbitals such as the *KLL, LMM, MNN, NOO,* and *OOO* families. These are the principal Auger electron energies for the direct and differentiated spectra, respectively (Davis *et al.,* 1976; Sekine *et al.,* 1982). Examination of published principal transition charts shows that Auger transitions usually involve electron binding energies less than 2.4 eV, a region in which the probability of finding an x ray is negligible. Since the Auger process involves three electrons and at least two energy levels, the direct detection of either hydrogen or helium is not possible. The dominant Auger energy transitions can be categorized by the atomic number of the atom analyzed:

$$3 < Z < 14 \quad KLL \text{ transitions}$$

$$14 < Z < 40 \quad LMM \text{ transitions}$$

$$40 < Z < 82 \quad MNN \text{ transitions}$$

$$82 < Z \quad NOO \text{ transitions}$$

Note that for transitions which involve outermost or valence bands for the emission of the Auger electrons, the notation "V" is often used. For example, an NOO transition is usually written NVV, designating that valence-band transitions are involved. For $N(E)$ spectra, the Auger kinetic energy position is indicated by the peak in the distribution. For differentiated spectra, the Auger kinetic energy position is indicated by the maximum negative excursion of the $dN(E)/dE$ peak. Therefore, the energy location defining each of these distributions is slightly different.

4.2.2. Auger Electron Escape Depths

The Auger electrons created by the incident energetic electron beam must escape from the sample in order to be detected and analyzed. The physics of the escape of the Auger electron is the basis for the surface sensitivity of AES (as well as the other electron spectroscopies). After the Auger electrons escape from their host atoms, they undergo energy losses through collisions, plasmon losses, core excitations, and interband transitions. Thus, the density of emitted Auger electrons depends on the mean free path of those carriers in the solid. Typically, mean free paths are small in semiconductors and metals, ranging from 4 to 50 Å (Seah and Dench, 1979; Penn, 1976; Fuggle, 1981). The escape depth of Auger electrons is not usually the measured mean free path for a bulk sample, since the Auger electrons are influenced significantly by the surface of the material—a region that is usually *electronically and chemically* distinct from the bulk. Thus, surface composition, lattice geometry, surface roughness and incident angle of the primary beam can profoundly affect escape depth.

Figure 4.3 shows the Auger electron escape depths for numerous materials. These parameters are generally determined empirically by depositing atomically-uniform films on metal substrates, simultaneously monitoring the Auger peak heights, or performing angular-resolved studies on layers of known thickness (Idzerda *et al.*, 1989; Tanuma *et al.*, 1990). The intensities usually decay exponentially, according to

$$N(z) = N_0 \exp(-z/\lambda) \qquad (4.2)$$

where λ is the Auger electron escape depth, $N(z)$ is the density of Auger electrons originating at a depth z, and N_0 is the electron surface density ($z = 0$). Thus, the Auger peaks decay exponentially with overlayer coverage, as predicted by the escape probability expressed in Eq. (4.2).

Escape depth is a perturbation of the value for the inelastic mean free path (IMFP) of an Auger electron in a host solid. The IMFP is primarily governed by valence-band excitations (about 10% of the total inelastic cross section involves core electrons). The primary beam diameter and escape depth determine the analysis volume. The number of atoms-analyzed is determined when atomic density is considered. Penn (1976) has formulated the dependence for λ upon Auger electron energy, E, given by

$$\lambda = \frac{E}{a[\ln E - b]} \qquad (4.3)$$

where a and b are constants that depend on the material being analyzed, and have been tabulated for the elements. Examination of Fig. 4.3a,b shows the

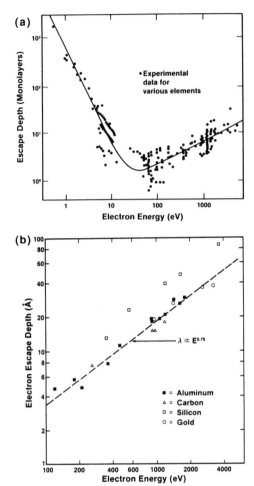

Figure 4.3. The dependence of the escape depth of the Auger electrons on electron energy: (a) experimental data compared with the calculations (solid line) of Seah and Dench (1979); (b) the functional $E^{0.75}$ dependence for the high-kinetic-energy regime. (Adapted with permission from Davis *et al.*, 1976.)

validity of this expression, but only for *E greater* than 100 eV. The fact that the escape depth is greater for Auger electrons having higher energies is logical. However, what happens in the lower Auger energy range departs from what is expected.

The understanding of the escape depth increasing with decreasing Auger energy in the region below ~100 eV is based on quantum mechanics, the electron physics in this region, and the difference in surface electronic structure. The probability that an event will occur, such as a collision associated with the limitation of the IMFP of the electron, is provided by the ionization cross section $\sigma_x(E)$. The larger the cross section, the more likely the event is. The early calculations of $\sigma_x(E)$ were provided by Bishop and Rivière (1969), and were based on previous x-ray microanalysis calculations by Worthington and Tomlin (1956). Their results showed

$$\sigma_x(E) = 1.3 \times 10^{-13}\ bC/E^2 \qquad (4.4)$$

where b is a function of the shell from which the Auger electron originates (e.g., 0.35 for the K and 0.25 for the L level) and C is a constant that depends on the incident (ionizing) beam energy. Gryzinski later modified this expression, recognizing that there is an additional contribution from backscattered electrons (Gryzinski, 1965). Therefore, the total ionization at a specific level is given by

$$\sigma(E) = \sigma_x(E)[1 + r_M(E, E_p, \theta)] \qquad (4.5)$$

where r_M is the backscattering term, which depends on the matrix in which the atoms are embedded, and θ is the angle the incoming electron beam makes to the surface. The evaluations of r_M involve both empirical approaches (such as Reuter, 1972, and Szajam and Leckey, 1975) and Monte Carlo calculations (e.g., Ichimura and Shimizu, 1981, 1983). The backscattering term is found to increase for lower E (Fig. 4.4) and with increasing atomic number (Fig. 4.5). The Gryzinski cross section is shown in Fig. 4.6. The ionized core level decays with a probability of Auger electron emission through the given transition. The probability that the created Auger electron will escape is inversely proportional to $\sigma(E)$. Therefore, the enhancement of λ for lower E is the result of the quantum mechanical effects which include backscattering.

The Auger electrons that do not escape from the sample contribute to the almost uniform background noise upon which the AES peaks are superimposed. Since the phonon losses are really very small with respect to the natural width of the Auger energy peaks, they do not significantly affect the Auger electron yield. Thus, from this consideration, low-temperature mea-

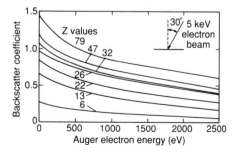

Figure 4.4 The backscatter coefficient as a function of electron energy for various atomic number values, for 5-keV incident electron beam, 30° from normal. (Adapted with permission from Ichimura and Shimizu, 1983.)

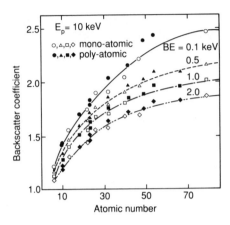

Figure 4.5. The backscatter coefficient as a function of the atomic number for various values of Auger electron energy. (Adapted with permission from Ichimura and Shimizu, 1981.)

surements provide no advantage. The use of low temperatures in Auger measurements is primarily to minimize the alteration of the materials being analyzed [e.g., to avoid the loss or evaporation of elemental compounds of high-vapor-pressure materials such as HgCdTe (Massopust *et al.*, 1985)].

The variation in escape depth with Auger electron energy, shown for several elements in Fig. 4.3a,b, can be used for analytical advantage. For example, the escape depth of the Si_{LMM} (92 eV) electrons is about 20% of that for the higher-energy (1609 eV) Si_{KLL} transition (Holloway, 1980; Ibach *et al.*, 1982). Thus, the monitoring of the *LMM* transition provides enhanced surface sensitivity, since the electrons originate from nearer the surface. The effect can be noted in the apparent interface width difference in the SiO_x/Si structure in Fig. 4.7.

4.2.3. Peak Shape, Fine Structure, and Matrix Effects

There are several major contributions to the peak shape and peak energy position of an AES transition (Finello and Marcus, 1977). In general, there

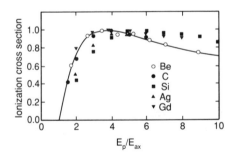

Figure 4.6. The dependence of the cross section on energy for an ionizing level, *x*. Experimental values for various elements are included. (Adapted with permission from Gryzinski, 1965.)

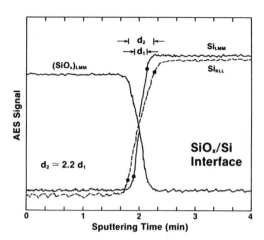

Figure 4.7. AES depth-compositional profile of SiO$_x$/Si interface, indicating difference in escape depths using the Si$_{KLL}$ and Si$_{LMM}$ transitions.

are three origins of the fine structure of Auger transitions, attributed to chemical and/or final-state effects: (1) the splitting of the levels, (2) chemical effects, and (3) losses. Fine structure is observed very often in both metals and nonmetals. The natural linewidth of an Auger peak is several electron volts, because of the short initial ionization state lifetime of about 10^{-16} s. Since electron emission in the Auger process always involves the binding energy of core levels and usually involves valence (outer shell) electrons, the line shape and the probable peak energy position are dominated by the chemical environment or matrix. Because the fine structure in AES transitions involves energy losses or transitions with lower energy, the effects are usually observed on the lower-energy side of the major Auger peaks.

The transition metal series Ru through Sn best illustrates fine structure relating to level splitting. These metals have a characteristic sharp, closely spaced doublet, associated with the transitions $M_{4,5}N_{4,5}N_{4,5}$. The energy separation (i.e., 5 eV for Ag, 6 eV for In) between the M_4 and M_5 levels is large enough, in most instances, to provide for resolution of the doublet (Briggs and Seah, 1983; Wagner, 1983). This fine structure is illustrated in the series of these transition metals presented in Fig. 4.8.

Peak width and shape are influenced by the density of states in the valence band, where a high density of states at a particular energy level will enhance the Auger yield from that energy level (Finello and Marcus, 1977). This will cause fine structure within the side Auger transition. The shape of an Auger electron transmission line is not a direct measure of the density of states, but reflects the transition probability density. Various loss mechanisms that affect the electron escaping the matrix (e.g., plasmon losses, interband excitations, core-level ionization) can provide structure in the low-energy side of the Auger

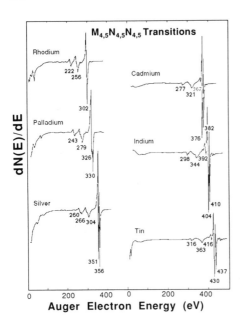

Figure 4.8. AES spectra for transition metal series showing $M_{4,5}N_{4,5}N_{4,5}$ doublet formation. The doublet is a result of the separation between the M_4 and M_5 levels, having sufficient difference in energy to be resolved by the spectrometer.

transition peak. These spectral differences make it possible to distinguish between atoms that are absorbed on the surface from those that are bound in the bulk. Plasmons are created by energetic electrons losing discrete amounts of energy to excite collected oscillations of valence electrons under the influence of the positive cores. Thus, because plasmon energy loss is specific or characteristic to a certain solid, chemical changes, such as oxidation, can be noted by differences in the fine structure of certain peaks. An electron that has given up an amount of energy equal to one of these characteristic energies, in the course of excitation, is said to have suffered plasmon loss. Within the bulk, the process involves bulk plasmons. At the surface, the regular atomic lattice of the solid terminates, and the conditions for setting up oscillations like bulk plasmons do not exist. Instead, a localized type of collective oscillation can be excited, and the surface plasmon has a frequency that is less than that of the bulk (i.e., it would have an energy closer to that of the primary Auger transition energy). The fundamental or first plasmon loss will always be visible. Reducing the incoming beam energy to 50–500 eV (low penetration) is usually required to observe the plasmon peaks. The surface loss process will dominate, and such is the case shown for the aluminum transitions (Massignon *et al.,* 1980). Depending on the material and experimental conditions, several multiple plasmon losses of decreasing intensity, but regularly located in energy position, are visible. Such is the case for the magnesium transition observed for a very clean surface.

The intensity of AES transitions and fine structure have been shown to be a function of the crystalline nature of the sample. Three mechanisms affect the nature of the AES signal:

- Density of atomic planes and their composition, parallel to the analysis surface
- Anisotropic emission of Auger electrons
- Channeling effects caused by diffraction of the incoming electron beam

The density/composition effects are observed most often in layered (graphitelike) compounds, in which the composition of the atomic planes alternates between sequential layers. Anisotropic emission effects involve the Auger emission process or diffraction of the Auger electrons as they are emitted. The effect is observed most often for low-energy transitions. If the analyzer has a high acceptance angle, these anisotropic effects are not significant. Angular-resolved techniques or use of movable analyzers are effective in measuring these distributions. Bishop *et al.* (1984) reported these effects on the Al-*KLL* and *LVV* transitions for (110) *ordered* surfaces with controlled oxygen coverages. In normal AES analyses, these effects do not overwhelm the spectra. Samples are either not usually sufficiently ordered, clean, or the experimental technique is not aimed at specific measurement of such low-order effects.

4.3. Experimental Methods and Instrumentation

Four basic components are required for AES: (1) a vacuum system and analysis chamber; (2) an electron gun (or other ionizing source) for excitation; (3) an electron spectrometer for energy analysis of the emitted electrons; and (4) some mechanism for data acquisition and analysis. Because AES is surface sensitive, an ultrahigh vacuum (UHV) is required to minimize sample surface contamination from the environment. Such contamination (e.g., oxygen, carbon) can result from adsorption or possibly from electron-beam-induced processes. The cost of providing clean ($\sim 10^{-10}$ torr) vacuum using ion or other oil-less pumps is certainly not prohibitive, and such vacuums are extremely reliable. The electron sources have evolved from simple tungsten filaments to the currently used high-brightness emitters. The typical electron source has incorporated a lanthanum hexaboride (LaB_6) emitter that can support relatively high current densities at very low beam sizes. Recently, electron guns with long-lifetime, Schottky thermal field emitters have been introduced (Perkin–Elmer, 1983, 1984, 1990). These can be coaxially mounted in a cylindrical mirror analyzer (CMA), and are capable of providing 1-nA beam currents at beam diameters below 150 Å at 20 kV and better than 250 Å at 10 kV. These specifications have obvious benefits for obtaining ultimate AES microanalysis conditions. The type, configuration, and sophistication of any of the basic AES system elements vary. Systems almost always support ion guns for sputter-etching and cleaning, and sample manipulation components for precisely locating the analysis area. Additional, sample introduction systems that do not disrupt the analysis chamber vacuum environment are now routine. Sample heating/cooling, fracture/breaking, electrical and optical biasing, surface treatment/deposition, and stressing attachments are also available for the *in situ* analysis under a variety of conditions. The availability of quality commercial equipment has brought AES to its current status as a reliable and useful analysis technique that can be operating in a laboratory within days after delivery. This section provides a generic summary of the major components, and is not meant to be an exhaustive treatment of the hardware. The reader can consult manufacturers for more details.

4.3.1. Cylindrical Mirror Analyzer

One basic configuration for the AES system is shown in Fig. 4.9. The spectrometer used in this application is a CMA, and the electron gun is coaxial with the CMA. This popular configuration has the advantage of higher transmission. However, other configurations and analyzer types are used, each with its own advantages and applications. Of importance to any detection system is the energy resolution of the analyzer. This is defined in terms of the pass energy, ΔE, of the analyzer and the range of electron energies that are transmitted, E, giving (Perkin–Elmer, 1983, 1984, 1990)

$$R = \Delta E/E \qquad (4.6)$$

For commercial CMAs, R is typically better than 0.005. The types and properties of a variety of electron analyzers used in AES are presented in Table 4.2. Note that in order to increase the energy resolution of a CMA, another configuration is possible—the double-pass CMA. This analyzer, shown in Fig. 4.9, provides for two passes of the Auger electrons before they are counted by the analyzer. This improves the pass energy factor for the analyzer, and the typical resolution is better than 0.002. However, the transmission of the CMA (and the ability to resolve signals from low-concentration elements) is decreased in this double-pass configuration. In most applications and for routine analysis, the resolution provided by the single-pass CMA is sufficient, especially since the higher transmission is desirable.

4.3.2. Detection Methods

An important and sometimes confusing aspect of the instrumentation is the type of detection used in obtaining the Auger signal. Four basic types (and the output) are:

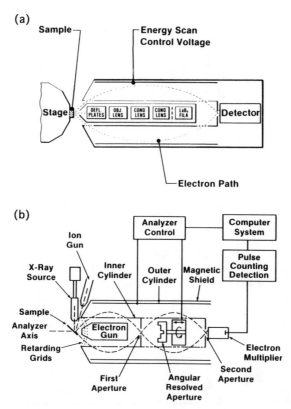

Figure 4.9. Schematic representations of cylindrical mirror analyzers: (a) single-pass; (b) double-pass configurations. (Adapted with permission from Perkin–Elmer, Physical Electronic Division.)

- Pulse-counting, PC $[N(E) \cdot E]$ (Perkin–Elmer, 1983, 1984, 1990)
- Beam-blanking modulation, BBM $[N(E)]$ (Sekine *et al.*, 1982)
- Voltage-to-frequency conversion, V/f $[N(E) \cdot E]$ (Perkin–Elmer, 1983, 1984, 1990)
- Lock-in amplifier $[dN(E)/dE]$ (Joshi *et al.*, 1975)

Each of these detection schemes can provide particular information. Mathematical differentiation or integration provides the mechanism of presenting the data in an identical format. However, the methods do differ, and the data (i.e., magnitude, peak shape) can differ subtly or significantly among them. Care must be exercised in the application of these techniques, and the analyst must have knowledge of the proper parameters to ensure correct results.

The first Auger spectra were reported by attempting to resolve the total number of Auger electrons at a given kinetic energy [i.e., the $N(E)$ versus E

Table 4.2. Summary and Comparison of Electron Analyzers

	Concentric hemispherical analyzer (CHA)	Retarding field analyzer (RFA)	Spherical sector analyzer (SSA)	Cylindrical mirror analyzer (CMA)	Double-pass cylindrical mirror analyzer (DCMA)
Lens	Retarding	Retarding	Retarding field	Retarding field	Retarding field
Scan mode	Retarding and hemisphere potential	Retarding potential	Retarding potential	Mirror potential	Mirror potential
Approximate energy resolution $\Delta E/E = Aw/2 + B(\Delta\alpha)^n/4$	0.2–2.0% of analyzer pass energy	0.5–1.0%	0.5–1.0%	0.3–5.0% of analyzer pass energy	0.15–3.0% of analyzer pass energy
Entrance angle	180°			42.3°	42.3°
A	$1/R_o$			$0.36/r_1$	$0.36/r_1$
B	1			5.55	5.55
n	2			3	3
Transmission	<10%	10%		10–12%	6–9%
Comments	Highest resolution; primarily for XPS; lower transmission than for CMS	Used in LEED; not commonly used in AES	XPS primarily; geometry used in SIMS	Most common in AES; adequate energy resolution, highest transmission with coaxial electron source	High energy resolution for XPS; good for EELS; lower transmission than for single-pass CMA

spectrum]. Without the availability of background subtraction methods and computer data handling, the ability to resolve an AES peak from the background and noise is difficult. Of course, the introduction of the methods to obtain the derivative spectra [i.e., $dN(E)/dE$ versus E] made the identification of the Auger transitions easier and more exact. The detection of Auger electrons in CMAs has evolved considerably since the mid-1970s. Currently, one or more of the four detection schemes, or modifications of them, are utilized.

Until the mid-1980s, the most common technique was *lock-in amplifier detection.* This method produces the first derivative of the energy spectrum electronically. A small AC modulation signal is superimposed on the Auger energy-discriminating DC voltage applied to the outer cylinder of the CMA. The modulated AES signal is compared to the AC reference, and the lock-in amplifier is used for synchronous detection of the output of an electron multiplier. The output from the lock-in amplifier produces the $dN(E)/dE$ signal directly. Signal-to-noise (S/N) characteristics are poor for lock-in amplification of low signal levels. These can be improved by using an isolation amplifier for pulse counting (PC). In PC, the AES signal from the electron multiplier is capacitively coupled to an amplifier discriminator. The output of the amplifier is an ECL pulse train with frequency proportional to signal intensity. Thus, the data acquired are the actual number of Auger electrons at a given energy. As a result of the development of algorithms with varying effects on energy resolution, lower-cost direct $N(E)$ detection schemes have been developed for high-level signal amplifications. These methods replace the costly lock-in amplifier with less expensive electronics. One such method is the voltage-to-frequency (V/f) amplifier method. The amplifier, actually a current-to-frequency amplifier, is optically coupled to the output of the electron multiplier, and the observed signal provides the desired $N(E)$ versus E information. A fourth detection scheme, beam-brightness modulation (BBM), was introduced to provide $N(E)$ data for all signal levels. The primary electron beam is blanked electronically at a given frequency (e.g., 10 Hz). The resultant AES signal is processed by a lock-in amplifier with a reference signal at the blanking modulation frequency. This approach has advantages for acquiring data at low primary beam currents during simultaneous ion sputtering. The appreciable ion-induced signal is not measured. One disadvantage is that under high-spatial-resolution conditions, the primary beam current is halved for a given probe diameter. Comparative electronic representations of these detection methods have been reviewed previously (Kazmerski, 1992).

Each of these detection schemes has its advantages and limitations. All can be implemented on a given AES system (either by the manufacturer or by the user), allowing the exploitation of the best method for the specific application. Recently, commercial microanalysis systems have been introduced that utilize high-brightness, Schottky thermal field emitter sources with multichannel detector for nanoprobe analysis. These systems provide for the highest spatial resolution with corresponding high current densities. The benefits include the ability to detect and analyze small features and to map compositions from nanoregions with excellent S/N and reproducibility.

4.3.3. Signal-to-Noise

An important figure of merit for AES instrumentation is the S/N ratio. This parameter indicates the ability to resolve a signal under given operating conditions. Higher S/N requires more time for data acquisition. The S/N is often used to evaluate the performance of AES instruments in the purchasing stage. The periodic monitoring of this important parameter can also indicate instrument problems such as changes or degradation in the electron multiplier, failures in the electronics, or contamination of the analyzer.

Several methods to evaluate the S/N exist, but consistency in the approach is important for proper monitoring or evaluation. It is common to use the Cu (914 eV) transition as a standard. In the direct $N(E)$ modes, the S/N can be evaluated from (Seah et $al.,$ 1983)

$$S/N = \frac{\text{peak signal(counts)} - \text{background signal(counts)}}{(\text{background signal})^{1/2}} \qquad (4.7)$$

The determination of the inputs to this equation involves the alignment of the sample, and cleaning of the surface (e.g., by sputter-etching or cleaving in $vacuo$). The voltage and current are set to the conditions required. The signal at peak is compared against the measured background at the higher energy side of the peak—usually some 40 eV from the peak signal, at a point at which the observed signal is constant. For dN/dE detection systems, an analogous relationship to Eq. (4.7) can be used, with the peak-to-peak intensities substituted for the pulse count currents. In any case, the S/N is a function of the detection system as well as the current (current density) regime used.

4.3.4. Electron Beam Effects

Various effects of electron, photon, and ion beams on the surfaces of solids have been reported in the literature. Specifically, electron and photon beam process have included surface oxidation (Madey and Yeats, 1971), dissociation (Madden and Ertl, 1973), gas adsorption (Brillson, 1976), contaminant or surface species desorption (Menzel, 1982), adsorbate dissociation (Coad et $al.,$ 1970), and atomic and/or ionic migration and evaporation (Chou et $al.,$ 1973). The mechanisms underlying these interactions vary from thermal and kinetic, to more complicated quantum-mechanical interactions (Shapira and Friedenberg, 1980). The literature contains examples of electron beam–semiconductor interactions, ranging from elemental semiconductors such as silicon (Knotek and Houston, 1982), through binaries such as GaAs (Corrallo *et al.*, 1986), to ternary and multinary compounds (Kazmerski, 1988). The potential changes in the surface chemistry by oxidation, desorption, and species removal are a concern for the utilization of analysis techniques based on energetic beams since it is possible for them to provide artifactual information instead of providing the actual mechanism or information involved.

The relationship between electron beam exposure and possible alteration of surface chemistry has been demonstrated by both AES and electron-beam-induced current investigations. Both the common beam-enhanced oxidation of silicon and the quite opposite beam desorption of oxygen from silicon surfaces have been reported, and which occurs depends critically on the characteristics of the incident electron beam. The question of the effect(s) of exposure to an electron beam has been critical to the interpretation of confusing EBIC results for thin-film, polycrystalline $CdS/CuInSe_2$ solar cells (Matson *et al.*, 1986, 1989; Noufi *et al.*, 1986; Kazmerski and Nelson, 1990). It was observed that the junction location indicated by the spatial response of the EBIC signal changed or shifted from the first electron beam sweep to subsequent sweeps. It was proposed that the electron beam might dislodge inherent oxygen or an oxide which resided on grain boundary or internal surfaces, or perhaps within the grain itself, and thus alter the measured electrical response. That is, the oxygen was desorbed during the electron beam interaction, similar to some particular cases for silicon. The ability to confirm this fact had been limited by the fact that any of the existing oxygen at the internal interfaces or within the bulk of the grains is at a concentration below the detectability of AES.

The chemistry of the surface of $CuInSe_2$ can change with exposure to an electron beam (Kazmerski, 1988). Similar results have been reported for the surface produced by fracturing *in vacuo* or by mild sputter cleaning the surface prior to analysis to remove the inherent 20- to 30-Å oxide and carbon layer. Even with the background pressure in the 10^{-10} torr range, sufficient oxygen is present in the system from the CO and/or originating from the ionization filament to serve as a source. The rate at which the oxygen layer grows can be inhibited by turning off the ionization gauge during analysis. The critical influence of the electron beam is indicated in Fig. 4.10, which presents similar data during which the electron beam is incident on the sample surface only during the AES analysis (approximately 0.78 min for each sweep). Initially (less than 30-s exposure time), the oxygen is physisorbed on the surface as established through the position of the O-$1s$ XPS peak. Thereafter, the oxygen becomes bonded to the indium and/or copper. The O_{KLL} intensity after 30

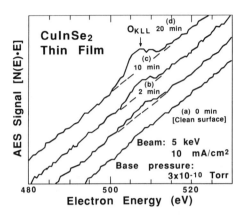

Figure 4.10. Evolution of the AES O_{KLL} signal from a CuInSe$_2$ polycrystalline thin film surface exposed to ultrahigh vacuum ambient with the electron beam incident only during the analysis (approximately 0.75 s per sweep). (Adapted with permission from Kazmerski (1988).)

min in the vacuum system is about the same as the 2-min exposure in Fig. 4.10, consistent with the proposition that electron exposure time is the controlling parameter. During this process, the selenium AES intensity decreases, and the copper and indium intensities increase. However, a significant increase in the indium signal is not distinguishable until after about 4–5 min of electron beam exposure. From this time, the copper signal intensity diminishes under the electron beam, indicating that the indium and not the copper is diffusing out. During the exposure, the physically small oxygen is replacing the selenium atom, providing sufficient space for the outward movement of the indium under the thermal and/or electrical fields originating from the electron beam. The carbon concentrations within the beam-exposed area and outside it are small (less than 4 at. %) and identical, similar to the results for Si(111) surfaces. This indicates that O$_2$ might dominate this beam-induced oxidation rather than CO.

4.3.5. Sample-Related Analysis Effects

Several specimen-related effects provide limitations for reproducible and accurate AES analysis. Since this is a vacuum-confined technique, the first limitation is that the sample must be stable in that environment. This would exclude the direct analysis of liquids or materials that might sublime or evaporate at pressures ranging from 10^{-8} to 10^{-11} torr. It is extremely important to know as much about the sample as possible before introducing it into a clean analysis chamber, and certainly before exposing it to an energetic electron beam. More than a few chambers and analyzers have suffered contamination because the investigators were unaware of the volatility of Zn, Se, Te, or Hg, for instance, even at low concentrations in another material matrix. An example is that of Hg$_x$Cd$_{1-x}$Te having higher Hg ($x > 0.3$) concentrations for

IR-detector applications. Such analysis is frequently performed at low temperatures using sample specimen stages that are cooled to liquid nitrogen (or lower) temperatures to inhibit the loss of the volatile component.

Another troublesome and major limitation is that of the analysis of insulating (low-conductivity) specimens. The incoming electron beam (current) must be removed from the sample at a rate high enough to prevent the accumulation of charge. If the total current entering the sample is larger than the current leaving via secondary emission and backscattering, the sample will charge until the incoming and outgoing current fluxes balance. There are several experimental adjustments that can overcome charging problems and permit adequate analysis. The first method, and one that is used in a routine analysis mode perhaps unconsciously by most operators, is to change the angle of incidence of the incoming electron beam with respect to the sample surface. This increases the secondary emission, as indicated in Fig. 4.11 (Bishop, 1989), provided the surface is sufficiently flat. A second approach is also provided from the representation in Fig. 4.11. This predicts that the secondary electron yield can also be increased by using lower primary beam voltages. Since many insulating materials charge only a few volts positive before equilibrium is reached, an AES spectrum can be obtained under proper incoming E_p conditions. Sometimes the charging is only suppressed, and the spectrum looks normal except that the energy locations of the peaks are shifted. Although using lower E_p conditions is the first logical approach, a third method can also be used if the nonconductive layer is thin. By increasing the potential of the incoming beam to a higher level, the penetrating path provided by the incoming beam can produce a higher conductivity region in the layer, with the z-direction penetrating the entire insulator thickness (punch through). This can result in a conductive path through the layer, eliminating or minimizing the accumulation of charge. Certainly other methods for charge removal can be used, including coating the sample with a thin metallic or carbon layer, similar to that used by scanning electron microscopists, then opening

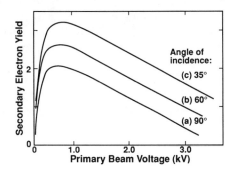

Figure 4.11. Dependence of the secondary electron yield on the primary beam energy for several angles of incidence.

a small hole by sputtering. The material in this aperture region is then analyzed, with the charge removed at the periphery of the mask.

Sample topography and form are important. Since both electron back-scattering and Auger electron emission are affected by the topology and roughness of the sample surface, such effects must be considered. A number of computer algorithms exist to correct for surface roughness (Wittmaack, 1990). Powders and granular materials can be prepared for analysis by compacting them or by embedding them in conducting matrices. One example is to embed the powder or granular pieces in a foil of indium. Manufacturers of commercial equipment not only have available a selection of sample holders to meet some of the difficult analysis conditions, but also have a great deal of background with analysis problems and often can offer immediate solutions.

4.4. Scanning Auger Electron Spectroscopy

One of the strengths and distinguishing features of AES compared with other electron and ion spectroscopies is that the excitation electron beam can be readily focused and scanned across a surface, controlled accurately and reproducibly by electrostatic and/or electromagnetic lenses. This permits the production of high-spatial-resolution, one- and two-dimensional AES compositional images of the surface under investigation. In essence, the AES imaging is accomplished by the same methods used in scanning electron microscopy, with some additional constraints on topography and surface contamination to ensure proper results. Because secondary electron images can be obtained in the scanning AES system, the compositional images can be compared directly and simultaneously with the surface topography or topology.

4.4.1. Resolution and Beam Conditions

Most AES users and clients demand or expect ultimate spatial and elemental (signal) resolution. Since there are distinct interrelationships between the experimental conditions (e.g., beam current, beam diameter, background noise, sample surface conditions) and the ability to resolve features or to detect elemental species, it is not possible to provide the highest spatial resolution for the lowest detectable concentrations. The relationships between some of these parameters are presented in Fig. 4.12, a nomograph for AES minimum detectability (Clough, 1984). The analysis time required as a function of primary beam current is shown for various minimum concentration parameters, which are direct functions of the S/N ratios and inversely proportional to the concentrations and the elemental sensitivity factors. The re-

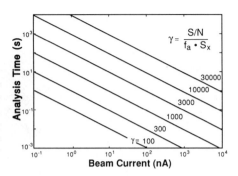

Figure 4.12. Nomograph for determining the minimum detectability for AES. The relationship among the normalized minimum detected concentration parameter, the analysis time per point, and the incident beam current is indicated. (Adapted with permission from Perkin–Elmer, 1984.)

lationship indicates that for a given beam current and a given parameter γ, a certain minimum analysis time is required. If the S/N is increased (and/or the concentration and/or the sensitivity factor of the element of interest decreases), the required analysis time deceases. Likewise, if a lower beam current is used, the minimum detectable concentration is less. Since the beam current density is inversely proportional to the area of the incoming beam, the minimum detectable concentration decreases as the beam size decreases, keeping the other parameters constant. Therefore, it is unlikely to provide maximum size and concentration resolution under the lowest current and smallest beam diameters. In turn, the resolution of the smallest features are limited somewhat by the attainable and used experimental conditions.

4.4.2. Limits of Spatial Resolution

The resolution of features is not determined solely or uniquely by the diameter of the incoming electron beam. The ability to resolve features is determined to an extent by incident probe area, but other probe characteristics (e.g., distribution, beam voltage, current, energy) are equally important. This is certainly evident in the tendency of the beam to spread on entering the sample, becoming pear-shaped at higher energies. Other experimental conditions also affect resolution, including the probe–sample geometries and feature size, the sample characteristics and topography, and instrument stability (i.e., mechanical vibrations as well as electronic drifting and variations).

AES may be able to detect $10^{20}/cm^3$ arsenic segregated at a 1000-Å-wide feature near a surface, but it would be impossible to provide this information on the $10^{18}/cm^3$ As concentrations because of limits in detectability. The limitations imposed by probe size to spatial resolution, and specifically to the relationships between the observed size and the actual feature dimensions when mapping impurities, are illustrated in the representation of Fig. 4.13. As the incident probe is scanned across the feature (e.g., a grain boundary),

Incident Probe: Integration Effects

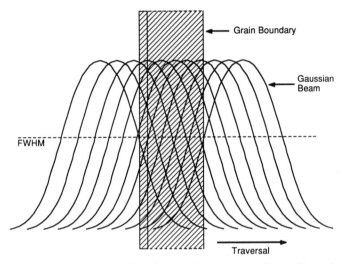

Figure 4.13. Analysis of input probe (Gaussian shape) transit across feature illustrating relative dimensional factors for image spatial resolution.

some integration of the detected species occurs, and the resulting imaged dimensions depend on the ensemble of conditions described in the previous paragraph. If the traversal of a Gaussian probe is exactly perpendicular to the feature, as shown in Fig. 4.13, the relative magnification factor can be predicted. In general, if the feature is relatively small with respect to the FWHM of the analysis beam, the beam dimension dominates the result. Conversely, if the beam size is much smaller than the feature, a more exact representation of the feature dimension is provided. The relationship among the probe and feature parameters is presented in Fig. 4.14, which gives both analytical and experimental data for grain boundaries in Si (Kazmerski, 1989a). The relative image magnification becomes significant at lower probe diameters and smaller feature widths. Spatial resolution is linked inherently to the experiment and analysis.

4.4.3. Mapping Techniques

The use of AES in the scanning mode provides the possibility of producing high-resolution images of surface compositions. Early Auger images were limited because of the relatively long times required to obtain the data and the noisy signal associated with the high background signals. The use of computers

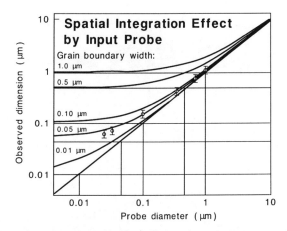

Figure 4.14. Image integration (magnification) effects by input probe, representing observed dimension as a function of the probe diameter for various feature widths. Experimental volume-indexed AES data are included for comparison with analytical calculations.

and image processing techniques has greatly enhanced the capabilities of obtaining high-quality images.

The general procedure in obtaining line maps and area maps of compositions on surfaces is to tune the analyzer for a selected energy range encompassing the peak (or peak–peak) energy of the element(s) of interest. Although such data specification and acquisition can be done in several manual modes, it is more common to utilize computer techniques for instrument control, data acquisition, and image processing. The output can be provided in gray-scale images or in color representations, with the color coding designated by a color palate selected by the operator. An example of a one-dimensional line scan is presented in Fig. 4.15. A two-dimensional area map is represented in Fig. 4.16.

4.4.4. Volume Mapping

The detection and mapping of subsurface features is often hampered by the inability of knowing exactly where these features are located. Fracturing the sample in order to expose the analysis regions has been shown to be difficult and uncertain. Conventional depth profiling using sputtering provides a method of detecting species essentially along a line from the surface through the feature. From a consideration of the analysis volume of the feature and size of the electron beam (Section 4.5.2), detection may be missed if the feature is small or the distribution of impurities is not uniform or of sufficient

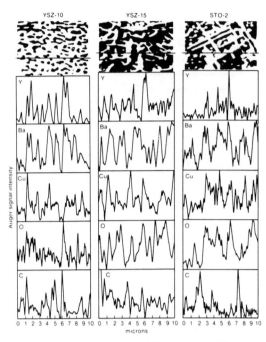

Figure 4.15. AES line scan (linear map) showing compositional differences on high-temperature superconductor surface. The levels of the Y, Ba, Cu, O, and C signals are compared with the features in the secondary electron images. Preferential accumulation of carbon at the grain boundaries is observed. The oxygen is not associated with the accumulated carbon, indicating free carbon and not carbonate. (Adapted with permission from Nelson *et al.,* 1988.)

concentration. Sputtering, of course, can be disruptive to such analysis (discussed later in this chapter), and is prone to produce artifacts for misinterpretation of the data (Kazmerski, 1988). Recently, three-dimensional or volume-mapping techniques have been introduced which provide for the optimization that subsurface compositional and chemical information can be obtained during a single analysis run (Kazmerski, 1989a). Volume mapping, schematically represented for a grain boundary in Fig. 4.17, provides the digital acquisition of AES (and SIMS) data throughout a selected microvolume, coding this information for elemental (or ionic) species, spatial location within the selected region, and concentration level. That is, this indexed information is obtained at each point in the chosen microvolume during a single depth-profile operation by control of the analyzer and beam, with spatial resolutions commensurate with the technique used. The best depth resolution is in the range 100–1000 Å. The computer software permits the operator to selectively display the information for any point, along any line, or on any plane within

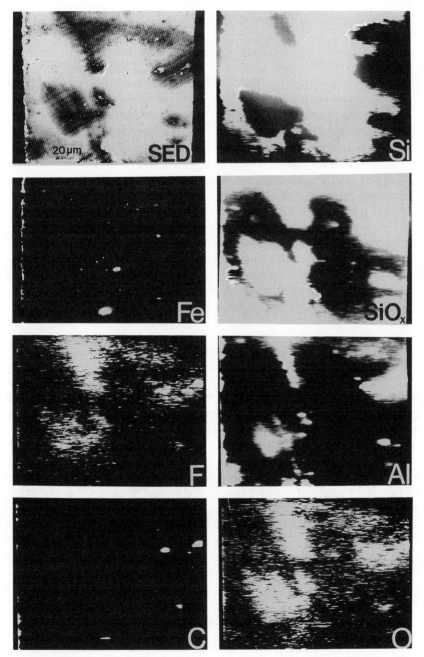

Figure 4.16. Scanning Auger map (two-dimensional) of impurities at Si grain boundary surface. The distribution of the elemental and chemical species is compared with the features of the secondary electron image.

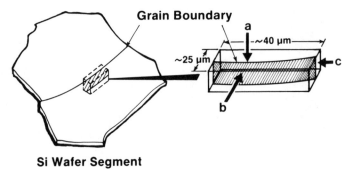

Figure 4.17. Schematic representation of volume-indexed or three-dimensional surface analysis mapping technique indicating three possible viewing directions.

the analysis microvolume after a single analytical operation. The technique minimizes the occurrences of missed or lost information encountered in the routine analysis or in fracturing or cleaving procedures. An example of a volume-indexed AES map is shown in Fig. 4.18 for a grain boundary in Si, having impurities segregated in that region.

4.5. Chemical Analysis

AES can be used to provide information about chemical environments and even bonding (Holloway, 1976, 1980). It has not been exploited in this area for two reasons: (1) the use of an energetic charged electron beam to provide the Auger excitation creates the real possibility of artifact generation, i.e., it is likely that the technique itself will be responsible for the chemistry observed, rather than the inherent property; and (2) the "competitive" techniques (e.g., XPS) are well established, involve input probes that are more gentle, can be used with a wider variety of chemical materials (e.g., polymers, insulators, powders), and provide a direct determination of the binding process. Although the energy shifts for a few AES cases can be many times larger than those observed in XPS, the uncertainties involved with damage or change make it far less utilized for this type of analysis. However, many specific chemical evaluations have been reported, and this area of AES remains one of interest and development.

4.5.1. Chemical Effects on Transitions

The basis for understanding the matrix and chemical effects on the resulting AES signals was provided in Section 4.3. In general, the valence-level

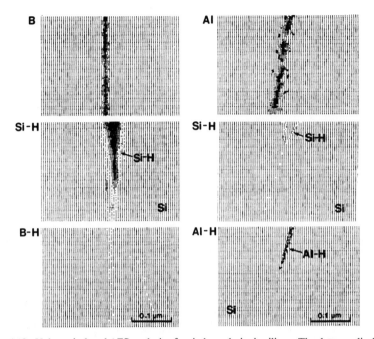

Figure 4.18. Volume-indexed AES analysis of grain boundaries in silicon. The data are displayed for the grain boundary edge (view c in Fig. 4.17), showing the segregation of arsenic to silicon grain boundaries as a function of the processing time at a set elevated temperature.

transitions show the major chemical effects as a result of their intrinsic relaxation, screening, and chemical shifts. Changes in the valence levels are reflected in the energy position and the shape of the AES transitions, and the majority of chemical shift examples in the literature involve transitions with at least one valence level and frequently two. A commonly cited and used example of chemical effects is for the Si_{LVV} low-energy transition presented in Fig. 4.19 (Hiraki *et al.*, 1979). This elemental $L_{2,3}VV$ peak from a clean surface occurs at 92 eV. The same valence-band transition in which oxygen is bonded to the Si (i.e., with the formation of SiO_2) is changed both in shape and in energy position (72 eV) as depicted in this same figure. Such large changes in energy are not common. Matrix effects are evident in other cases including those shown for the In_{MNN} in Fig. 4.20 (Kazmerski *et al.*, 1980). The usefulness of AES chemical identification is illustrated in Fig. 4.21, which shows the C_{KVV} transition for various carbon bonding mechanisms (Haas *et al.*, 1972). Such identifications are important in the identification layers used for hard and durable coatings (Meyer *et al.*, 1988). The differences between SiC, graphite, and diamond are important in this technology area, and AES provides the

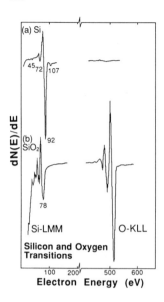

Figure 4.19. Chemical effects on Si-LVV Auger spectra for (a) elemental and (b) SiO$_2$ species. Oxygen-KLL spectra are included for reference.

mechanism to realize this identification to the first order on a high spatial resolution level.

An interesting example of how chemical effects can be used in AES is that relating to hydrogen. AES cannot detect hydrogen *directly*. However, its presence can be evaluated through the chemical effects on the shape and energy position of the Si-$L_{2,3}VV$ transition (Madden, 1981; Burnham et al., 1987), as presented in Fig. 4.22. The elemental spectrum is compared to one from hydrogenated amorphous Si. The elemental transition contains only the silicon p–p bonding. The hydrogen-treated case shows the addition Si-H s–p and p-p bonding. (A small, about 1.8 eV, shift in the elemental p–p signal is apparent.) Such data have been used to analyze hydrogenated a-Si, including information on hydrogen concentrations for 5–30 at. % levels, and on the presence of hydrogen at grain boundaries in polycrystalline Si material.

4.5.2. Auger Difference Spectroscopy

This method provides for the resolution of compositions by a careful analysis of the changes in the energy position and shape of a given peak, usually a valence transition (Coad et al., 1970). The method takes and records the differences between the measured transition and the corresponding elemental transition. The comparison (difference) between the control or elemental transition and the measured one assesses the presence and effects of the impurity and its bonding with respect to the host species. If the difference-

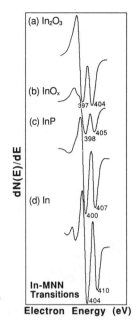

Figure 4.20. Chemical effects on In-*MNN* Auger spectra for (a) In_2O_3, (b) InO_x, (c) InP, and (d) elemental In.

result is a straight line, the detected species is elemental. If the difference-characteristic is structured, some other bonding or elemental transition is present. Such a difference spectrum is presented for the Si-$L_{2,3}VV$ transition in Fig. 4.23 for the hydrogenated and untreated amorphous Si, similar to those shown in Fig. 4.22. The structure in the difference spectrum for the hydrogenated Si enhances the transitions relating to s–p and p–p bonding. Such data can also be obtained by analytical deconvolution routines. The advantage of difference techniques is that the calculation times are significantly less complex and are more rapid. This can enhance the use of difference spectra to provide chemical maps, for example.

4.6. Sputtering and Depth Profiles

Ion beams are commonly used to clean surfaces prior to analysis and to provide depth-compositional data with surface analysis techniques (Morabito and Lewis, 1975). Ion beams can have major effects on material surfaces, even at very low incidence energies (Cheng *et al.*, 1987). Ion beam interactions can result in preferential removal of one or more species, implantation or knock-on, surface roughening, and redeposition of inherent elemental species (Wehner, 1975). Even hydrogen impinging at energies in the range 100–200

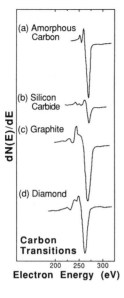

dN(E)/dE

(a) Amorphous
    Carbon

(b) Silicon
    Carbide

(c) Graphite

(d) Diamond

**Carbon
Transitions**

200   250   300
Electron Energy (eV)

Figure 4.21. Comparison of C-*KVV* spectra for several bonding configurations showing chemical effects on the valence transition: (a) amorphous carbon, (b) SiC, (c) graphite, and (d) diamond.

eV can cause significant damage to semiconductors to depths of micrometers. The reaction of the ion beam with a surface is one of the more difficult processes to control, and is one that is likely to cause severe changes in the chemistry of the region (Wehner, 1975). Some geometrical and etching-inherent problems associated with sputtering are represented in Fig. 4.24. Therefore, extreme care must be employed when utilizing any ion-beam techniques.

4.6.1. Ion Beam Effects and Depth Profiles

Certainly the most common application of sputter-etching is in combination with AES in order to provide information about composition as a

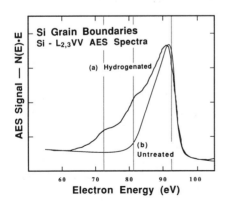

AES Signal — N(E)·E

Si Grain Boundaries
Si - $L_{2,3}VV$ AES Spectra

(a) Hydrogenated

(b)
Untreated

60   70   80   90   100
Electron Energy (eV)

Figure 4.22. Si-*LVV* Auger transition for (a) hydrogenated Si and (b) elemental Si.

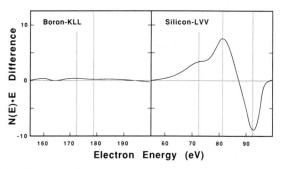

Figure 4.23. AES difference spectra for B-*KLL* and Si-*LVV* transitions.

function of depth or to investigate subsurface features. Depth-compositional profiles obtained by simultaneous AES analysis abound in the scientific literature. An example illustrating the significance of the operational parameters is provided in Fig. 4.25. CuInSe$_2$ is a ternary semiconductor that is sensitive to preferential removal of its elemental species. This has been shown to be a problem especially with the incident ion beam at an angle greater than 30°

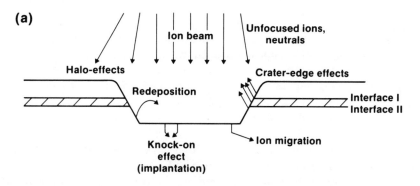

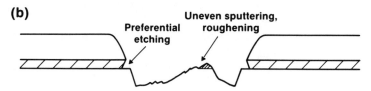

Figure 4.24. Indications of problems and artifacts encountered in use of sputter-etching with surface analysis techniques.

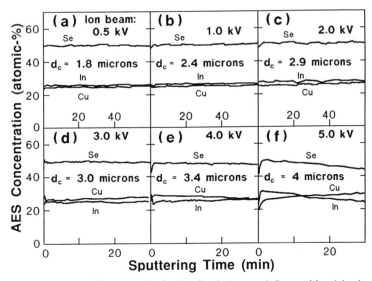

Figure 4.25. Ion beam artifact generation for CuInSe₂ single crystal. Compositional depth profiles are shown as a function of increasing ion beam energy from (a) through (f). At higher beam energies, preferential removal of indium can be observed, giving the impression that the near-surface region of the uniform single crystal has a copper-deficient, then copper-rich layer.

to the surface. Figure 4.25 presents a series of sputter depth profiles for CuInSe₂ crystals, in which the composition is known to be uniform for the primary elemental species (Matson *et al.,* 1989). Panels (a) to (f) show the data as a function of increasing ion beam energies. At low beam energies, below 1 kV, the crystal compositional profile appears to be uniform, with stoichiometric elemental levels. As the energy of the ion beam is increased, the depth profile appears to be nonuniform, with increasing deviations from uniformity as the ion beam voltage increases. In Fig. 4.25c, it appears that two distinct layers are present in this crystal, one copper-rich and one copper-deficient. However, this is the result of the preferential removal of the indium (and selenium), leaving the copper behind to be detected and to dominate the AES spectrum as the sputtering proceeds. Thus, if the proper sputtering conditions are not implemented in the analyses of these complex material systems, it is possible to interpret results that are actually artifacts of the sputtering process rather than inherent in the material or device analyzed.

Caution should be exercised in using sputtering with any analysis. Operator familiarity with the material under analysis is helpful in bringing attention to possible problems or artifacts. Alternatives to sputtering include the use of Auger line scans (using scanning Auger analysis). It is possible to fracture a sample *in situ,* and examine the cross section to obtain a profile. Other special techniques to obtain depth-compositional data include crater edge profiling and ball cratering with scanning AES (Walls *et al.,* 1979), and variation of electron emission angle [angular-resolved AES (Armitage *et al.,* 1980; Idzerda *et al.,* 1989)] and electron energy. The reader is referred to several review articles and references (see Bibliography) that provide more detail on the physics of the sputtering process, and a more expansive treatment of depth profiling.

4.6.2. Zalar Rotation

The accuracy and reproducibility of depth-profile analysis are sensitive functions of incident ion-beam properties, analytical instrumentation parameters, experimental geometry, and inherent sample surface and chemical characteristics. Multicomponent, polycrystalline thin-film semiconductor systems [such as the $(Cd,Zn)S/CuInSe_2$ heterostructure and its more complicated solar cell relatives] are especially prone to inaccuracies in interpretation of depth-compositional investigations. Problems associated with ion beam energies, sample-beam angle alignments, as well as electron beam current and energy densities have been cited in the previous section. This section extends these previous results to include potential artifacts associated with nonuniform or preferential sputtering that can lead to misinterpretation of depth resolutions and/or interface widths in topologically rough or textured samples or for junction regions several microns below the cell surface. Included are representations of experimental techniques that can provide more accurate representations of interfacial widths and properties for such common analysis cases.

Figure 4.26 presents AES depth-compositional profiles for two thin-film heterointerfaces, differing only in the thickness of the top or window layer. For each case, the sputtering rate was calibrated to be the same, approximately 220 Å/min. The first (Fig. 4.26a) represents a relatively shallow junction, with the deposited CdS layer about 0.19 μm in thickness. The second profile is that for a 2.2-μm CdS layer deposited on the same $CuInSe_2$ film, under identical conditions. The relative intralayer compositions are the same, but significant differences in the interface widths of the junctions between the binary and ternary semiconductors are observed. The deeper junction appears to be much wider, and can be interpreted as representing some extensive interdiffusion over this interface. However, the dark current–voltage relationships of the two junctions are nearly identical, leading to the unexpected result that the chemically interdiffused interface region, apparently spread over some 4500 Å in this case, has little effect on the minority carrier transport. These conflicting results can be resolved by consideration of the effects of surface to-

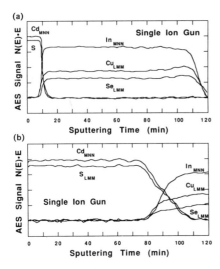

Figure 4.26. AES depth-compositional profiles of CdS/CuInSe$_2$ thin-film solar cell structures: (a) profile with thin CdS window layer; (b) profile with thick CdS window layer. Data were taken under identical sputter-etching and instrument parameters.

pography and angular dependence of ion sputtering on the depth resolution. The artifactual results can be minimized by techniques first reported by Zalar (1985) for improving depth resolution during AES depth profiling.

   The degradation of depth resolution during the analysis of samples results from the angular distribution of microplanes. These microplanes are a result of either the initial sample roughness and/or can be produced via preferential etching over extensive sputtering times (Zalar and Hofmann, 1987). The re-

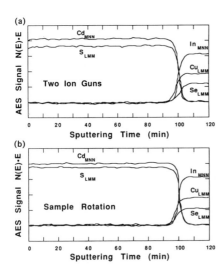

Figure 4.27. AES depth-compositional profiles of CdS/CuInSe$_2$ thin-film solar cell structures, for the same device presented in Fig. 4.26b: (a) profile using two ion sources, focused on same film region from opposite positions (approximately 120° apart); (b) profile using substrate or Zalar rotation (1 rev/min) during sputter etching. (Adapted with permission from Kazmerski and Nelson, 1990.)

sulting AES signals are significantly altered because of angular amplitude and shadowing effects. To diminish these artifacts, the ion etching can be made more uniform by eroding the surface from two (or more) different directions using multiple ion sources (Sykes *et al.,* 1980). Figure 4.27a presents such data using two ion guns. The film structure is exactly the same as that in Fig. 4.26b but the deep interface can now be observed to be more abrupt—with a transition region more that of the shallow junction and more representative of the comparative electrical characteristics.

Problems exist with the use of multiple ion sources, including cost, maintaining the focus for extended periods of time in investigations of thick layers, and keeping the ion currents and emissions constant and at the same level so that one source does not dominate. Zalar (1984) has developed an improved technique that provides uniform etching results through the slow rotation of the sample during the combined etching and analysis process. In this technique, the sample is turned at approximately 1 rev/min (continuous, on a modified sample positioning stage), while maintaining the analysis position relative to the AES electron beam. Zalar reported increases in depth resolution of more than an order of magnitude for Cr/Ni thin-film multilayers, and similar results have been reported recently for $Si/SiO_2/Ti/Al/Ti-N$ semiconductor structures. Figure 4.27b presents data for the $CdS/CuInSe_2$ interface, using Zalar rotation. During these analyses, the angle between the ion source and the electron beam was maintained at $65°$, and the angle between the incident electron beam and the sample normal at $30°$. The rastered (2×2 mm) ion beam energy was 2 keV, and the electron beam parameters were: $V_b = 5$ kV, $I_b = 1$ nA (pulse counting mode). The expected sharpness of the transition layer results from this analysis technique, consistent with the two-ion-source analysis. These results support the Zalar rotation method for accurate analysis of deep interfaces or semiconductors having rough surfaces, and imply the need for further caution in the interpretation of analytical characterizations utilizing ion beams.

4.7. Quantitative Analysis

Early work with AES concerned gaining qualitative and quantitative information from the analysis technique. Efforts have been toward advancing and improving accuracy in quantification, and the recent review by Powell and Seah (1990) provides an excellent updated treatment of quantitative analysis. AES is really a *semiquantitative* technique, with accuracy tied to the experimental and data reduction techniques employed. Quantification remains one of the more difficult areas of surface analysis, primarily because the interpretation of the quantification process is critical and, many times, the major inexactness associated with the analysis. Part of the problem is that users of techniques such as AES want *numbers,* sometimes without regard to what these mean. Because the strengths of AES and related methods are in their surface sensitivity, the numbers generated relate to the compositions of the near-surface regions and not the bulk. Bulk representations of the compositions can be obtained by depth-profile methods, but the cautions involving ion beam effects must be considered, especially for preferential etching, implantation, surface migration, and geometrical effects. Quantitative measurements should always be considered with some error analysis, which depends on the method utilized, and the instrumentation and operator expertise (Seah, 1991).

The signals in Auger (and photoemission) spectra are quantified through evaluation of their intensities, the detected electron currents. The implementation of this evaluation provides two approaches to quantification. The first involves calculations from *basic principles,* and the analysis is predictive and meant to involve minimal empirical input. The second approach includes several parameter-based methods, using analytical modeling, published data bases, and calibration from standards.

4.7.1. Calculations from Basic Principles

It is very difficult to provide a complete, accurate analytical representation of quantification based solely on basic physical principles. This approach, however, provides important insights into the processes for understanding accuracy and limitations of all methodologies for quantifying AES data. In this section, the total Auger current will be derived based on an ensemble of mechanisms that govern the Auger electron generation and emission.

Auger yield. Consider the surface that is exposed to an incident beam of monoenergetic, high-energy electrons. The incident electron beam ionizes atoms, producing Auger electrons with energy E_{ABC}. In AES, the Auger electrons (detected signals) are generated in *electron–specimen* interactions. In most real cases, the electron beam enters the sample at some angle, θ. The total number of Auger electrons generated by the primary electron beam, the *Auger electron yield,* can be expressed as

$$N_{X,ABC} = (1/2\pi) \iiint \gamma_{ABC} \Phi(z, E) \sigma_A(E) \gamma_{ABC}$$

$$\times \exp[-z/\lambda(E_{ABC}) \cos\phi] \cdot n_x(z) \, dz dE d\Omega \quad (4.8)$$

where $\Phi(z, E)$ is the number of electrons with kinetic energy E at depth z; $\sigma_A(E)$ is the ionization cross section of core level A of an atom of element x;

γ_{ABC} is the probability of Auger electron emission through transition ABC following the ionization process; $\lambda(E_{ABC})$ is the escape depth of the Auger electron of the ABC transition in the matrix; $n_x(z)$ is the atomic density of the element at depth z; and ϕ is the emission or takeoff angle of the Auger electron, measured from the surface normal. The atomic density

$$n_x(z) = \int \gamma_{ABC}\Phi(z, E)\sigma_A(E)\,dE \qquad (4.9)$$

corresponds to the in-depth distribution of Auger electrons.

The ionization cross section was discussed in Section 4.3, and the representation of $\sigma_A(E)$ is given by Eq. (4.5) and shown in Fig. 4.6. This parameter depends on the backscattering term, r_M, which increases with increasing atomic number of the elements and decreases with the depth of the core level E_A.

Total Auger signal. The ionized core level A decays with a probability γ_{ABC} of Auger electron emission, through the ABC transition. The created Auger electron has a probability $1/e$ of traveling a distance characterized by the IMFP in the matrix before being inelastically (diffuse) scattered and no longer appearing in the Auger signal. To the first order, the escape depth is

$$\lambda = \mathrm{IMFP}(E_{ABC})\cos\phi \qquad (4.10)$$

The emitted Auger electron is detected by an electron spectrometer with transmission efficiency $T(E_{ABC})$, and an electron detector of efficiency $D(E_{ABC})$. These are experimental variables that must be considered when analyzing the detected Auger signal. The Auger current is then expressed (Powell and Seah, 1990) as

$$I = I_p \sigma TD \int N_A(z)\exp[-z/\lambda]\,dz \qquad (4.11)$$

Substituting the expressions for σ into Eq. (4.11),

$$I = I_p \sigma_A(E_p)[1 + r_M(E_A, \theta)]\,T(E)D(E)\int N_A(z)\exp[-z/\lambda]\cos\phi\,dz \qquad (4.12)$$

This expression for the Auger current is still too complex for use for direct quantification of AES results, and further simplifications are required. For a homogeneous, binary system, the integral in Eq. (4.12) simplifies to $N_A\lambda_M\cos\phi$, and

$$I = I_p \sigma_A(E_p)[1 + r_M(E_A, \theta)] \, T(E)D(E) \, N_A \lambda_M \cos\phi \qquad (4.13)$$

The use of Eqs. (4.11)–(4.13) for the calculation of the concentrations has been discussed in the literature. Values of λ can be predicted from Seah and Dench (1979) and from Penn (1976), and the cross sections (Gryzinski, 1965) and backscattering terms (Reuter, 1972) can be analytically derived. The instrument factors, T and D, however, are very specific to the measurement system. Some simplification is gained since most current spectrometers operate in the constant resolution $\Delta E/E$ mode, and the transmission is proportional to the Auger energy. The expressions for the concentrations involve ratios for the T and D terms, and the transmission coefficients simply cancel for data taken on the same instrument. The case for D is not as convenient. However, by use of proper detector biasing, errors can be limited to less than 5% over the usual 2000 eV Auger energy range.

4.7.2. Practical Parameter-Based Methods

The signals in the AES spectra are quantified through their intensities, with the Auger current derived in the last section giving insight into the important parameter. These methods originate in early Auger quantitative work when Auger currents were related to calibrated, submonolayer coverages of an element on a dissimilar substrate. Coverage calibration was achieved with LEED (Palmberg and Rhodin, 1968), ellipsometry (Vrakking and Meyer, 1971), and quartz crystal oscillator (Penn, 1976) measurements. Later, determinations for absolute atomic concentrations of an element i in an unknown (C_x^{unk}) were attempted by comparing the Auger signals of the constituent elements (I_x^{unk}) with those from pure elemental standards (I_x^{std}). That is,

$$\frac{C_x^{unk}}{C_x^{std}} = \frac{I_x^{unk}}{I_x^{std}} \qquad (4.14)$$

Accuracy of this first approximation is poor. Certainly, the expression of Eq. (4.14) predicts that the material and electronic properties between the unknown samples and the standards would provide nonproportional Auger currents. If $C_x^{unk} = C_x^{std}$, and the host matrices are identical, the properties will be the same, and Eq. (4.14) is most accurate. However, such standards are usually difficult to identify and obtain for practical Auger analysis. Atomic density, escape depth, emission angles, and backscatter coefficient, predicted by the expression for the Auger current in Eq. (4.13), are the material/electronic properties of consequence for which correction may be necessary in quantitative Auger analyses. In addition, experimental conditions such as primary beam current and energy, incidence angle of electron probe, analyzer type and configuration, electronic system parameters (e.g., modulation voltage, amplifier gain), and sample surface properties (roughness, cleanliness, purity, and integrity) must be considered.

Consider the parameters from Eq. (4.13) and the corrections required to the simple, first-order representation of Eq. (4.14); several alternatives for quantitative analysis are available. All require standards for comparisons, and identical instrument measurement conditions, for obtaining the Auger current. Correcting the unknown concentration for the two major parameters, the backscatter coefficient and the Auger electron escape depth, one derives using ratios from Eqs. (4.13) and (4.14) (Hall and Morabito, 1979):

$$C_x^{*\text{unk}} = \frac{I_x^{\text{unk}}}{I_x^{\text{std}}} \frac{R_x^{\text{std}}}{R_x^{\text{unk}}} \frac{\lambda_x^{\text{std}}}{\lambda_x^{\text{unk}}} \tag{4.15}$$

Density differences for a multicomponent system can be included, providing the more complete expression:

$$C_x^{\text{unk}} = \frac{d_x C_x^{*\text{unk}}}{S_i d_i C_i^{*\text{unk}}} \tag{4.16}$$

The application of Eq. (4.16) has two inherent limitations. First, any inaccuracy in the determination of a given $C_x^{*\text{unk}}$ is transmitted to all $C_i^{*\text{unk}}$ results. Second, this equation is valid when the densities of constituent atoms in a compound are the same as in elemental form. This is usually true for alloys but not for crystalline structures. An alternative to Eq. (4.16) is (Gries, 1989)

$$C_x^{*\text{unk}} = C_x^{*\text{std}} \frac{f_x^{\text{unk}}}{f^{\text{std}}} \tag{4.17}$$

The factor f contains information about the backscatter intensity, and is minimally sensitive to surface topography effects. Equation (4.17) is valid for direct $N(E)$ data only.

Combining Eqs. (4.13), (4.15)–(4.17) for the case of crystalline materials (Masspust et al., 1984),

$$C_x^{\text{unk}} = \frac{I_x^{\text{unk}}}{I_x^{\text{std}}} \frac{d_x^{\text{std}}}{d_x^{\text{unk}}} \frac{R_x^{\text{std}}}{R_x^{\text{unk}}} \frac{\lambda_x^{\text{std}}}{\lambda_x^{\text{unk}}} \tag{4.18}$$

In applying this, pure elemental standards are used, and d_x^{std} is readily obtained from one of the handbook values. The value for the density of the unknown is necessary in this method. This is available when analyzing near-stoichiometric samples for which densities have been determined. Otherwise, total density can be found experimentally, and then d_x^{std} similar to the technique proposed by Sekine *et al.* (1982). This method provides an independent calculation for detectable elements in an unknown sample based on fundamental considerations and is usable even if there is an inability to obtain an I_x^{std} for H, He, gases, insulators, and low-vapor-pressure elements.

The most common and a very practical method for quantitative Auger analysis involves using a single standard and a given system of relative sensitivity factors (S) for the elements. Atomic concentrations are calculated by (Davis *et al.*, 1976)

$$C_x^{unk} = \frac{I_x^{unk}}{I_x^{std}} \frac{S_x^{std}}{S_x^{unk}} \qquad (4.19)$$

This method can be extended for analysis involving no standards, provided that Auger currents are determined for all i constituent elements in the unknown. Equation (4.19) becomes

$$C_x^{unk} = \frac{\dfrac{I_x}{S_x}}{\displaystyle\sum_i \dfrac{I_i}{S_i}} \qquad (4.20)$$

Other than this last relationship, the accuracy of any of the above methods depends on the ability to reproduce all instrumental conditions for measurements on the standards and unknowns.

The Auger current can be determined in several ways from $N(E)$ and $dN(E)/dE$ data, as shown in Fig. 4.28. The Auger current is deduced from $dN(E)/dE$ spectra by measuring the peak-to-peak excursions of the AES peaks. Auger electron detection for this determination is realized with analyzer energy modulation (AEM) and lock-in detection. Derivative data may also be obtained by mathematically differentiating $N(E)$ data, which is obtained by any of the methods described in Section 4.3. The term I is defined as the difference between peak Auger intensity and neighboring background intensity for a given Auger transition, or as the area of the peak for the transition. The latter method is desirable for cases in which Auger peak shape differences exist between the unknowns and the standards (Gries, 1989), because of valence-band density shape differences. Hardware effects on the Auger current include the primary excitation conditions, analyzer collection efficiency, gain of the detection system, and the instrumental linewidth.

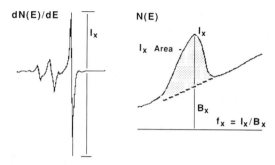

Figure 4.28. Definition of signal intensities for differentiated and direct spectra for electron spectroscopies, used for quantification.

Auger emission will occur during ion, photon (x-ray), or electron irradiation. Primary electron beams of 1–15 keV are usually employed. Collection currents are typically in the 1–50 pA range. Auger yields are weakly dependent on the primary beam voltage for $E_p > 3\ E$, and are directly proportional to the primary beam current, I_b. Constant Auger emission is energy analyzed using a hemispherical configuration or a CMA. Sample-to-analyzer distance affects collection, while analyzer geometry affects transmission. Transmission is normally fixed for a CMA, but collection efficiency has been found to be strongly dependent on sample positioning. Repeatable sample positioning may be accomplished by energy analyzing the backscattered primary electrons. Best accuracy occurs for smaller values of E_p.

The gain of the detection system must be constant during data acquisition. The electron multiplier is assumed to have constant gain if the applied voltage is kept constant and all pertinent measurements are conducted in a time period during which no electron multiplier degradation occurs. In addition, all subsequent amplifiers should maintain equal gain for all measurements. The instrumental linewidth is dominated by one of the following: (1) the analyzer resolution ($R = \Delta E/E$); (2) system bandwidth; or (3) the eV/channel interval for digital data acquisition. With good R, D, and I correction, the ability to measure reproducible Auger currents is the primary task. It should be noted that fundamental relationships exist among beam current, beam size, and detectability (recall Fig. 4.12).

4.8. Summary

The microanalysis of materials using AES has been presented. The physics, operation, and applications of AES have been covered, with some comparisons with other materials and device characterization methods. Examples incorporating other micron-level to atomic-level analysis techniques were included to demonstrate the evolving nature and power of modern microanalysis. The purpose of this chapter has been to provide a detailed introduction to the analysis method, presenting strengths and limitations and sufficient references for the reader to further supplement the topics.

ACKNOWLEDGMENTS. The author gratefully acknowledges the significant contributions and considerable support and assistance of A. Nelson, A. Swartzlander-Franz, S. Asher, H. Moutinho, D. Niles of the Measurements and Characterization Branch at the National Renewable Energy Laboratory, and Thomas Massopust of Rocky Mountain Laboratories, Golden, Colorado, in the preparation of this chapter, particularly in providing much of the data. This work was supported by the U.S. Department of Energy through Contract Number DE-AC01-83CH10093 with the National Renewable Energy Laboratory.

References

Armitage, A. F., Woodruff, D. P., and Johnson, P. D. (1980). *Surf. Sci.* **100**, L483.

Auger, P. (1925). *J. Phys. Radium* **6**, 205.

Bergstrom, I., and Nordlind, C. (1965). In *Alpha, Beta and Gamma Ray Spectroscopy* (K. Siegbahn, ed.), North-Holland, Amsterdam, p. 11.

Bishop, H. E. (1989). In *Methods of Surface Analysis* (J. M. Walls, ed.), Cambridge University Press, London, pp. 87–126.

Bishop, H. E., and Rivière, J. C. (1969). *J. Appl. Phys.* **40**, 1740.

Bishop, H. E., Chornik, B., LeGressus, C., and LeMoel, A. (1984). *Surf. Interface Anal.* **5**, 116.

Briggs, D., and Seah, M. P. (eds.) (1983). *Practical Surface Analysis by Auger and X-Ray Photoelectron Spectroscopy,* Wiley, New York, Appendix 6, pp. 514–518.

Brillson, L. J. (1976). *J. Vac. Sci. Technol.* **13**, 325.

Burnham, N. A., Fisher, R. F., Asher, S. E., and Kazmerski, L. L. (1987). *J. Vac. Sci. Technol. A* **5**, 2016.

Carlson, T. A. (1975). *Photoelectron and Auger Spectroscopy,* Plenum Press, New York.

Cheng, Y.-T., Dow, A. A., Clemens, B. M., and Cirlin, E.-H. (1989). *J. Vac. Sci. Technol. A* **7**, 1641.

Chou, N. J., Osburn, C. M., Van der Meulen, Y. J., and Hammer, R. (1973). *Appl. Phys. Lett.* **22**, 380.

Clough, S. (1984). Perkin–Elmer, Physical Electronics Division, Technical Bulletin, No. T8401.

Coad, J. P., Bishop, H. E., and Rivière, J. C. (1970). *Surf. Sci.* **21**, 253.

Corrallo, C. F., Asbury, D. A., Pipkin, M. A., Anderson, T. J., and Hoflund, G. B. (1986). *Thin Solid Films* **139**, 299.

Davis, L. E., MacDonald, N. C., Palmberg, P. W., Riach, G. E., and Weber, R. E. (1976). *Handbook of Auger Electron Spectroscopy,* Perkin–Elmer, Physical Electronics, Eden Prairie, Minn.

Finello, D., and Marcus, H. L. (1977). In *Handbook of X-Ray and Photoelectron Spectroscopy* (D. Briggs, ed.), Heyden, London, pp. 138–140.

Fuggle, J. C. (1981). In *Electron Spectroscopy* (C. R. Brundle and A. D. Baker, eds.), Academic Press, New York, p. 5.
Gries, W. H. (1989). *J. Vac. Sci. Technol. A* 7, 1639.
Gryzinski, M. (1965). *Phys. Rev. A* 138, 305.
Haas, T. W., Grant, J. T., and Dooley, G. J. (1972). *J. Appl. Phys.* 43, 1853.
Hall, P. M., and Morabito, J. R. (1979). *Surf. Sci.* 83, 391.
Hiraki, A., Kim, S., Iwamura, T., and Iwamai, M. (1979). *Jpn. J. Appl. Phys.* 18, 1767.
Holloway, P. H. (1976). *Surf. Sci.* 54, 506.
Holloway, P. H. (1980). *Adv. Electron. Electron Phys.* 54, 241.
Ibach, H., Bachmann, H. D., and Wagner, H. (1982). *Appl. Phys. A* 29, 113.
Ichimura, S., and Shimizu, R. (1981). *Surf. Sci.* 112, 386.
Ichimura, S., and Shimizu, R. (1983). *Surf. Sci.* 124, L449.
Idzerda, Y. U., Lind, D. M., and Prinz, G. A. (1989). *J. Vac. Sci. Technol. A* 7, 1341.
Joshi, A., Davis, L. E., and Palmberg, P. W. (1975). In *Methods of Surface Analysis* (A. W. Czanderna, ed.), Elsevier, Amsterdam, pp. 159–222.
Kazmerski, L. L. (1988). *Sol. Cells* 24, 387.
Kazmerski, L. L. In *Physical Methods of Chemistry, Vol. 1XB* (J. Wiley and Sons, New York).
Kazmerski, L. L. (1989a). In *Polycrystalline Semiconductors* (J. H. Werner, H. J. Möller, and H. P. Struck, eds.), Springer-Verlag, Berlin, pp. 96–107.
Kazmerski, L. L. (1989b). U.S. Patent: Volume-Indexed Secondary Ion Mass Spectrometry, Patent No. 4,874,946, October 17, 1989.
Kazmerski, L. L., and Nelson, A. J. (1990). *Sol. Cells* 28, 273.
Kazmerski, L. L., Ireland, P. J., Sheldon, P., Chu, T. L., Chu, S. S., and Lin, C. L. (1980). *J. Vac. Sci. Technol.* 17, 1061.
Knotek, M. L., and Houston, J. E. (1982). *J. Vac. Sci. Technol.* 20, 544.
Madden, H. H. (1981). *J. Vac. Sci. Technol.* 18, 677.
Madden, H. H., and Ertl G. (1973). *Surf. Sci.* 5, 211.
Madey, T. E., and Yates, J. T. (1971). *J. Vac. Sci. Technol.* 8, 525.
Massignon, D., Pellerin, F., Fontaine, J. M., LeGressus, C., and Inchinokawa, T. (1980). *J. Appl. Phys.* 51, 808.
Massopust, T. P., Ireland, P. J., and Kazmerski, L. L. (1984). *J. Vac. Sci. Technol. A* 2, 1123.
Massopust, T. P., Kazmerski, L. L., Whitney, R. L., and Starr, R. (1985). *J. Vac. Sci. Technol. A* 3, 955 (1985).
Matson, R. J., Noufi, R., Ahrenkiel, R. K., Powell, R. C., and Cahen, D. (1986). *Sol. Cells* 16, 495.
Matson, R. J., Kazmerski, L. L., Noufi, R., and Cahen, D. (1989). *J. Vac. Sci. Technol. A* 7, 230.
Menzel, E. (1982). *J. Vac. Sci. Technol.* 20, 538.
Meyer, D. E., Ianno, N. J., Woollam, J. A., Swartzlander, A. B., and Nelson, A. J. (1988). In *Diamond Optics*, SPIE Vol. 969, pp. 66–69.
Morabito, J. M., and Lewis, R. K. (1975). In *Methods of Surface Analysis* (A. W. Czanderna, ed.), Elsevier, Amsterdam, pp. 279–328.
Muller, R. O., and Kiel, K. (1972). *Spectrochemical Analysis by X-Ray Fluorescence*, Plenum Press, New York.
Nelson, A. J., Swartzlander, A., Kazmerski, L. L., Kang, J. H., Kampwirth, R. T., and Gray, K. E. (1988). In *High T_c Superconducting Thin Films, Devices and Applications* (G. Margaritondo, R. Joynt, and M. Onellion, eds.), American Institute of Physics, New York.
Noufi, R., Matson, R. J., Powell, R. C., and Herrington, C. R. (1986). *Sol. Cells*, Vol. 16, 479.
Palmberg, P. W., and Rhodin, T. N. (1968). *J. Appl. Phys.* 39, 2425.
Penn, D. R. (1976). *J. Electron Spectrosc. Relat. Phenom.* 9, 29.
Perkin–Elmer (1983, 1984, 1990). PHI Data Sheet 1064 8-82 13M, PHI Microprobe 600, Perkin–

Elmer, Physical Electronics Division, Eden Prairie, Minn. (1983). For nanoprobe and Schottky thermal emission source, see Perkin–Elmer PHI Model 670 Auger Nanoprobe Product Specifications, September 15, 1990, Physical Electronics Division; also, PHI Technical Bulletin No. 8404, May, 1984.
Powell, C. J., and Seah, M. (1990). *J. Vac. Sci. Technol. A* **8,** 735.
Reuter, W. (1972). *Proc. Sixth Int. Conf. X-Ray Optics and Microanalysis* (G. Shinoda, K. Kohra, and T. Ichiokawa, eds.), University of Tokyo Press, p. 121.
Seah, M. P. (1991). *J. Vac. Sci. Technol. A* **9,** 1227.
Seah, M. P., Anthony, M. T., and Dench, W. A. (1983) *Journal of Phys. E. Sci. Instrum.* **16,** 848.
Seah, M. P., and Dench, W. A. (1979). *Surf. Interface Anal.* **1,** 2.
Sekine, T., Nagasawa, Y., Kudoh, M., Sakai, Y., Parkes, A. S., Geller, J. E., Mogami, A., and Hirata, K. (1982). *Handbook of Auger Electron Spectroscopy,* JEOL Ltd., Tokyo.
Shapira, Y., and Friedenberg, A. (1980). *J. Appl. Phys.* **51,** 710.
Shirley, D. A. (1973). *Phys. Rev. A* **7,** 1520.
Sykes, D. E., Hall, D. D., Thurstans, R. E., and Walls, J. M. (1980). *Appl. Surf. Sci.* **5,** 103.
Szajam, J., and Leckey, R. C. G. (1975). *J. Electron Spectrosc. Relat. Phenom.* **23,** 83.
Tanuma, S., Powell, C. J., and Penn, D. R. (1990). *J. Vac. Sci. Technol. A* **8,** 2213.
Vrakking, J. J., and Meyer, F. (1971). *Appl. Phys. Lett.* **18,** 226.
Wagner, C. D. (1983). In *Practical Surface Analysis by Auger and X-Ray Photoelectron Spectroscopy* (D. Briggs and M. P. Seah, eds., Wiley, New York, Appendix 8, pp. 521–526.
Walls, J. M., Hall, D. D., and Sykes, D. E. (1979). *Surf. Interface Anal.* **1,** 29.
Wehner, G. K. (1975). In *Methods of Surface Analysis* (A. W. Czanderna, ed.), Elsevier, Amsterdam, pp. 5–38.
Wittmaack, K. (1990). *J. Vac. Sci. Technol. A* **8,** 2246.
Worthington, C. R., and Tomlin, S. G. (1956). *Proc. Phys. Soc. London Sect. A* **69,** 401.
Zalar, A. (1984). *Abstr. Proc. Sixth Int. Conf. Thin Films,* Stockholm.
Zalar, A. (1985). *Thin Solid Films* **124,** 223.
Zalar, A., and Hofmann, S. (1987). *Vaccum* **37,** 169.

Selected Bibliography

Glauert, A. M. (ed.) (1972). *Practical Methods in Electron Microscopy,* North-Holland, Amsterdam.
Roy, D., and Carette, J.-D. (1972). *Current Topics in Physics,* Vol. 4, Springer-Verlag, Berlin.
Sevier, K. D. (ed.) (1972). *Low Energy Electron Spectrometry,* Wiley–Interscience, New York.
Kane, P. F., and Larrabee, G. B. (eds.). (1974). *Characterization of Solid Surfaces,* Plenum Press, New York.
Czanderna, A. W. (ed.) (1975). *Methods of Surface Analysis,* Elsevier, Amsterdam.
Wißmann, P., and Müller, K. (1975). *Surface Physics,* Springer-Verlag, Berlin.
Carlson, T. A. (1976). *Photoelectron and Auger Spectroscopy,* Plenum Press, New York.
Ibach, H. (1977). *Electron Spectroscopy for Surface Analysis,* Springer-Verlag, Berlin.
Davis, L. E., MacDonald, N. C., Palmberg, P. W., Riach, G. E., and Weber, R. E. (1978). *Handbook of Auger Electron Spectroscopy,* Perkin–Elmer, Physical Electronics Division, Eden Prairie, Minn.
Fiermans, L., Vennick, J., and Dekeyser, W. (eds.) (1978). *Electron and Ion Spectroscopy of Solids,* Plenum Press, New York.
McIntyre, N. S. (ed.) (1978). *Quantitative Surface Analysis of Materials,* ASTM STP 643, American Society for Testing and Materials, Philadelphia.
Buch, O., Tien, J. K., and Marcus, H. L. (eds.) (1979). *Electron and Positron Spectroscopies in Materials Science and Engineering,* Academic Press, New York.

McGuire, G. E. (1979). *Auger Electron Spectroscopy Reference Manual,* Plenum Press, New York.

Brundle, C. R., and Baker, A. J. (eds.) (1979–1983). *Electron Spectroscopy: Theory, Techniques and Applications,* Vols. I–V, Academic Press, New York.

Barr, T. L., and Davis, L. E. (eds.) (1980). *Applied Surface Analysis,* ASTM STP 699, American Society for Testing and Materials, Philadelphia.

Langeron, J. (ed.) (1980). *Spectrometrie Auger,* Soc. Francaise du Vide, Paris.

Dwight, D. W., Fabish, T. J., and Thomas, H. R. (eds.) (1981). *Photon, Electron and Ion Probes of Polymer Structure and Properties,* American Chemical Society, Washington, D.C.

Robinson, J. W. (1981). *Handbook of Spectroscopy,* CRC Press, Boca Raton, Fla.

Sekine, T., Nagasawa, Y., Kudoh, M., Sakai, Y., Parkes, A. S., Geller, J. D., Mogami, A., and Hirata, K. (1982). *Handbook of Auger Electron Spectroscopy,* JEOL Ltd., Tokyo.

Windawi, H., and Ho, F. (eds.) (1982). *Applied Electron Spectroscopy for Chemical Analysis,* 2nd ed. (1993), Wiley, New York.

Prutton, M. (1983). *Surface Physics,* Oxford University Press, London.

Briggs, D., and Seah, M. P. (eds.) (1983). *Practical Surface Analysis by Auger and X-Ray Photoelectron Spectroscopy,* Wiley, New York.

Czanderna, A. W. (ed.) (1984). *Crucial Role of Surface Analysis in Studying Surfaces,* Florida Atlantic University, Boca Raton, Fla.

Oechsner, H. (ed.) (1984). *Topics in Current Physics, Thin Film and Depth Profile Analysis,* Vol. 37, Springer-Verlag, Berlin.

Thompson, M., Baker, M. D., Christie, A., and Tyson, J. F. (1985). *Auger Electron Spectroscopy,* Wiley, New York.

ASTM Standards on Surface Analysis (1st ed.) (1986). American Society for Testing and Materials, Philadelphia.

Feldman, L. C., and Mayer, J. W. (1986). *Fundamentals of Surface and Thin Film Analysis,* North-Holland, Amsterdam.

Thompson, M., Baker, M. D., Christie, A., and Tyson, J. F. (1986). *Auger Electron Spectroscopy,* Wiley-Interscience, New York.

Kemeny, G. (1988). *Surface Analysis of High Temperature Materials—Chemistry and Topography,* Applied Science, London.

Tu, K. N., and Rosenberg, R. (eds.) (1988). *Analytical Techniques for Thin Films,* Academic Press, New York.

Sibilia, J. (1988). *A Guide to Materials Characterization and Chemical Analysis,* VCH Publishers, New York.

Zangwill, A. (1988). *Physics at Surfaces,* Cambridge University Press, New York.

Hoffmann, R. (1988). *Solids and Surfaces,* VCH Publishers, New York. Woodruff, D. P., and Delchar, T. A. (1988). *Modern Techniques of Surface Science,* Cambridge University Press, London.

Cubiotti, C., Mondio, D., and Wandelt, K. (eds.) (1989). *Auger Spectroscopy and Electronic Structure: Proc. of the First International Workshop,* Springer-Verlag, Berlin.

McGuire, ed. (1989). *Characterization of Semiconductor Metals, Volume 1,* Noyes Publications, New Jersey.

Walls, J. M. (1989–1990). *Methods of Surface Analysis,* Cambridge University Press, London.

Ferguson, I. F. (1989). *Auger Microprobe Analysis,* Adam Hilger, IOP Publ. Co., Bristol.

Krakow, W., Ponce, F. A., and Smith, D. J. (eds.) (1989). *High Resolution Microscopy of Materials,* Materials Research Society, Pittsburgh.

Coyne, L., McKeever, S., and Blake, D. (eds.) (1990). *Spectroscopic Characterization of Minerals and Their Surfaces,* American Chemical Society, Washington, D.C.

Rivière, J. C. (1990). *Surface Analysis Techniques, Monographs on the Physics and Chemistry of Materials,* Oxford University Press, London.

MacRitchie, F. (1990). *Chemistry at Interfaces,* Academic Press, New York.

Watts, J. F. (1990). *An Introduction to Surface Analysis by Electron Spectroscopy,* Oxford University Press, U.K.

Vvensky, D. D., ed. (1992). *Graphics and Animation in Surface Science,* IOP Publishing Inc., Philadelphia.

Lifshin, E., ed. (1992). *Characterization of Materials, Part 1,* VCH, New York.

Rossiter, B. W. and Baetzold, R. C., eds. (1993). *Physical Methods of Chemistry, Volume 1XB,* "Investigations of Surfaces and Interfaces", John Wiley and Sons, New York.

Brundle, C. R. *et al.,* eds. (1993) *Encyclopedia of Materials Characterization: Surfaces, Interfaces, Thin Films,* Butterworth–Heineman, Stoneham, Massachussetts.

Journals: *Journal of Vacuum Science and Technology* (AVS)
Journal of Electron Spectroscopy and Related Phenomena
Surface Science Spectra (AVS) [New introduction 1992]
Surface Science
Surface and Interface Analysis
Vacuum (Pergamon)

Part III

Ion Beam Techniques

5

Secondary Ion Mass Spectrometry

S. E. Asher

5.1. Introduction

Secondary ion mass spectrometry (SIMS) has become an established method for the analysis of trace elements in semiconductor and other materials. The flexibility of this technique has led to its use in a wide variety of materials analysis problems. It is routinely used to determine dopant and trace contaminant levels in semiconductor devices of micron and submicron dimensions. It is used to determine isotopic abundances in mineral phases and it is used to study the localization of trace elements in biological tissues. Time-of-flight SIMS is becoming a standard tool for the analysis of polymeric and other high-mass materials. SIMS is a complementary technique for methods such as Auger electron spectroscopy and photoemission spectroscopy, discussed in Chapters 4 and 9 in this volume, respectively, and for the technique of laser ionization mass spectrometry, discussed in Chapter 10. This chapter will outline basic principles of the SIMS technique, the equipment used for the technique, and several closely related techniques (SIMS variants). Information which can be obtained by SIMS will be discussed and several examples presented.

Several excellent references exist for the reader needing more extensive information about this technique than can be provided in this chapter. The review by McHugh (1975), although several years old, still provides a well-

S. E. Asher • National Renewable Energy Laboratory, Golden, Colorado 80401.

Microanalysis of Solids, edited by B. G. Yacobi *et al.* Plenum Press, New York, 1994.

written basic reference for the technique of SIMS. More recently, two books have been published which cover all aspects of the SIMS technique, from the ion emission process to practical applications and related techniques (Benninghoven *et al.*, 1987; Vickerman *et al.*, 1990). The practical aspects of SIMS analysis, as applied primarily to semiconductor materials, have been thoroughly covered in an excellent text by Wilson *et al.* (1989a). A handbook of static SIMS has also been published which provides an introduction to the technique together with a library of reference spectra (Briggs *et al.*, 1989a). Other good information sources about the current and past status of SIMS and related techniques are found in the published proceedings from the biennial international conferences on SIMS, SIMSII–SIMSVIII (Benninghoven *et al.*, 1979, 1982, 1984, 1986, 1988, 1990, 1992).

5.2. Technique

5.2.1. Outline of Physical Process

The standard configuration of SIMS uses an ion beam (called the primary ions) to remove material from a solid surface. The primary ion beam is incident on the sample surface with energies between 0.5 and 20 keV at an angle between 0 and $\sim 70°$, as measured from the surface normal. The energy of the primary ions is transferred into the near-surface region of the sample through a series of binary collisions with atoms in the target. The process by which energy is transferred has been termed the collision cascade. Several processes occur as a result of the bombardment process. First, some of the primary ions are implanted into the near-surface region of the sample where they can affect the chemistry of the sample. Second, energy transferred from the primary ions to the target results in disruption of the matrix. This is known as cascade mixing. The depth of cascade mixing primarily depends on the primary ion penetration depth (function of mass, kinetic energy, and angle of incidence of the primary ion). Third, the bombardment process results in the sputter removal of material from the sample surface. The escape depth of secondary ions is two to three atomic layers. The sputtering process is shown schematically in Fig. 5.1.

The principal components of the sputtered material are uncharged atoms and molecules, although a small fraction (10^{-1}–10^{-5}) are ejected as positive and negative (secondary) ions (McHugh, 1975). These ions are both monoatomic elements and polyatomic clusters. Secondary ions, which are representative of the original composition of the target, comprise the analytical signal used for conventional SIMS. The proportion of secondary ions generated during the sputtering process varies greatly between different elements and

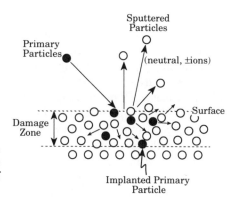

Figure 5.1. Schematic diagram of the sputtering process showing the initiation of the collision cascade and the sputter removal of material from the sample surface.

compounds. This leads to difficulties in converting secondary ion intensities into concentrations. Quantitative evaluation of secondary ion signals will be discussed in Section 5.3.

The sputtered secondary ions are collected by an electric field, biased to extract either positive or negative ions, and focused into a mass spectrometer. Measurements of secondary ion energy distributions show that the majority of ions are ejected with energies between 0 and 10 eV. Atomic ions, however, are observed to have high-energy tails extending to hundreds of electron volts. Because of the large energy spread, secondary ions are usually energy and mass filtered by the mass spectrometer (double focused) to obtain better resolution. After exiting the mass spectrometer, the mass-separated ions are detected by an ion-to-electron converter. All SIMS experiments are performed under high or ultrahigh vacuum conditions. Data acquisition is computer controlled to facilitate handling the large amounts of data obtained during analysis.

5.2.2. Advantages and Disadvantages of SIMS

SIMS is a microanalytical technique that combines high sensitivity with good elemental selectivity. It is able to analyze all elements in the periodic table including isotopes (from H to U) with ppm sensitivities for most all elements, and ppb for a few.* SIMS also has excellent dynamic range, $>10^5$ for typical analyses. The lateral resolution, which can be as good as 2–3 nm (depending on the primary ion source), is typically 2–10 μm. Depth resolutions

---

* Concentration values in SIMS are generally specified in terms of atoms per cubic centimeter. For example, the atomic density of silicon is 5.0×10^{22} at/cm^3. This translates to 1 ppm = 5 $\times 10^{16}$ at/cm^3, and 1 ppb = 5×10^{13} at/cm^3.

for SIMS analyses are typically 10–20 nm, although resolutions as good as 2–3 nm have been achieved (Grasserbaüer *et al.*, 1986). SIMS does require a conducting surface; however, conductive overlayers and charge neutralization allow the analysis of insulating materials. A wide variety of materials systems have been analyzed by SIMS, including semiconductors (Magee and Botnick, 1985), insulators (Van den Berg, 1986), metals (Degreve *et al.*, 1988), minerals (Reed, 1989), polymers (Briggs, 1989), organic solids (Benninghoven, 1985), and biological materials (Burns, 1988). Finally, quantitative results can be obtained with standards.

To its detriment, SIMS is based on a complicated physical phenomenon, the production of ions during the sputter removal of material from a surface. The proportion of ions in the sputtered flux varies considerably depending on the specific element/matrix combination studied. The SIMS "matrix effect" (Section 5.3.1) is well known, and matrix effects often make quantitative analysis tedious because of the need for complicated correction procedures. SIMS is a destructive technique; thus, each analysis must be performed on a new area of the sample. Detailed chemical bonding information is lost during the sputtering process. Finally, to obtain accurate profiles it is important that the sample surface be (locally) flat and conducting.

5.2.3. Static versus Dynamic SIMS

There are two regimes in which SIMS analyses are generally conducted, called static and dynamic. The term static SIMS (SSIMS) was first used by Benninghoven (1973) to describe a method for studying organic molecules. SSIMS uses low total primary ion doses (10^{12}–10^{13} cm^{-2}) to limit the amount of damage to the sample surface. For true SSIMS analyses the total primary ion dose is kept low enough that every primary ion impact is on a virgin area of the surface. This results in the quasistatic analysis of the surface monolayer. Signal levels for SSIMS are low because of the low primary ion dose, and thus it is desirable to use a mass analyzer which has a high transmission. In addition, since SSIMS is often used to study high-weight organic matrices and polymers, it is also desirable to use a mass analyzer with a large mass range. The second regime, called dynamic SIMS, uses a high primary ion dose ($>10^{13}$ cm^{-2}) with the intent being to maximize signal intensity for trace element analysis. One consequence of the high primary ion dose is that the sample is eroded, allowing rapid access to subsurface layers of the sample. In dynamic SIMS, primary ion doses are large enough to cause significant disruption of the near-surface region of the sample, and most molecular information is thus lost.

5.2.4. Other Techniques Related to SIMS

There are numerous variations of the SIMS technique now in use. These techniques utilize a variety of methods to generate secondary ions, mass analyze them, and detect them. They can be broken into two categories. In the first, secondary ions are generated directly by the incident primary particle (ion, atom, or photon). These techniques include static and dynamic SIMS, fast atom bombardment mass spectrometry (FABMS), and laser ionization mass spectrometry (LIMS). The second set of techniques makes use of the large proportion of neutral atoms and molecules which are sputtered from the sample, the so-called sputtered neutral mass spectrometries (SNMS). In these techniques, secondary ions are generated indirectly by ionizing the atoms and molecules of the sputtered neutral flux. Various methods such as electron impact, plasma sources, and laser ionization with resonant and nonresonant photon transitions [e.g., surface analysis by laser ionization (SALI)] are used to perform the postionization. Postionization techniques have the potential to surpass the sensitivity of conventional dynamic SIMS with fewer problems for quantitative analysis. They will be discussed at the end of this chapter. The instrumentation used to perform SIMS and related techniques is discussed in the following section.

5.2.5. Equipment

5.2.5.1. Primary Ion Sources

Commercial dynamic SIMS machines are typically equipped with at least two, and often three, ion sources. This allows the analyst to take full advantage of the chemical enhancement in secondary ion yields which can be obtained by using a reactive species for the primary ion. SIMS instruments are equipped with a gas source, generally a duoplasmatron, which is used to produce ion beams of O_2^+, O^-, Ar^+, Xe^+, or other gases. Surface ionization or liquid metal ion sources (LMIS) are used to produce Cs^+ ion beams. Gallium LMISs provide small spot sizes (typically 20–100 nm) for high-lateral-resolution ion microprobe analyses. In general, oxygen primary beams are used to enhance the ion yields from electropositive elements such as the group IA and IIIA elements. Negative oxygen primary beams have been shown to reduce surface charging on insulating samples (Anderson et al., 1969). Cesium primary beams are used to enhance the yields from electronegative elements such as the group VIA and VIIA elements (Storms et al., 1977). Inert gas primary beams (e.g., Ar^+) are primarily used for SSIMS and as particle beams for FABMS.

The development of ion sources which can provide high currents in small spot sizes (e.g., >25–50 mA/cm^2 in a 50-μm-diameter spot) has made it possible to improve sputter rates, thereby improving detection limits and allowing smaller areas to be analyzed. A relatively recent addition to SIMS instruments are the LMISs, particularly the high-brightness liquid Ga source. Liquid metal sources are capable of providing reasonable working currents (10 pA) with spot sizes down to 20–30 nm. For instruments operating in the microprobe mode, the use of an LMIS yields a dramatic increase in the lateral resolution (Levi-Setti, 1990).

5.2.5.2. Mass Analyzer

The mass analyzer is used to separate the secondary ions according to their mass-to-charge ratio, m/z. There are three types of mass analyzers generally used for SIMS: the quadrupole, the magnetic sector, and the time-of-flight. Quadrupoles are probably the most common type of mass analyzer because they are inexpensive and easily adaptable to different machines. They have low transmission because of a small acceptance energy and limited high-mass-resolution capabilities. The second type of analyzer is the magnetic sector mass spectrometer, which is of the same type used for more traditional mass spectrometry. Magnetic sector instruments are characterized by high transmissions and the ability to operate with high mass resolution ($M/\Delta M \approx 10,000$). The mass range of these analyzers is limited by the magnetic field strength, generally to the range of elemental masses ~300 amu. In addition, magnet hysteresis limits the rate at which the masses can be scanned during an analysis. The final type of mass analyzer used for SIMS is the time-of-flight (TOF) analyzer. These mass analyzers are ideally suited for SSIMS analysis because they have virtually unlimited mass ranges, the capability for high mass resolution, and high transmission. These characteristics make TOF analyzers well suited for the analysis of polymeric and biological materials. TOF analyzers require a pulsed ion source and sophisticated multichannel analyzers to determine the exact arrival time of ions at the detector. Because of the low primary ion currents in the pulsed ion source, depth profiling is cumbersome with this type of analyzer.

5.2.5.3. Ion Microprobe versus Ion Microscope

There are two different kinds of SIMS instruments which are primarily differentiated by their approach to forming a secondary ion image. In the first type, called the ion microprobe, image formation is similar to that used for a scanning electron microscope. A finely focused ion beam is scanned (rastered) across the sample surface and the secondary ion signal is measured as a function of position. The scan signal used for rastering the primary beam is synchronized with a display CRT monitor. Variation in intensity on the CRT is caused by variations of the secondary ion signal from that point on the sample surface. The image is then generated point by point and the lateral resolution is determined by the focusing of the primary ion beam, and ultimately by the emission volume of the secondary ions. The second type, called the direct imaging approach or ion microscope, uses specialized ion optics to preserve the spatial relationship of the ions from where they were removed from the surface. The real, mass-filtered image is observed simultaneously with a position-sensitive detector. For ion microscopes the ion beam can be larger and does not necessarily need to be scanned across the surface to generate the image. The image resolution is determined by focusing of and aberrations in the secondary ion optics and by the resolution of the detector (typically 0.5–1 μm). One other important difference between these two philosophies of operation is the method of limiting the analysis area. In the microprobe mode the analyzed area is determined by the size of the rastered area. To limit this area it is necessary to use electronic gating to select a portion out of the entire rastered area from which data will be collected. For the microscope mode, the analyzed area is limited by physically aperturing an area out of the center of the imaged area. Both types of machines are commercially manufactured and in common use for SIMS analysis. The microprobe-type machines are used in combination with quadrupoles, magnetic sectors, and TOF mass analyzers (discussed in Section 5.2.5.2). The direct imaging machines are used in combination with either a magnetic sector or TOF mass analyzer.

5.2.6. Data Types

In the following section it is worthwhile to point out that SIMS data are almost always plotted with a logarithmic scale along the y-axis. This is because the dynamic range of the technique covers more than six orders of magnitude. Within a single analysis it is not uncommon to find that the secondary ion signal from a matrix element is greater than 10^5 counts/s, whereas at the same time the signal from an element of interest is below 10 counts/s. The three commonly used methods of acquiring and displaying SIMS data are described in the following sections.

5.2.6.1. Mass Spectrum

A mass spectrum provides a graph of secondary ion signal as a function of mass. To acquire a mass spectrum, the desired range of ion masses must be scanned by the mass analyzer and the secondary ion signal collected at every mass. The upper limit of the mass range is analyzer dependent. In a mass spectrum, the x-axis is displayed as mass (atomic mass units or amu) and the y-axis is displayed as the number of ion counts. Mass spectra are used to (1) qualitatively determine the elements of interest in the sample, (2) determine potential mass interferences which could obscure the signal from the element(s) of interest, and (3) make qualitative comparisons between samples. The range for commonly found commercial magnetic sector machines varies from 0 to ∼300 amu, quadrupoles to ∼500 amu, and TOF > 5000 amu. Figure 5.2 shows the positive ion spectrum from a clean piece of Si. In this spectrum, peaks are observed for the three silicon isotopes at masses 28, 29 and 30, and also for the doubly charged ions (Si^{2+}), the dimer, trimer, etc. peaks (Si_n^+), and the oxide peaks SiO^+, SiO_2^+, etc. There is different information obtained from the negative secondary ion mass spectrum of the same sample, as shown in Fig. 5.3. Here the oxide peaks are found to be more intense than in the positive ion spectrum. Although a mass spectrum provides information about the composition of a sample, it does not provide this information as a function of depth.

5.2.6.2. Depth Profile

Depth profiles are a natural display of SIMS data. All SIMS analyses employ sputtering to generate the secondary ion signals used for analysis, and this leads to the removal of material as the analysis progresses. To obtain a depth profile, the element(s) of interest are identified and the analyzer is set to collect data from each element sequentially. Thus, each cycle contains a point from each element at close to the same depth in the sample. The resulting spectrum shows the change in ion signal for a particular species as a function

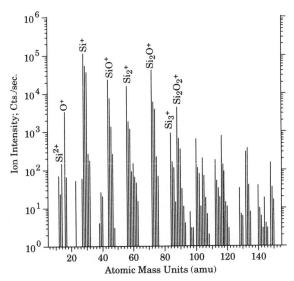

Figure 5.2. Positive secondary ion mass spectrum from crystalline silicon, obtained with a 12.5-keV O_2^+ primary beam. The vertical axis gives the number of secondary ion counts recorded at each mass (horizontal axis). Compare with Fig. 5.3.

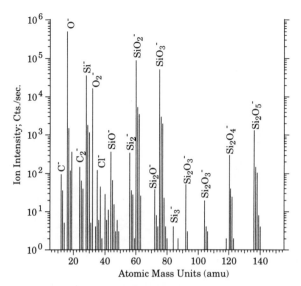

Figure 5.3. Negative secondary ion mass spectrum from crystalline silicon obtained with a 12.5-keV O_2^+ primary beam. Compare with Fig. 5.2.

of the time sputtered into the sample. The time axis is converted into sputtered depth by measuring the crater depth after analysis by stylus profilometry or interferometry (smooth surfaces). The y-axis is converted to concentration via sensitivity factors obtained from standards. Figure 5.4 shows a depth profile of a vanadium implant into Si. In this graph the x-axis has been converted to sputtered depth in microns and the y-axis is shown as concentration on the left and raw ion counts on the right. Depth profiles are used to provide information about the change in concentration of the element(s) of interest as a function of depth. They are also used to study changes in a sample as a function of treatment, e.g., interface sharpness and diffusion broadening after annealing.

5.2.6.3. Secondary Ion Image

The third method for acquiring SIMS data is as a secondary ion image. The ion image yields information about the lateral position of a chosen element in the sample. By collecting images for the different elements in a sample, it is possible to find the lateral distribution of all species. Figure 5.5 shows ion images of Si^+ and Al^+, respectively, obtained from an integrated circuit. Light areas in each of the images indicate areas where secondary ions were generated

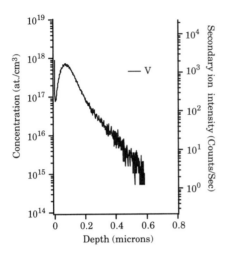

Figure 5.4. Depth profile of a vanadium implant into crystalline silicon, obtained with a 12.5-keV O_2^+ primary beam. The profile has been quantified from the dose of the implant.

from the sample surface. Note that for this device, areas which emit Si do not emit Al, and vice versa. Ion images can be collected electronically and stored for postacquisition processing.

Ion images provide a powerful method for depth profiling. If a single image gives information about the lateral distribution of an element, then it is possible to obtain three-dimensional information by collecting a series of ion images at different depths. Electronically stored images can be used to reconstruct separate depth profiles for the element(s) studied in any area of

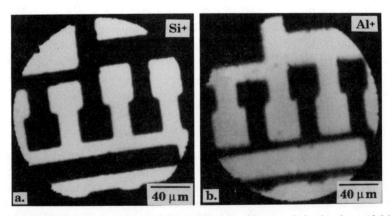

Figure 5.5. Secondary ion images obtained from a piece of integrated circuit using a 12.5-keV O_2^+ primary beam. (a) Si^+ image and (b) Al^+ image.

the image (Rüdenauer, 1984; Bryan *et al.*, 1985; Brown, 1988; Lee *et al.*, 1992). Image depth profiling is an extremely powerful technique which is beginning to find more widespread applications. Applications of image depth profiling have been limited by the computer-intensive data processing and by the large storage requirements for the image data.

5.3. Quantitative Analysis

An important aspect of any analytical technique is its ability to obtain reproducible and accurate quantitative information about the analyte(s) from the sample of interest. The analyte is defined as the elemental or molecular species of interest in the sample. In order to quantify SIMS data, the measured secondary ion intensity from the analyte must be converted to a concentration and the analysis time must be converted to a sputtered depth. One of the major problems with SIMS, however, is the difficulty in assigning concentrations to secondary ion signals. The difficulty arises primarily from the large variation in the proportion of ions formed during the sputtering of different elements and materials. The following sections discuss the parameters affecting quantitative SIMS depth profile analysis.

5.3.1. Secondary Ion Yield

The ion yield is the most important parameter to be considered in any SIMS analysis because it determines the conditions to be used during the analysis. Unfortunately, the ion yield is also the most difficult parameter to control during a SIMS analysis.

The ion yield is defined as the number of ions of a particular element, *i*, formed for each atom of *i* sputtered from the surface. The positive or negative ion yield can be written as

$$Y_i^{\pm} = \gamma_i^{\pm} \cdot S_i \qquad (5.1)$$

where $\gamma_i^{\pm}$ is the positive or negative ionization efficiency of the sputtering process for element *i* and S_i is the sputter yield of element *i* (sputter yield is defined as the number of atoms and ions of element *i* sputtered for each incident primary ion). The ionization efficiency is affected by the chemical environment in the matrix of interest and by changes in that chemical environment which result from the sputtering process (such as implantation of the primary species during analysis). Experimentally, ion yields have been found to correlate with the ionization potential (IP) of the analyte for positive secondary ions, and with the electron affinity (EA) of the analyte for negative secondary ions (Deline *et al.*, 1979a,b). Ion yields for different elements in the same matrix can vary by a factor of $>10^6$, and ion yields for the same element in different matrices also vary widely (Storms *et al.*, 1977; Wilson *et al.*, 1989b). Variation in secondary ion yield for different elements and matrices is known as the "SIMS matrix effect."

Secondary ion yields are extremely sensitive to the amount of oxygen contained within the matrix, and heavily oxidized matrices are found to enhance the ion yield of electropositive elements. Enhancements are also observed when oxygen is used as the primary ion or when the surface of a sample is flooded with oxygen during sputtering. A similar enhancement is observed for negative secondary ion yields if an electropositive species, such as cesium, is used for bombardment. The chemical enhancement of secondary ion yields by sputtering with a reactive primary species is important for trace element analysis. Higher ion yields are critical to obtaining better sensitivities and lower detection limits needed for the analysis of small volumes of material.

Secondary ion yields are also affected by changes in the sputter rate. The sputter rate is the amount of material removed from the sample for a given unit of time, e.g., nm/s. For constant primary beam conditions, sputter rate changes will be the result of disparate sputter yields or different atomic densities between materials. Ion yield differences of this type are often observed at the interfaces between materials during depth profile analysis or in different grains during secondary ion imaging. The effects can be magnified when a reactive primary particle (e.g., O_2^+ or Cs^+) is used for sputtering because of enhancement or depletion of the reactive species in the near surface region of the sample. Matrix effects make it difficult to quantify changes in secondary ion intensity at interfaces between materials.

5.3.2. Standards and Relative Sensitivity Factors

From the discussion of secondary ion yields in Section 5.3.1 it becomes evident why SIMS has garnered a reputation as a semiquantitative technique at best. Although there has been progress in attempts to model the sputter ionization process, there is still no theoretical model which accurately describes all materials systems (Williams, 1990). To overcome this problem, SIMS analysts rely on the use of empirical standards for quantitative analysis. The sensitive dependence of $Y_i^\pm$ on the matrix composition, however, necessitates that SIMS standards closely match the element and matrix combination of interest.

The secondary ion intensity measured by the detector can be written as the following function,

$$I_i^\pm = I_p \cdot S \cdot a(i) \cdot \gamma_i^\pm \cdot f_i^\pm \cdot C(i) \qquad (5.2)$$

where I_p is the primary ion current, S is the sputter yield (material sputtered from the sample per primary ion), $a(i)$ is the abundance of the isotope chosen for analysis, $\gamma_i^{\pm}$ is the ionization efficiency of the sputtering process for element i, $f_i^{\pm}$ is an instrumental transmission factor which includes contributions from the efficiency of the extraction lens, the spectrometer transmission, and the detector efficiency, and finally, $C(i)$ is the concentration of element i in the matrix (Werner, 1980). If the intensity of the analyte is normalized to a reference element in the same matrix (generally a matrix element signal), then any variations in I_p will be removed and the remaining terms will be constant for a given element/matrix combination. This constant is called the relative sensitivity factor (RSF). After an RSF has been determined for a sample of known concentration, it can be used to calculate the concentration of an unknown sample, provided the element of interest and the matrix are matched to the standard sample. Quantitative analysis with RSFs provide accuracies of better than 3% for carefully controlled analyses (Deng and Williams, 1989), and 10% on a regular basis (Spiller and Ambridge, 1986).

As stated in Section 5.3.1, early experimental observations had showed that positive and negative secondary ion yields were correlated to the ionization potential and electron affinity of the analyte, respectively (Deline *et al.*, 1979a,b). Recent work by several groups has shown that not only are the IP and EA correlated with the positive and negative secondary ion yields, respectively, but that there is also a systematic dependence of the RSF on the IP and EA as well (Wilson and Novak, 1988; Stevie *et al.*, 1988a; Wilson *et al.*, 1992). There is also evidence that RSF values do not change significantly between different instruments if experimental conditions are controlled (Simons *et al.*, 1990). This large body of observations has led to an empirical procedure for the prediction of RSFs for uncharacterized elements and matrices (Wilson *et al.*, 1989c). This is a major advance for SIMS because it allows quantitative analysis of new matrices and elements with a significantly smaller set of standards.

The most common method used to fabricate standards for SIMS analysis is ion implantation. Ion implantation introduces a known amount (implanted dose) of an element of interest directly into the matrix of interest. SIMS is then used to depth profile the implanted element. The RSF is calculated directly from the implant fluence, the measured SIMS depth profile, and the sputtered crater depth. Implant peaks must be sufficiently deep to avoid surface effects during depth profiling. Bulk doped standards can also be used for quantitative analysis provided the concentration of the element of interest can be accurately determined by some other technique, and mass interferences are known. RSF values for bulk standards are calculated directly from the known concentration and the intensities of the reference and analyte signals.

RSFs are found to be applicable in the concentration range where the element of interest can be considered to be a "trace" element in the matrix of interest, i.e., approximately ≤1 at. %. In this case the matrix composition remains essentially constant and the secondary ion intensity is a linear function of the concentration. At levels greater than 1 at. % the graph of secondary ion intensity versus concentration is not necessarily linear and the same RSF may not apply to the entire curve. If a matrix contains more than one minor element, then the sum of all of the trace elemental concentrations should be ≤1 at. % for accurate determinations with RSFs.

5.3.3. Sputtered Depth

The final step in quantifying a SIMS depth profile is the conversion of sputtered time into sputtered depth. The depth of a particular feature or the thickness of a layer is an important value needed from a depth profile; thus, care must be taken in rescaling this axis. The most accurate method of determining sputtered depth is to measure the sputtered crater depth after analysis. Crater depth measurement is commonly performed with a stylus profilometer. If the sputter rate is constant throughout the analysis, then the depth scale is a linear function of the sputtering time. In practice, however, instabilities in the primary beam and changes in the matrix can both result in sputter rates which vary significantly through the course of a profile.

5.4. Sensitivity and Resolution

There are four parameters which measure the quality of a SIMS analysis: depth resolution, sensitivity, dynamic range, and lateral resolution. These four factors are intimately linked together by the conditions chosen for the analysis. In any SIMS analysis, the "best" conditions to acquire data are determined by balancing the requirements for sensitivity and dynamic range with the desired depth and lateral resolutions.

5.4.1. Depth Resolution

The depth resolution is a measure of the sharpness of an interface in a depth profile. It is calculated by taking the interval over which the signal drops from 84% to 16% of the maximum signal level (Magee and Honig, 1982). The main factors which affect the depth resolution are (1) the roughness of the sample surface (both native and sputter induced), (2) ion beam mixing (randomization of subsurface layers because of collisions between implanted primary ions and atoms of the target) including radiation-enhanced diffusion, and (3) the sputter rate. The first two factors are determined by the physical properties of the sample, and by the interaction of the primary beam particles with the sample matrix, while the third factor is determined by the analysis conditions. The depth resolution degrades as a function of depth into the sample, i.e., worsens as the analysis progresses, generally because of sputter-induced topography or nonuniformities in the sputtered crater. Depth resolutions of ~2 nm have been obtained for delta-doped spikes in GaAs (Clegg and Gale, 1991), although depth resolutions of 10–20 nm are more usual. Collisional mixing tends to broaden the trailing edge of any feature more than the leading edge, resulting in asymmetric profiles (Clegg and Gale, 1991). The depth resolution is also affected by the migration of elements segregating toward or away from the implanted primary particles. This effect is usually associated with oxygen and can cause distortions on the order of several microns in extreme cases (Vriezema and Zalm, 1991). Finally, the depth resolution is never any better than the fastest rate of data collection. For example, if the distance sputtered between successive points in a depth profile is 10 nm (not uncommon), then the depth resolution will be limited to this level at best.

In a SIMS analysis the primary ion beam is generally rastered across the surface of the sample to produce a flat-bottomed sputtered area, referred to as the sputtered crater. A flat sputtered crater is imperative for good depth resolution. So-called "crater edge effects" are distortions in a depth profile which are caused by contributions to the analytical signal from the edges of the sputtered crater. As the analysis progresses into the sample, all of the layers of previously sputtered material are exposed to the primary beam at the edges of the crater. If ions from the crater walls are allowed to pass to the detector with the ions from the bottom of the analysis crater, they will contribute signals from different depths. This degrades the depth resolution and can also affect the detection limit of an analysis (Magee and Honig, 1982). Crater edge effects are alleviated by illuminating an area which is large relative to the area from which data are collected. In a microprobe-type machine, this is accomplished by rastering the focused ion beam over a large area and electronically gating the data collection from only the center of the crater. In an ion microscope, the beam is either defocused to cover a large area of the sample or focused and rastered across the surface similarly to the ion microprobe case. The analyzed area is then limited by a physical aperture centered over the cratered area.

An example of depth resolution distortion caused by a rough sample surface and sputter-induced topography is shown in Fig. 5.6a–c. The sample is a highly textured polycrystalline CdTe film grown on CdS. The surface of the sample before and after sputtering is shown in Fig. 5.6a and b, respectively.

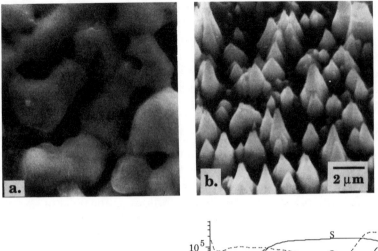

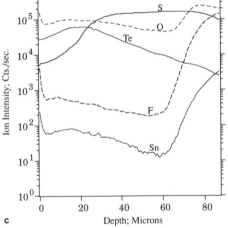

Figure 5.6. Scanning electron microscope (SEM) images and depth profile of textured polycrystalline CdTe deposited on CdS. (a) SEM image of surface before sputtering; (b) SEM image of surface after sputtering with 10-keV Cs$^+$ ion beam showing development of sputter cones; (c) SIMS depth profile from this sample showing extremely poor depth resolution.

The evolution of sputter cones is clearly seen after bombarding the surface with 12.5-keV O$_2^+$ primary ions. Figure 5.6c shows the SIMS depth profile obtained from this sample. There is no clear definition of the layers in the depth profile, even though cross-sectional SEM analysis clearly shows two distinct layers. Sputter roughening is worst for polycrystalline materials, e.g., metal films, although roughening is also observed in semiconductor materials (Stevie *et al.*, 1988b). In some cases, sputter-induced topography can be reduced by changing the angle of incidence of the primary ion (Wittmaack, 1990) or by rotating the sample under the ion beam during analysis (Cirlin *et al.*, 1991). When analyzing a material for the first time by SIMS, it is important to examine the sputtered crater after analysis to ensure that no significant roughness has formed.

5.4.2. Sensitivity, Detection Limit, and Dynamic Range

The detection limit is the minimum detectable amount of an element in a given matrix. It is determined by the SIMS sensitivity of the element (primarily a function of the ion yield and the analysis area), the spectral interferences at the mass of interest from the sample or the instrumental background, and the conditions used for analysis. Even though the sensitivity for an element is high, the detection limit can be poor if there are interferences or background contributions at the mass of interest, i.e., the dynamic range is small. For favorable elements the dynamic range in a SIMS profile can be 10^5. An example of an element with good sensitivity but poor dynamic range is observed for the analysis of O in Si. Figure 5.7 shows the depth profile from an O implant into crystalline silicon. Although the SIMS sensitivity to O is good (high count rate at the implant peak), the detection limit is poor because of the presence of O-containing contaminants in the residual gases in the vacuum chamber. These gases adsorb onto the sample from the vacuum and are subsequently sputtered by the ion beam. They can then contribute to the total O signal observed from the sample. Another cause of poor detection limits are interferences at the same nominal mass of the element of interest. An example of this is the analysis of P in Si, where the ^{31}P signal is obscured by the ^{30}SiH signal at the same nominal mass. For instruments which have ion optics in the region near the sample, contamination on the ion extraction lenses in the vicinity of the sample can contribute to background signals, thereby raising the detection limit. Finally, the detection limit can be affected by crater edge effects discussed in Section 5.4.1.

There are several methods for removing mass interferences. In some cases it is possible to use a different isotope of the element to avoid or minimize

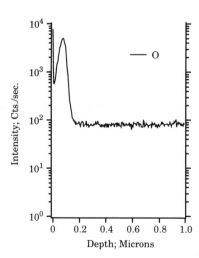

Figure 5.7. Depth profile of O implanted into crystalline silicon. The poor detection limit is caused by the high residual gas level in the vacuum chamber.

the interference. Other common methods used to remove interferences when another isotope is not available are either high mass resolution or energy filtering (Wilson et al., 1989d). High mass resolution is possible with magnetic sector and TOF instruments. Energy filtering takes advantages of differences between the energy spectra for molecular ions versus atomic ions; the former has a spectrum which is sharply peaked at low energies, the latter has a broader spectrum which can have significant intensity at energies > 75 eV. Both of these methods restrict the transmission of the instrument and result in a loss in the sensitivity. Sensitivity can be increased by analyzing a larger area (at the expense of lateral resolution) or by sputtering faster (at the expense of depth resolution).

5.4.3. Lateral Resolution

The ultimate lateral resolution of a SIMS analysis is determined by the type of instrument (Section 5.2.5.3) and by the sensitivity needed from the analysis. In the imaging mode of analysis the resolution can be 20–30 nm for microprobe machines with LMIS, and 0.5–1 μm for the ion microscope. For practical purposes, however, the lateral resolution is often determined by the sensitivity needed from the analysis. A small analyzed area provides fewer total ion counts, which consequently raises the detection limit. Sensitivity can be increased by increasing the sputter rate, although this decreases the depth resolution.

5.5. Experimental Examples

The data shown in the following examples were obtained with a direct-imaging, double-focusing magnetic sector SIMS instrument. Primary ion beams of O_2^+ and Cs^+ were used, with detection of positive and negative secondary ions, respectively.

5.5.1. Dopant Profiling

Depth profile analysis of dopants in semiconductor materials and devices is the most common use of SIMS. SIMS is able to provide good lateral and depth resolution, together with high sensitivity and low detection limits necessary for the measurements of dopant profiles. The most basic requirement for photovoltaic (PV) devices is the need for a p–n junction within the device. High-efficiency (30%) tandem PV devices are fabricated from III–V materials. The formation of a sharp p–n junction with the proper doping levels is important for proper device operation. In this example, the n-type dopant Se was studied in GaInP$_2$/GaAs cells. Figure 5.8 shows the depth profile obtained for Se and O in the device. The depth profile showed a Se contamination spike at the beginning of the p-doped GaInP$_2$ layer growth (nominally Zn doped). In addition, the depth profile indicated O contamination in the Al-containing window layer of the device.

5.5.2. Distribution of an Analyte following Processing

The ability to successfully getter detrimental impurities during photovoltaic device fabrication is a desirable characteristic of any processing step. Various transition metal atoms can be gettered during the diffusion anneal which creates the p–n junction to form the device. In this example, the gettering of Cr during high-temperature P diffusion is observed by SIMS depth profiling. Figure 5.9 shows the depth profile from a Cr implant into silicon. Figure 5.9 also shows the Cr profile from the same sample following diffusion of P from a solid source at 900°C for 30 min. The profile from the annealed sample shows that the Cr has been gettered from the silicon into the oxide layer formed on the surface during the anneal. The gettered Cr is removed with the oxide layer during a subsequent processing step.

5.5.3. Stable Isotope Tracer Analysis

The ability of SIMS to study stable isotopes is useful for the study of diffusion properties of major elements in materials. Hydrogenated amorphous

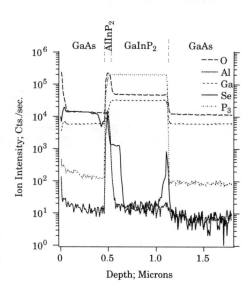

Figure 5.8. Se dopant profile from a GaAs/GaInP$_2$ photovoltaic device, showing Se contamination at the back GaInP$_2$/GaAs interface.

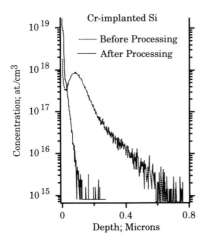

Figure 5.9. Depth profiles of Cr implant into silicon before (-----) and after (———) solid source diffusion of P at 900°C for 30 min. The Cr is gettered by the phosphorosilicate glass formed on the silicon surface during the diffusion step.

silicon (a-Si:H) is an attractive material for production of low-cost PV modules, although loss of conversion efficiency during illumination has limited its use. The random motion of H within amorphous silicon is thought to contribute to the observed degradation of this material. SIMS is the only surface analytical technique which is capable of directly observing H in materials. In addition, the ability of SIMS to distinguish between H and D allows the use of D as a stable isotope tracer to study the self-diffusion of H in a-Si:H.

In this example, amorphous silicon films were grown with a layer intentionally doped with D at a level of $\sim 10^{20}$ atoms/cm^3 (nominal H concentration is ~ 10 at. %, or 2–3×10^{21} atoms/cm^3). The deuterated layer was sandwiched between two layers of hydrogenated amorphous silicon, all grown on crystalline Si, a-Si:H/a-Si:H:D/a-Si:H/c-Si. The films were annealed at different temperatures and SIMS was used to measure the D profile in each film. Depth profiles from the as-deposited film and pieces of the same film annealed at 180 and 210°C are shown in Fig. 5.10. The depth profiles from the annealed films show the diffusion of D caused by the high-temperature annealing. From the SIMS profiles it is possible to determine the diffusion coefficient of D in a-Si:H. Figure 5.10 also shows the ability of SIMS to discern small amounts of diffusion in materials (the small "wings" on the 180°C profile).

5.5.4. Multilayer Analysis

SIMS can be used to determine the structure of multilayer materials. In this case, 30 periods of a GaInP$_2$ (12.2 nm) and GaAs (9.1 nm) superlattice were deposited on a GaAs substrate with GaInP$_2$ capping layer. Figure 5.11 shows the depth profile from the major matrix elements (with the exception

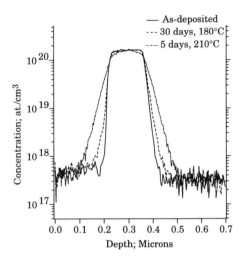

— As-deposited
--- 30 days, 180°C
····· 5 days, 210°C

Figure 5.10. Depth profiles of the D distribution in an amorphous silicon sandwich structure containing a D-doped layer (a-Si: H/a-Si:H:D/a-Si:H), annealed at different temperatures.

of P) in this sample. All of the layers are clearly visible, although the thin GaAs layers are not completely resolved because of the depth resolution limits of the analysis. Gao (1988, 1990) has shown that it is possible to use SIMS to provide atomic compositions from thin layers by using molecular ions of Cs^+ $(M + Cs)^+$ for the analysis.

5.5.5. Image Analysis

The segregation of dopants and impurities to the grain boundaries can greatly affect material quality. In this example, ion images were used in conjunction with light microscopy to identify lanthanum in the grain boundaries of Si. Figure 5.12a shows an optical micrograph of a grain boundary in heavily La-doped silicon. Figure 5.12b shows the La secondary ion image from the same grain boundary region. The La image clearly shows that the majority of emission came from the grain boundary rather than from inside the grain. This indicated that most of the La had segregated to the boundary during growth rather than being incorporated into the crystal. Subsequent depth profiles of an intragranular area relative to an intergranular area confirmed this finding.

5.6. Related Techniques

Dynamic SIMS has enjoyed widespread use as a microcharacterization technique in the semiconductor field because of its high sensitivity, excellent

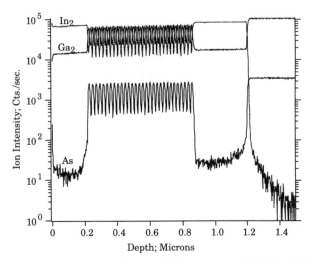

Figure 5.11. Depth profile of a $GaInP_2$/GaAs multiple-quantum-well device with 30 periods of a $GaInP_2$ (12.2 nm) and GaAs (9.1 nm) superlattice. All 30 periods of the superlattice structure are visible in the profile.

depth resolution, and good lateral resolution. Semiconductor samples, however, are ideally suited for study by this technique because they are flat, (mostly) conducting, and are generally well characterized prior to SIMS analysis (i.e., the composition is known and impurities are generally at trace levels). In spite of this success, SIMS analysis of semiconductor and other materials suffers from several problems which are all related to the variability of the ion yield. Primarily the difficulties are found during quantitative analysis of unknown matrices, in matrix effects at interfaces between different materials, and in poor ion yields for some elements. In addition, dynamic SIMS is not able to analyze organic or polymer molecules without causing significant damage to the molecule in the process. The related techniques discussed in the following sections (static SIMS, TOF SIMS, and sputtered neutral techniques) are designed to augment and/or improve on the capabilities of dynamic SIMS. Although it is unlikely that either will replace dynamic SIMS in the near future, each has some strengths which are unmatched by dynamic SIMS for particular analyses.

5.6.1. Static, Molecular, and Time-of-Flight SIMS

Benninghoven and co-workers (1973) were the first to use the term *static SIMS* (SSIMS) to describe the analysis of organic solids with low-density

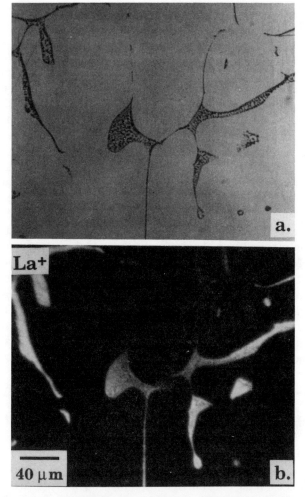

Figure 5.12. Comparative images of grain boundaries in polycrystalline silicon: (a) optical micrograph of a distinct grain boundary region; (b) La⁺ secondary ion image of the same boundary region as in (a).

primary ion beams (total dose $< 10^{13}$ cm^{-2}). In the intervening years, SSIMS has become a popular technique for the analysis of surfaces (top monolayers of a sample) and complicated molecular structures. Because of its widespread use to characterize molecular fragments from organic and polymer species, SSIMS is often called *molecular SIMS*. SSIMS allows high-mass molecules to be removed from the surface and ionized intact. There is some fragmentation, but this is used to advantage in the generation of fingerprint spectra (Briggs *et al.*, 1989). Quantitative analysis of molecular SIMS information is also plagued by variations in the ion yield; thus, as discussed in the section on quantitative analysis (5.5.3), standards are required. In general, it is not possible to perform depth profile analyses of polymers and other organic materials because of the requirement that each molecule sputtered come from a virgin area of the surface. The primary particle damages the surface and the near-surface region; thus, the molecular information obtained from the surface may be lost farther into the sample. SSIMS measurements can be made with any mass analyzer, although they are particularly suited to the TOF mass analyzer.

Application of SSIMS to a broader range of problems was made possible by coupling the low-dose primary ion sources required for SSIMS to TOF mass spectrometers. This technique is now referred to as TOF-SIMS. TOF analyzers are well matched to the SSIMS process because they have the high transmission required to make use of the low signal levels generated during SSIMS analyses, and they have the high mass range (in excess of 10,000 amu) needed for the study of high-mass materials. In addition, TOF analyzers are able to operate with high mass resolutions (unlike quadrupoles) allowing complicated molecular structures to be accurately determined. Finally, even the low current densities needed for SSIMS are compatible with the requirement that the ion source be pulsed for the TOF analyzer. TOF analyzers have extended the range of organic and polymer species which can be studied, and the number of applications have grown accordingly.

There are two areas of research for molecular SIMS: trace molecular analysis of mixtures and direct analysis of surfaces. The direct analysis of polymer sample surfaces is illustrated by the work of Mawn *et al.* (1991). In this work, TOF-SIMS was used to study the polymer additives present in polyethylene. Spectra taken from an extract of the polyethylene were compared with spectra taken directly from the bulk polymer. The additives could be analyzed directly in the bulk polymer by comparing the spectra from the bulk with spectra from the extract. Quantitative analysis of mixed fatty acid Langmuir–Blodgett films was studied by Cornelio and Gardella (1990). This study illustrates the difficulties in trying to quantify molecular secondary ion intensities.

5.6.2. Sputtered Neutral Mass Spectrometries (SNMS)

As discussed in Section 5.2.1, most of the material sputtered from the sample during a SIMS analysis is in the form of neutral atoms and molecules. This material is lost in conventional SIMS machines because, being uncharged, it cannot be collected by the mass spectrometer. Sputtered neutral techniques discard the small fraction of true secondary ions, and instead make use of the larger sputtered neutral flux. Since the sputtered neutral flux is not as affected by changes in matrix, these techniques have fewer matrix effects and simpler quantification schemes. In addition, SNMS is not affected by charging at the sample surface so that analysis of bulk insulators is simplified.

There are several methods used for postionization of sputtered neutrals. Secondary ions can be generated in an Ar plasma (which is also used to sputter the sample surface), by electron impact sources and by lasers (multiphoton resonant and nonresonant). Material is removed from the surface by ion beams or in the case of the plasma ionization by the plasma itself. In the case of nonresonant multiphoton ionization [surface analysis by laser ionization (SALI)], pulsing the laser when the ion beam is in the center of a rastered crater has resulted in excellent depth resolutions for multilayer semiconductor films (Becker, 1991). A good review of the SALI technique is that presented by Becker (1991). A review of electron and plasma ionization methods is given by Jede (1990). An example of matrix quantification of metal overlayers on p-HgCdTe is found in Storm et al. (1991). Most of the examples of SNMS have been more demonstrative than routine, but there is little doubt that this will change as the technique becomes more widely available.

5.7. Future Trends

With the decreasing dimensions of semiconductor devices, the challenge to dynamic SIMS and SNMS is to make maximum use of each and every atom sputtered from a sample surface. For dynamic SIMS and SSIMS this will mean further optimizing instrumental transmission and detector sensitivity and improving direct secondary ion yields. For SNMS this means increasing the useful yield of postionized species, either through more efficient postionization or through optimization of the collection optics to gain higher transmissions. The sputtered neutral techniques hold the most promise for improved detection capabilities because they take advantage of the large fraction of material which is not ionized directly during the sputtering process. Although most work with these techniques has been of the demonstrative nature, availability of commercial instrumentation will undoubtedly improve access and application of these techniques to a larger variety of problems. In addition, the optimization of the collection optics of SNMS instruments will lead to higher transmissions and higher useful secondary ion yields.

Further improvements in ion source design (high brightness in small spots) will also aid both conventional and SNMS analysis. Better postionization techniques will improve both depth and lateral resolution of these techniques.

It seems likely that the optimization of SNMS ion collection schemes and secondary ion optics will improve the useful yield of this technique significantly.

SIMS and related techniques have all gained from the proliferation of large, rapid, and low-cost computer storage media. This has made it possible to save large amounts of data rapidly which is important for imaging SIMS and related techniques. There is a tremendous amount of information contained in one secondary ion image, but until recently it was difficult to store enough data to realize the idea of image depth profiling by SIMS. In addition, most laboratory computers were inadequate to process the data once they were obtained. Individual graphics workstations are making it possible to perform complicated three-dimensional reconstructions in the laboratory (Lee *et al.*, 1992). Secondary ion image analysis has the potential to solve many problems of lateral localization in small areas during depth profiling, but still suffers from the loss of sensitivity and dynamic range through the detectors (for imaging machines) and low primary ion currents (for microprobe machines). For the ion microscope the loss occurs because of the detector sensitivity and dynamic range. For the ion microprobe the loss of primary ion current because of extremely small beam spot sizes limits signals and the ability to depth profile. Also, the use of a Ga LMIS to obtain small primary beam spot sizes results in the loss of chemical enhancement normally gained with O_2^+ and Cs^+ ion beams.

TOF-SIMS has become an important tool for the study of polymers and nonvolatile organic molecules. The ability to desorb large molecules and fragments intact and mass analyze them accurately at high mass range is an important development for organic mass spectrometry. A major problem with traditional organic mass spectrometry has been difficulty in volatilizing large, nonthermally labile molecules intact. One of the strengths of SSIMS is its ability to postionize large molecules. It is possible that (similarly to dynamic SIMS) the ability to decouple the desorption process from the ionization process will also be important for improving the capabilities of this process.

Materials systems with multiple components, e.g., conducting polymers and inorganic materials, are becoming important in manufacturing of many products. SIMS, SSIMS, and related techniques offer the ability to analyze surface and interface processes with excellent sensitivity, high depth and lateral resolution. SIMS is already finding use as a trace element analysis method for the study of biological materials and it may also have applications in the study of interfaces in biomedical materials. The broad family of SIMS characterization techniques have overcome or found ways to alleviate many limitations experienced at the beginning. Although there are still formidable problems for quantitative analysis, dynamic SIMS has found widespread acceptance as a trace element analysis technique. TOF-SIMS is fast becoming a standard method for organic solid and polymer analysis. Finally, SNMS offers a way to avoid some of the problems with quantification of standard SIMS results and will undoubtedly find more applications in the future.

References

Anderson, C. A., Roden, H. J., and Robinson, C. F. (1969). Negative ion bombardment of insulators to alleviate surface charge-up, *J. Appl. Phys.* **40**, 3419–3420.

Becker, C. H. (1991). Laser resonant and nonresonant photoionization of sputtered neutrals, in *Ion Spectroscopies for Surface Analysis* (A. W. Czanderna and D. M. Hercules, eds.), Plenum Press, New York, pp. 273–310.

Benninghoven, A. (1973). Surface investigation of solids by the statical method of secondary ion mass spectroscopy (SIMS), *Surf. Sci.* **35**, 427–457.

Benninghoven, A. (1985). Static SIMS applications—from silicon single crystal oxidation to DNA sequencing, *J. Vac. Sci. Technol.* **A3**, 451–460.

Benninghoven, A., Evans, C. A., Powell, R. A., Shimizu, R., and Storms, H. A. (eds.) (1979). *Secondary Ion Mass Spectrometry (SIMS II), Springer Series in Chemical Physics* Vol. 9, Springer-Verlag, Berlin.

Benninghoven, A., Giber, J., Laszlo, J., Riedel, M., and Werner, H. W. (eds.) (1982). *Secondary Ion Mass Spectrometry (SIMS III), Springer Series in Chemical Physics* Vol. 19, Springer-Verlag, Berlin.

Benninghoven, A., Okano, J., Shimizu, R., and Werner, H. W. (eds.) (1984). *Secondary Ion Mass Spectrometry (SIMS IV), Springer Series in Chemical Physics* Vol. 36, Springer-Verlag, Berlin.

Benninghoven, A., Colton, R. J., Simons, D. S., and Werner, H. W. (eds.) (1986). *Secondary Ion Mass Spectrometry (SIMS V), Springer Series in Chemical Physics* Vol. 44, Springer-Verlag, Berlin.

Benninghoven, A., Rüdenauer, F. G., and Werner, H. W. (1987). *Secondary Ion Mass Spectrometry,* Wiley, New York.

Benninghoven, A., Huber, A. M., and Werner, H. W. (eds.) (1988). *Secondary Ion Mass Spectrometry (SIMS VI),* Wiley, New York.

Benninghoven, A., Evans, C. A., McKeegan, K. D., Storms, H. A., and Werner, H. W. (eds.) (1990). *Secondary Ion Mass Spectrometry (SIMS VII),* Wiley, New York.

Benninghoven, A., Janssen, K. T. F., Tümpner, J., and Werner, H. W. (eds.) (1992). *Secondary Ion Mass Spectrometry (SIMS VIII),* Wiley, New York.

Briggs, D. (1989). Surface analysis, *Encycl. Polym. Sci. Eng.* **16**, 399–442.

Briggs, D., Brown, A., and Vickerman, J. C. (1989). *Handbook of Static Secondary Ion Mass Spectrometry,* Wiley, New York.

Brown, J. D. (1988). Three dimensional secondary ion mass spectrometry, *Scanning Microsc.* **2**, 653–662.

Bryan, S. R., Woodward, W. S., Linton, R. W., and Griffis, D. P. (1985). Secondary ion mass spectrometry/digital imaging for the three-dimensional chemical characterization of solid state devices, *J. Vac. Sci. Technol.* **A3**, 2102–2107.

Burns, M. S. (1988). Biological microanalysis by secondary ion mass spectrometry: Status and prospects, *Ultramicroscopy* **24**, 269–281.

Cirlin, E.-H., Vajo, J. J., Doty, R. E., and Hasenberg, T. C. (1991). Ion-induced topography, depth resolution, and ion yield during secondary ion mass spectrometry depth profiling of GaAs/AlGaAs superlattice: Effects of sample rotation, *J. Vac. Sci. Technol.* **A9**, 1395–1401.

Clegg, J. B., and Gale, I. G. (1991). SIMS profile simulation using delta function distributions, *Surf. Interface Anal.* **17**, 190–196.

Cornelio, P. A., and Gardella, J. A., Jr. (1990). Quantitative aspects of secondary ion mass spectrometry analysis of pure and mixed fatty acid Langmuir–Blodgett films, *J. Vac. Sci. Technol.* **A8**, 2283–2286.

Degreve, F., Thorne, N. A., and Lang, J. M. (1988). Metallurgical applications of secondary ion mass spectrometry (SIMS), *J. Mater. Sci.* **23**, 4181–4208.

Deline, V. R., Evans, C. A., and Williams, P. (1979a). A unified explanation for secondary ion yields, *Appl. Phys. Lett.* **33**, 578–580.

Deline, V. R., Katz, W., Evans, C. A., and Williams, P. (1979b). Mechanism of the SIMS matrix effect, *Appl. Phys. Lett.* **33**, 832–835.

Deng, R.-C., and Williams, P. (1989). Factors affecting precision and accuracy in quantitative analysis by secondary ion mass spectrometry, *Anal. Chem.* **61**, 1946–1948.

Gao, Y. (1988). A new secondary ion mass spectrometry technique for III–V semiconductor compounds using the molecular ions CsM^+, *J. Appl. Phys.* **64**, 3760–3762.

Gao, Y. (1990). Cs-SIMS quantitative analysis using secondary negative molecular ions for GaAs, InP and related compounds: Matrix effects, in *Secondary Ion Mass Spectrometry (SIMS VII)* (A. Benninghoven, C. A. Evans, K. D. McKeegan, H. A. Storms, and H. W. Werner, eds.), Wiley, New York, pp. 155–158.

Grasserbaüer, M., Stingeder, G., Pötzl, H., and Guerrero, E. (1986). Analytical science for the development of microelectronic devices, *Fresenius Z. Anal. Chem.* **323**, 421–449.

Jede, R. (1990). Postionization with electrons, plasma and beams (invited), in *Secondary Ion Mass Spectrometry (SIMS VII)* (A. Benninghoven, C. A. Evans, K. D. McKeegan, H. A. Storms, and H. W. Werner, eds.), Wiley, New York, pp. 169–177.

Lee, J. J., Gray, K. H., Lin, W. J., Hunter, J. L., and Linton, R. W. (1992). Three-dimensional visualization of secondary ion images, in *Secondary Ion Mass Spectrometry (SIMS VIII)* (A. Benninghoven, K. T. F. Janssen, J. Tümpner, and H. W. Werner, eds.), Wiley, New York, pp. 505–508.

Levi-Setti, R. (1990). Recent applications of high resolution secondary-ion-mass-spectrometry imaging, *Vacuum* **41**, 1598–1600.

McHugh, J. A. (1975). Secondary ion mass spectrometry, in *Methods of Surface Analysis* (A. W. Czanderna, ed.), Elsevier, Amsterdam, pp. 225–278.

Magee, C. W., and Botnick, E. M. (1985). On the use of secondary ion mass spectrometry in semiconductor device materials and process development, in *Proceedings of Applied Materials Characterization* (W. Katz and P. Williams, eds.), Materials Research Society, Pittsburgh, pp. 229–240.

Magee, C. W., and Honig, R. E. (1982). Depth profiling by SIMS—Depth resolution, dynamic range and sensitivity, *Surf. Interface Anal.* **4**, 35–41.

Mawn, M. P., Linton, R. W., Bryan, S. R., Hagenhoff, B., Jürgens, U., and Benninghoven, A. (1991). Time-of-flight static secondary ion mass spectrometry of additives of polymer surfaces, *J. Vac. Sci. Technol.* **A9**, 1307–1311.

Reed, S. J. B. (1989). Ion microprobe analysis; a review of geological applications, *Mineral. Mag.* **53**, 3–24.

Rüdenauer, F. G. (1984). Spatially multidimensional SIMS analysis, *Surf. Interface Anal.* **6**, 132–139.

Simons, D. S., Chi, P. H., Kahora, P. M., Lux, G. E., Moore, J. L., Novak, S. W., Schwartz, C., Schwartz, S. A., Stevie, F. A., and Wilson, R. G. (1990). Are relative sensitivity factors transferable among SIMS instruments? in *Secondary Ion Mass Spectrometry (SIMS VII)* (A. Benninghoven, C. A. Evans, K. D. McKeegan, H. A. Storms, and H. W. Werner, eds.), Wiley, New York, pp. 111–114.

Spiller, G. D. T., and Ambridge, T. (1986). Reproducible quantitative SIMS analysis of semiconductors in the Cameca IMS 3f, in *Secondary Ion Mass Spectrometry (SIMS V)* (A. Benninghoven, R. J. Colton, D. S. Simons, and H. W. Werner, eds.), Springer-Verlag, Berlin, pp. 127–129.

Stevie, F. A., Kahora, P. M., Singh, S., and Kroko, L. (1988a). Atomic and molecular relative secondary ion yields of 46 elements in Si for O_2^+ and Cs^+ bombardment, in *Secondary Ion Mass Spectrometry (SIMS VI)* (A. Benninghoven, A. M. Huber, and H. W. Werner, eds.), Wiley, New York, pp. 319–322.

Stevie, F. A., Kahora, P. M., Simons, D. S., and Chi, P. (1988b). Secondary ion yield changes in Si and GaAs due to topography changes during O_2^+ or Cs^+ ion bombardment, *J. Vac. Sci. Technol.* **A6**, 76–80.

Storm, W., Altebockwinkel, M., Wiedmann, L., Benninghoven, A., Ziegler, J., and Bauer, A. (1991). Depth profile analysis of Pt, Cu, and Au overlayers on p-$Hg_{1-x}Cd_xTe$, *J. Vac. Sci. Technol.* **A9**, 14–20.

Storms, H. A., Brown, K. F., and Stein, J. D. (1977). Evaluation of a cesium positive ion source for secondary ion mass spectrometry, *Anal. Chem.* **49**, 2023–2030.

Van den Berg, J. A. (1986). Neutral and ion beam SIMS of non-conducting materials, *Vacuum* **36**, 981–989.

Vickerman, J. C., Brown, A., and Reed, N. M. (1990). *Secondary Ion Mass Spectrometry, Int. Ser. Monogr. Chem.* **17**, Oxford University Press, London.

Vriezema, C. J., and Zalm, P. C. (1991). Impurity migration during SIMS depth profiling, *Surf. Interface Anal.* **17**, 875–887.

Werner, H. W. (1980). Quantitative secondary ion mass spectrometry: A review, *Surf. Interface Anal.* **2**, 56–74.

Williams, P. (1990). Sputtered ion formation, in *Secondary Ion Mass Spectrometry (SIMS VII)* (A. Benninghoven, C. A. Evans, K. D. McKeegan, H. A. Storms, and H. W. Werner, eds.), Wiley, New York, pp. 15–24.

Wilson, R. G., and Novak, S. W. (1988). Systematics of SIMS relative sensitivity factors versus electron affinity and ionization potential for Si, Ge, GaAs, GaP, InP, and HgCdTe determined from implant calibration standards for about 50 elements, in *Secondary Ion Mass Spectrometry (SIMS VI)* (A. Benninghoven, A. M. Huber, and H. W. Werner, eds.), Wiley, New York, pp. 57–61.

Wilson, R. G., Stevie, F. A., and Magee, C. W. (1989a). *Secondary Ion Mass Spectrometry*, Wiley, New York.

Wilson, R. G., Stevie, F. A., and Magee, C. W. (1989b). *Secondary Ion Mass Spectrometry, Appendix E,* Wiley, New York.

Wilson, R. G., Stevie, F. A., and Magee, C. W. (1989c). *Secondary Ion Mass Spectrometry*, Wiley, New York, pp. 3.3-1–3.3-4.

Wilson, R. G., Stevie, F. A., and Magee, C. W. (1989d). *Secondary Ion Mass Spectrometry*, Wiley, New York, pp. 2.11-1–2.11-4.

Wilson, R. G., Stevie, F. A., and Kahora, P. M. (1992). SIMS depth profiling and relative sensitivity factors/relative ion yields and systematics for elements from hydrogen to uranium implanted in metal matrices (Be, Al, Te, Ni, W, and Au), in *Secondary Ion Mass Spectrometry (SIMS VIII)* (A. Benninghoven, K. T. F. Janssen, J. Tümpner, and H. W. Werner, eds.), Wiley, New York, pp. 487–490.

Wittmaack, K. (1990). Effect of surface roughening on secondary ion yields and erosion rates of silicon subject to oblique oxygen bombardment, *J. Vac. Sci. Technol.* **A8**, 2246–2250.

6

Applications of Megaelectron-Volt Ion Beams in Materials Analysis

P. Revesz and J. Li

6.1. Introduction to Rutherford Backscattering

Rutherford backscattering spectrometry dates back to the classical Rutherford experiment done in the early 1900s. Naturally, at that time no one considered the Rutherford experiment as the beginning of a new analytical technique. It took many years and advances in a number of fields for it to become a widely used analytical tool.

In the original experiment by Rutherford, Geiger, and Marsden, a radioactive isotope was used as a source of energetic charged particles. On the one hand, a radioactive source would be the easiest solution for a Rutherford backscattering measurement because of its compactness and simplicity of use. On the other hand, the use of a radioactive source has two basic drawbacks: the radiation takes place in all directions resulting in a low flux at the position of the target, and the energy of the particles is fixed. The energy of the emitted particles from a radioactive isotope is not ideal for most Rutherford backscattering measurements.

P. Revesz and J. Li • Department of Materials Science and Engineering, Cornell University, Ithaca, New York 14853. *Present address of J. L.:* Intel Corporation, Santa Clara, California 95052.

Microanalysis of Solids, edited by B. G. Yacobi *et al.* Plenum Press, New York, 1994.

Rapid developments in the field of nuclear technology have made particle accelerators increasingly common in laboratories throughout the world. The advantages offered by Rutherford backscattering even in those early times were quite obvious. Analysis of the backscattering energy spectrum was a known method used by nuclear physicists to identify various contaminants of the target.

The Rutherford spectrometry first attracted the attention of solid state physicists at the beginning of the 1960s when the channeling effect was discovered. The channeling effect offered a relatively easy way to determine the depth location of lattice defects.

The backscattering technique had to overcome different instrumental difficulties to become a widespread analytical tool. One of the largest problems was the development of a reasonably fast backscattering measurement. This pointed at the need for a suitable particle detector. Classically, investigators used magnetic analyzers for energy measurement. Although such equipment provided superior energy resolution for charged particles, their low efficiency and large size prevented the wider use of backscattering. In the 1960s solid-state particle detectors became an inexpensive solution to this problem. With the advancements of nuclear electronics combined with solid-state detectors, backscattering matured to a relatively simple and fast analytical technique. Application of personal computers and computational techniques have made data reduction procedures routine and user-friendly.

Two books give a detailed description of the backscattering spectrometry. Chu et al. (1976) deal with Rutherford backscattering and related techniques, whereas Feldman and Mayer (1986) discuss other analytical techniques as well.

All analytical techniques share a common principle, i.e., incoming radiation causes an elastic or inelastic transition in the material under study, as a result of which radiation emerges from the target material. It is natural to expect that the characteristics of the outgoing radiation are determined by the incoming radiation and various properties of the target material. During the process of radiation–target interaction, various properties of the material may change substantially (destructive methods) or the change may be negligible (nondestructive methods). The scale of these changes depends not only on the method of study (radiation) but also on the characteristics of the material itself.

There are a number of ways to classify the analytical methods. It is convenient to categorize them by the difference in the probing and the outgoing radiation. Basically, the type of incoming and outgoing radiation may be the same or different. In the first case we have:

Incoming particle	Outgoing particle	Method	
Ion	Ion	Rutherford backscattering	RBS
Electron	Electron	Auther electron spectrosopy	AES
X ray	X ray	X-ray fluorescence spectroscopy	XRF

For different types of incoming and outgoing radiation:

Incoming particle	Outgoing particle	Method	
Ion	(Other) ion	Secondary ion mass spectroscopy	SIMS
Electron	X ray	Electron microprobe analysis	EMA
X ray	Electron	X-ray photoelectron spectroscopy	XPS
Ion	X ray	Particle-induced x-ray emission	PIXE
Ion	(Lighter) ion	Forward recoil elastic scattering	FRES
Ion	Particle/photon	Nuclear reaction analysis	NRA

Analytical techniques provide at least one of the following pieces of information about the target material. They are: identification of the constituents, concentrations, depth distributions, and structure. The choice of a particular analytical technique is always very problem-dependent. In a number of cases relating to ambiguous results or different "artifacts" it is necessary to use more than one measurement technique. It is interesting to note that with the use of a particle accelerator, four different types of measurement (RBS, FRES, PIXE, and NRA) can be accomplished with one apparatus.

6.2. Physical Principles of Backscattering

During the event of backscattering of a charged particle, there are four basic physical principles that provide the foundation of quantitative analysis of materials. All four phenomena describe different aspects of Rutherford backscattering spectrometry and each can be viewed as separate entities, representing different concepts.

As mentioned earlier, the main attraction of Rutherford backscattering spectrometry is that it provides four independent pieces of information encoded in one backscattering spectrum. The concepts of backscattering spectrometry together with their corresponding analytical results can be summarized as follows:

Concepts			Result
Kinematic factor:	k	⇨	Mass identification
Scattering cross section:	σ	⇨	Concentration
Stopping cross section:	ϵ	⇨	Depth scales
Minimum (channeling) yield: χ_{min}		⇨	Lattice defects

In the course of the mathematical description of the above-mentioned concepts, one can use different levels of approximation. It is always very important to know the limits imposed by the particular physical model used. In the case of Rutherford backscattering spectrometry, even the zeroth order of approximation ("eyeballing") can give useful information on the qualitative level. The model using first order of approximation is quite simple and enables the analyst to derive reliable quantitative results with sufficient accuracy in most cases.

In the following sections we will discuss the four basic concepts of backscattering.

6.2.1. The Concept of Kinematic Factor

Rutherford backscattering deals with elastic scattering of charged particles with the nuclei of the target material. During elastic scattering events the interaction between the colliding particles can be fully described by the principles of conservation of energy and momentum.

What are the basic guidelines that ensure the validity of this assumption?

- During the scattering events the energy spent on nonmechanical interactions (e.g., chemical bond breaking, electronic shell excitations) must be negligible.
- The scattering processes must not be distorted by the presence of nuclear phenomena like reactions and inelastic resonances.

It is relatively easy to fulfill the first condition. Kinetic energy spent on the excitation of electron shells during the event of backscattering is negligible for megaelectron-volt energy charged particles.

The question of possible nuclear reactions and resonances is somewhat more complex. As a rule of thumb, the likelihood of some kind of nuclear process increases as the energy of the projectile becomes comparable with the Coulomb barrier of the nucleus. For most cases no set threshold energy exist. It is obvious that the likelihood of nuclear interaction is greatest when proton beams are used to study targets with light elements. Nuclear processes are so intensive in this case that they become the primary interactions (RBS transforms into NRA). On the other hand, it is fair to say that for most of the nuclei the collisions of α particles with an energy of about 2 MeV can be considered as scatterings free of nuclear effects.

Let us consider a simple collision event when a projectile of mass M_1 and energy E_0 impinges on a target made of stationary atoms of mass M_2. During the Rutherford backscattering experiment the energy of the projectile after the collision is measured. This scattering problem can be solved explicitly by applying the principles of conservation of energy and momentum. An important constant for a pair of scattering particles is the kinematic factor defined as $k \equiv E_1/E_0$, where E_0 and E_1 are the energies of the projectile before and after collision, respectively.

The explicit form of the kinematic factor for the case $M_1 \geq M_2$ is a function of the masses and the scattering geometry:

$$k_{M_2} = \left[\frac{(M_2^2 - M_1^2 \sin^2\theta)^{1/2} + M_1 \cos\phi}{M_1 + M_2} \right]^2 \qquad (6.1)$$

The energy transfer to the target nucleus, i.e., the relative recoil energy, is given as

$$E_2/E_0 = \frac{4M_1 M_2}{(M_1 + M_2)^2} \cos^2\phi \qquad (6.1a)$$

where θ and ϕ are the scattering and recoil angles relative to the direction of the incident beam. It is clear that the kinematic factor depends only on two factors: the scattering geometry (θ) and the mass ratio $x = M_1/M_2$. The proper choice of the scattering geometry is important for the optimal performance of Rutherford backscattering spectrometry. Analysis of Eq. (6.1) for the k factor shows that at total backscattering ($\theta = 180°$) the mass resolution is the highest for a given pair of projectile and target atom. In practice it is not possible to place a particle detector at this position because it would obstruct the path of the incoming particle. In the case of small-angle scatterings, mass resolution deteriorates and certain deviations from the elastic nature of scattering may be observed. At small scattering angles the impact parameter of the collision (related to the closest approach of the projectile to the scattering center) may become comparable to the Bohr radius of the atom.

The meaning of the mass resolution is very important. Mass resolution is determined by the choice of energy, mass of the projectile, and the scattering geometry. However, the quality of the electronic system also plays an important role. During Rutherford backscattering measurements it is necessary to optimize the conditions by weighing different goals of the analysis. The optimization for either mass resolution, depth resolution, concentration sensitivity, or measurement time in most cases goes at the expense of the other.

In summary, with the use of the concept of kinematic factor the backscattering spectrum of an infinitesimally thin target composed of different constituents can be interpreted as a mass spectrum. Therefore, based on the Rutherford backscattering spectrum these masses can be determined.

6.2.2. The Concept of Scattering Cross Section

In the preceding section we discussed the relation between the energy E_0 of the incident particle of mass M_1 and its energy after an elastic scattering event at an angle of θ with an initially stationary target particle of mass M_2. The concept of the scattering cross section deals with the likelihood of this scattering event.

The differential scattering cross section, $d\sigma/d\Omega$, is defined as follows: at an angle θ from the direction of incidence an ideal detector is counting the number of particles scattered in the differential solid angle of $d\Omega$. Let us assume that Q is the total number of particles that have hit the target and dQ is the number of scattered particles recorded by the detector, then $d\sigma/d\Omega$ is defined as

$$d\sigma/d\Omega = (1/Nt) \cdot [(dQ/d\Omega)/Q] \tag{6.2}$$

where N is the volume density of the target atoms in atoms/cm^2 and t is its thickness. Nt is the number of target atoms per unit area (areal density). In practice, when the detector has a finite solid angle of acceptance the scattering probability is characterized by the integral scattering cross section:

$$\sigma\ (\theta) = 1/\Omega \int_\Omega (d\sigma/d\Omega)\ d\Omega \tag{6.3}$$

The number of backscattered particles is proportional to the scattering cross section, i.e.,

$$\begin{pmatrix} \text{number of} \\ \text{detected particles} \end{pmatrix} = \sigma\Omega \cdot \begin{pmatrix} \text{total number of} \\ \text{incident particles} \end{pmatrix} \cdot \begin{pmatrix} \text{number of target} \\ \text{atoms per unit area} \end{pmatrix}$$

i.e.,

$$A = \sigma\Omega \cdot Q \cdot Nt \tag{6.4}$$

This equation has an important meaning. It shows that if the scattering cross section and the detector solid angle are known, then based on the number of backscattered particles one can determine the number of target atoms per unit area.

So far we have not dealt with the physical nature of the collision event itself. To calculate the scattering cross section for an elastic collision, the principles of conservation of energy and momentum must be complemented by a specific model for the repulsive force that acts during the collision. In most cases Coulomb repulsion of the two colliding nuclei describes this force very well. The Coulomb potential acting between particle and nucleus has the simplest form when the following two basic conditions are met. First, the closest approach between the particle and the nucleus is much greater than the dimension of the nucleus. Second, this distance should be much less than the Bohr radius, $a_o = 0.53$ Å. For high-energy particles colliding with low-Z nuclei, the distance of closest encounter may approach dimensions of nuclear sizes. At this point short-range nuclear forces may start to play a role in the outcome of the scattering event. As a result, the scattering becomes inelastic. Consequently, the energy of the scattered particle cannot be given as kE_0. In this case, to determine the energy after collision the nature of the nuclear interaction must be taken into account. Nuclear reactions may also produce a different kind of particle than the incoming one. In other cases the scattering event remains elastic but the probability of the process departs from the Rutherford law (Rutherford, 1911).

The differential scattering cross section for pure Coulomb potential acting between a charged particle and a stationary nucleus in the laboratory frame of reference is given as

$$\frac{d\sigma}{d\Omega} = \left(\frac{Z_1 Z_2 e^2}{4E}\right)^2 \frac{4}{\sin^4\theta} \cdot \frac{\{[1-((M_1/M_2)\sin\theta)^2]^{1/2} + \cos\theta\}^2}{[1-((M_1/M_2)\sin\theta)^2]^{1/2}} \quad (6.5)$$

where e is the electron charge, E is the particle energy immediately before scattering. Z_1, M_1, Z_2, and M_2 are the atomic numbers and masses of nucleus and the particle, respectively. It is easy to see that at scattering angles close to total backscattering ($\theta \approx 180°$) the cross section does not change rapidly with the scattering angle. From a practical point of view this means that by using an average value for the acceptance angle (solid angle) of the detector the calculated value for the scattering cross section will remain accurate.

What are the most important functional properties of the Rutherford differential cross section?

1. $d\sigma/d\Omega \propto Z_1^2$. Therefore, a backscattering experiment done with a He beam gives four times the yield in comparison with a proton beam

for the same target material. Naturally, heavier particle beams give even more gain in the backscattering yield. On the other hand, the use of heavy ion beams creates a number of experimental difficulties, like shorter detector lifetime, lower energy resolution, and the need for a special ion source.

2. $d\sigma/d\Omega \propto Z_2^2$. This means that heavier target atoms are more efficient scatterers than light atoms. Consequently, the backscattering spectrometry is much more sensitive for heavier elements. At the same time, the analysis of the mass dependence of the kinematic factor shows that the mass resolution is the worst for the heaviest target atoms.

3. $d\sigma/d\Omega \propto E^{-2}$. The backscattering yield rapidly increases with decreasing beam energy. Also, at constant bombarding energy the backscattering yield from the deeper regions of a thick target is higher because of the energy loss of the penetrating particle beam.

4. $d\sigma/d\Omega \propto \sin^{-4}\theta$. The scattering yield is lowest at backscattering direction and highest at forward scattering direction.

At the angle of perfect forward scattering the cross section $d\sigma/d\Omega$ becomes infinity. In practice this never happens. Small scattering angles correspond to large flyby distances that may exceed the radius of the innermost electron shell. When this happens the potential of the nucleus will deviate from the pure Coulomb potential because of the perturbing effect of the electron shell. This situation is analogous to a low-energy light particle colliding with a much heavier nucleus. In such cases a model including the electron screening should be used (e.g., Everhart et al., 1955).

Instrumental reasons may also cause deviations from the expected Rutherford cross section. In some cases the energy spectrum of the backscattered particles is recorded by electrostatic or magnetic analyzers to achieve higher energy resolution. These instruments measure particles of one charge state at a time. As Marion and Young (1968), observed, the charge state of the backscattered particles strongly depends on the escape velocity of the bombarding particle. Therefore, in such experiments it is necessary to correct for the undetected fraction of the scattered particles.

6.2.3. The Concept of Energy Loss and Stopping Cross Section

The nuclear dimensions are much smaller than the interatomic distances. The consequence of this rather primitive geometrical consideration is that the most likely process during the interaction between a megaelectron-volt particle and the target material is the penetration of the particle into the target. If we observe a megaelectron-volt particle we will see that it will pass by rows and rows of atoms *almost* undisturbed before we would be able to see a single backscattering occur.

The energy loss, dE/dx, is defined as $\lim_{x \to 0} \Delta E/\Delta x \equiv (dE/dx)(E)$. There are extensive energy loss tables for a wide variety of particle/target pairs and energy range in the literature, e.g., by Ziegler and Chu (1974). Typical values for the energy loss for He particles in the energy range 0.5–3 MeV lie between 10 and 100 eV/Å.

The energy of the particle at any depth x below the surface can be expressed as

$$E(x) = E_0 - \int_0^x (dE/dx)\, dx \qquad (6.6)$$

This is not a very practical way to determine the energy of the particle at a depth x because the energy loss is usually known as a function of energy, not depth. The dependence of $x(E)$ is expressed as

$$x = \int_E^{E_0} (dE/dx)^{-1}\, dE \qquad (6.7)$$

In most cases, instead of the functional energy dependence of dE/dx some kind of approximation is used. The so-called surface approximation replaces dE/dx by its value at energy E_0. As a result for $x(E)$ we obtain:

$$E = E_0 - \left.\frac{dE}{dx}\right|_{E_0} \cdot x \quad \text{or} \quad x = (E_0 - E)\left.\left(\frac{dE}{dx}\right)\right|_{E_0} \qquad (6.8)$$

The surface energy approximation gives valid estimates only for the first several hundreds of nanometers of the sample.

Energy loss, dE/dx, is a result of numerous small-angle collisions between energetic particles with the nuclei and electrons of the target material. This process can be viewed as a series of small independent dissipative processes. The total number of atoms per unit area that take part in the energy loss process in a target with a thickness of Δx is $N\Delta x$, where N is the atom density of the target (atoms/cm^3). In such a way, the value of ΔE can be set proportional to $N\Delta x$. The proportionality factor between ΔE and $N\Delta x$ is the *stopping cross section* ϵ:

$$\epsilon = \frac{1}{N}\frac{dE}{dx} \qquad (6.9)$$

It is very important to understand the difference between dE/dx and ϵ. Let us assume different targets consisting of the same number of atoms per unit area but having different densities. It is evident that the total energy loss will be independent of how the atoms are "packed" together, closely or loosely. In the process of energy loss, what counts is the total number of atoms participating in the process of energy dissipation. Therefore, because the stopping cross section ϵ is invariant of the density the physical picture of the energy loss process is much better reflected by ϵ than dE/dx. It is worth mentioning that in some cases the normalization is done to the mass density ρ (g/cm^3) of the material. This approach is convenient when thickness units of μg/cm^2 are used.

We shall mention only the very basic physical aspects of energy loss. The theory of the energy loss is quite extensive and represents an active field of study even today. For a light particle traversing the target material there are two basic interactions responsible for the energy loss. Different types of interactions between moving ions, electrons, and nuclei take place independently, therefore their contributions to the energy loss can be separated as

$$\epsilon = \epsilon_e + \epsilon_n \qquad (6.10)$$

Here ϵ_e is the electronic stopping and ϵ_n is the so-called nuclear stopping. The term *nuclear stopping* is somewhat misleading. What we mean by nuclear stopping is transferring kinetic energy and momentum to the nucleus rather than the excitation of the nuclear state. The electronic stopping is analogous to the term *frictional resistance* as the projectile passes through the electron clouds around the atoms.

For energetic particles the energy transferred to the electrons is $(dE/dx)|_e \propto E^{-1}$. As the energy of the ion decreases, the ion will spend an increasingly longer time in the vicinity of electrons. Energy loss reaches its maximum when the speed of the particle becomes comparable to the Bohr velocity v_0. For He ions this velocity corresponds approximately to 0.1 MeV. As a result of slowing down, the probability for the ion capturing electrons along its passage increases too, thus the effective charge of the particle decreases. In addition, at lower energies excitation of the innermost electrons of the target atoms becomes less and less likely. These factors lead to deviation from the E^{-1} law and at lower energies result in decreasing energy loss with decreasing particle energy.

Contribution of the nuclear loss in total energy loss is almost negligible. For example, the electronic stopping component for protons is 3000 times greater than the nuclear stopping component. Nuclear stopping only starts to play a role during the last few collisions before the particle becomes fully at rest. Nuclear stopping is very important when low-energy heavy particles interact with materials (ion implantation, ion sputtering).

So far in our discussions we have not distinguished between different atoms making up the target material. Because the energy loss is considered to be a sum of uncorrelated collisions, the contributions of different atoms in this process can also be treated as independent events. Therefore, it is natural to assume that energy loss in a compound is the sum of energy losses attributed to constituent elements weighed by relative concentration (or abundance) of elements. This is known as Bragg's rule stating the stopping cross section $\epsilon^{A_mB_n}$ of a solid of composition A_mB_n is given as:

$$\epsilon^{A_mB_n} = m\epsilon^A + n\epsilon^B \tag{6.11}$$

where ϵ^A and ϵ^B are the stopping cross sections of the atomic constituents A and B. Let the volume density of the A_mB_n be $N_{A_mB_n}$, then for the energy loss of the compound material we get:

$$\left.\frac{dE}{dx}\right|_{A_mB_n} = N_{A_mB_n} \cdot \epsilon^{A_mB_n} \tag{6.12}$$

As an energetic particle moves through material it loses energy via numerous collisions. Such process has statistical characteristics. As a result, particles with the same initial energy passing through the same thickness of material will not have the same exact energy. The uncertainty introduced as a result of statistical fluctuations in energy loss is called energy straggling. Straggling was first derived by Bohr (1915) by assuming a Poisson distribution for the collision events. For a layer thickness t the Bohr straggling has a variance

$$\Omega_B^2 = 4\pi(Z_1e^2)^2Z_2Nt \tag{6.13}$$

According to the Bohr theory, straggling does not depend on the energy of the particles. Energy variation increases with the square root of the electron density per unit area, NZ_2t, of the target material.

6.2.4. Depth Scale of the Rutherford Backscattering Spectra

As shown in the previous section, the energy loss and the stopping cross section allow determination of the energy of the particle after traversing a thickness t of the target material. In the case of the backscattering event with a semi-infinite target material, the total energy loss of the particle is the sum of three energy loses:

1. Energy loss during penetration to depth x
2. Energy loss related to the kinematics of the scattering at depth x
3. Energy loss along the outgoing path of the particle

In Fig. 6.1 we show the schematics of the backscattering geometry and the symbols used in the description. E_0, E, and E_1 denote the particle energy of incidence, immediately before scattering at a depth x, and emerging from the target, respectively. The angle of incidence is θ_1 and of the scattering is θ_2 with respect to the target normal. In a number of cases we can assume a constant value for the energy loss along the inward and outward paths of the particle; therefore,

$$E = E_0 - \frac{x}{\cos\theta_1}\frac{dE}{dx}\bigg|_{E_{in}} \quad \text{and} \quad E_1 = KE - \frac{x}{\cos\theta_2}\frac{dE}{dx}\bigg|_{E_{out}} \quad (6.14)$$

The last expression for E_1 contains the kinematic factor K responsible for the energy loss during the collision. From the above two expressions we get:

$$\Delta E = KE_0 - E_1 \quad (6.15)$$

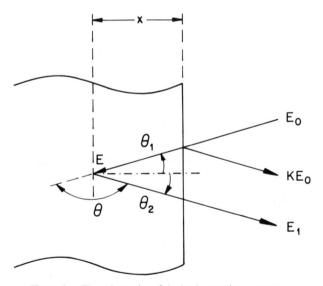

Figure 6.1. The schematics of the backscattering geometry.

The difference $KE_0 - E_1$ corresponds to the measured energy difference between the scatterings from the surface and the depth x of the material. Denoting this value

$$\Delta E = KE_0 - E_1 \quad \text{and} \quad [S] = \left[\frac{K}{\cos\theta_1} \frac{dE}{dx}\bigg|_{E_{in}} + \frac{1}{\cos\theta_2} \frac{dE}{dx}\bigg|_{E_{out}} \right] \quad (6.16)$$

we obtain:

$$\Delta E = [S]x \quad (6.17)$$

The term $[S]$ is called the energy loss factor of the S factor. Similarly, we can write a set of expressions in terms of stopping cross section:

$$\Delta E = [\epsilon]Nx, \quad \text{where } [\epsilon] = \left[\frac{K}{\cos\theta_1} \epsilon\big|_{E_{in}} + \frac{1}{\cos\theta_2} \epsilon\big|_{E_{out}} \right] \quad (6.18)$$

Here $[\epsilon]$ is the stopping cross section factor. This quantity is useful when converting the energy width into the number of atoms per unit area.

In most practical cases when regions near the surface are studied the surface approximation gives valid results. The surface approximation claims that the energy loss along the inward path is small, therefore for $(dE/dx)_{in}$ the value of $(dE/dx)_{E_0}$ can be used. Similarly, for the outgoing path the value of $(dE/dx)_{out}$ can be taken at KE.

So far we have dealt with targets made up of a single element. When the target material consists of two or more elements, then the contributions from all constituents should be taken into account when determining the energy loss of the particles. The energy width of different elements of the target is then expressed in terms of different stopping cross section factors of the compound. In the case of a binary compound target A_mB_n, analogously to Eq. (6.18) we get:

$$\Delta E_A = [\epsilon]_A^{AB} N^{AB} x$$

$$\Delta E_B = [\epsilon]_B^{AB} N^{AB} x \quad (6.18a)$$

Here N^{AB} denotes the atomic density of the compound. The stopping cross section factors for each element A and B are expressed in term of the compound stopping cross section ϵ^{AB} and the corresponding kinematic factors K_A and K_B:

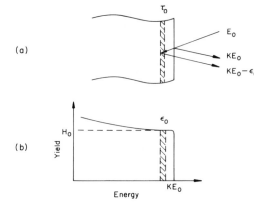

Figure 6.2. The schematic of the backscattering process for a homogeneous target material and the relationship between spectrum height and width of a channel and its corresponding imaginary slab.

$$[\epsilon]_A^{AB} = \frac{K_A}{\cos\theta_1} \epsilon_{in}^{AB} + \frac{1}{\cos\theta_2} \epsilon_{out}^{AB}$$

$$[\epsilon]_B^{AB} = \frac{K_B}{\cos\theta_1} \epsilon_{in}^{AB} + \frac{1}{\cos\theta_2} \epsilon_{out}^{AB} \qquad (6.18b)$$

To determine the compound stopping cross section ϵ^{AB}, it is necessary to use the Bragg rule given by Eq. (6.11).

6.2.5. The Height of the Backscattering Spectrum

During backscattering the measuring electronics records the backscattered particles emerging from different depths of the sample. The result of the data collection is the backscattering (energy) spectrum representing the number of detected backscattering particles as a function of their energy. The energy axis of the spectrum is split up into a number of discrete units called channels. There is a linear correspondence between the particle energy and the channel number; therefore, the energy-to-channel conversion can be given as:

$$E = \#_{ch} \cdot \mathscr{E} + \Delta \qquad (6.19)$$

where $\#_{ch}$, $\mathscr{E}$, and Δ are the channel number, the channel energy width (keV/channel), and the energy corresponding to the zeroth channel, respectively.

Particles having an energy variation within the limits of $\mathscr{E}$ will produce signals that correspond to the same channel. In the first approximation the origin of the energy variation $\mathscr{E}$ is that the particles emerge from slightly different depths of the sample. Therefore, every channel of a backscattering spectrum can be related to a thin imaginary slab within the sample with thickness τ_0 from where the particles emanate. Figure 6.2a,b shows the schematic of the backscattering process for a homogeneous target material and the relationship between spectrum height and width of a channel and its corresponding imaginary slab.

The number of backscattering counts in one channel (spectrum height) H_i is expressed analogously to (6.4), i.e., $H_i = \sigma \Omega Q N \tau_0$. This formula is correct for normal incidence. For $\theta_1 > 0$ the number of atoms in a slab seen by the incoming particle beam increases by $1/\cos\theta_1$, therefore for the ith channel of the spectrum we have:

$$H_i = \sigma(E_i)\Omega Q \frac{N\tau_i}{\cos\theta_i} \qquad (6.20)$$

The slab thickness τ_i can be easily expressed using Eq. (6.18) in terms of the stopping cross section factor:

$$\mathcal{E} = [\epsilon]N\tau \qquad (6.21)$$

For the height of the ith channel we get:

$$H_i = \sigma (E_i)\Omega Q \frac{\mathcal{E}}{[\epsilon]\cos \theta_i} \qquad (6.22)$$

It is important to note that in Eq. (6.22) the values for the scattering cross section σ and stopping cross section factor $[\epsilon]$ should be taken at E, the energy of the particle immediately before scattering. In the case of surface approximation these values are taken at the energy of incidence E_0.

The backscattering spectrometry is most frequently applied to the analysis of compound targets. As we have seen above, in the backscattering spectrum for each element a separate surface position exists determined by the kinematic factor K. For each constituent, separate depth scales should be calculated using the corresponding stopping cross section factor. The backscattering spectrum of a multielemental target can be viewed as a superposition of elemental backscattering spectra. Therefore, in general, the spectrum height at a particular emerging energy E_1 for a compound target $A_m B_n$ is given as:

$$H(E_1) = H_A(E_1) + H_B(E_1) \qquad (6.23)$$

In the case of surface approximation the elemental spectrum height for A and B can be expressed analogously to Eq. (6.22) as:

$$H_A = \sigma_A(E_0)\Omega Q \cdot m \cdot \frac{\mathcal{E}}{[\epsilon]_A^{AB} \cos\theta_1}$$

$$H_B = \sigma_B(E_0)\Omega Q \cdot n \cdot \frac{\mathcal{E}}{[\epsilon]_B^{AB} \cos\theta_1} \tag{6.22a}$$

where m and n are the fractional concentrations of A and B of the compound AB. The goal of the elemental analysis is to determine the relative concentrations of A and B based on the measured spectrum. These concentrations can be determined using Eq. (6.22a) as:

$$\frac{m}{n} = \frac{H_A}{H_B} \frac{\sigma_B}{\sigma_A} \frac{[\epsilon]_A^{AB}}{[\epsilon]_B^{AB}}\bigg|_{E_0} \tag{6.24}$$

The analytical expressions for a target with more than two constituents and/or for the case of nonsurface approximation are quite complicated. Most of the actual data reductions are performed with the use of computer programs capable of handling multilayer multielemental targets. For example, the RUMP program developed at Cornell University by Doolittle (1985) can handle basically all physical aspects of the backscattering phenomenon. Computer programs like RUMP made it possible to carry out fast and accurate analysis of large amounts of data in a very short time.

6.2.6. The Resolution and Sensitivity of Backscattering

All experimental techniques have limitations. Limits of a technique have two basic origins. On the one hand, they originate from the accuracy and specifics of the apparatus, and on the other hand they originate from the physical phenomena on which the measurement technique is based.

The most important parameters of the analysis are:

- The mass resolution
- The accessible depth
- The depth resolution
- The sensitivity

In the following we will briefly discuss these parameters and their optimization.

6.2.6.1. Mass Resolution

The mass spectroscopic feature of backscattering, as pointed out in Section 6.2.2., is achieved through the kinematic factor $K = E_1/E_0$. Different masses of the target material show up at different energies on the energy spectrum. Therefore, the mass resolution is linked to the energy resolution of the measuring system at the specified incident energy E_0. The energy resolution, δE_1, is the FWHM of the Gaussian response of the detecting system to a perfectly monoenergetic excitation. The mass resolution, δM_2, for near-perfect backscattering geometry ($\theta = 180°$) is given as:

$$\delta M_1 = \frac{\delta E_1}{E_0} \frac{(M_1 + M_2)^3}{4M_1(M_2 - M_1)} \tag{6.25}$$

From this equation it is obvious that the mass resolution rapidly deteriorates for heavier target elements. With ^{4}He particles, isotopic resolution is achieved for target atoms of phosphorus and lighter when the energy resolution is 16 keV. At heavier target masses the situation worsens; e.g., at $M_2 = 100$ amu, $\delta M_2 = 3$.

The obvious way to increase the mass resolution is to use a detection system with better energy resolution. Increasing the particle energy E_0 and mass M_2 according to Eq. (6.24) should result in better mass resolution too. In practice, however, the use of heavier particle beams for backscattering analysis in combination with silicon surface barrier detectors does not necessarily lead to the desired effect. In fact, the deterioration of the detector's energy resolution and lifetime for particles with higher masses offsets the expected gain from the kinematics.

6.2.6.2. Accessible Depth

Particles backscattered from different depths will emerge from the target material with different energies. Besides the kinematics of the scattering event, this emerging energy will be determined by the energy loss in the material. The depth from which the emerging particle can be detected by the detector will be accessible. A sensible value for the lower limit of the emerging energy E_1 is one fourth the energy of the particles scattered from the surface, i.e.,

$$\Delta E = KE_0 - \tfrac{1}{4}KE_0 = \tfrac{3}{4}KE_0 = [\bar{S}] \cdot x \tag{6.25}$$

The numerical evaluation of Eq. (6.25) shows that at 2 MeV for target masses between Al and Au, the accessible depth ranges between 1.7 and 1 μm for ^{4}He particles and between 15 and 6.4 μm for ^{1}H particles.

6.2.6.3. Depth Resolution

Perhaps the most often asked question about the features of the backscattering technique concerns its depth resolution. The question of depth sensitivity and optimization was studied by Williams (1978) in detail. The energy resolution of the measuring system δE can be transformed into depth resolution δx using Eq. (6.16):

$$\delta x = \frac{\delta E}{\left[\dfrac{K^2}{\cos\theta_1} S|_{in} + \dfrac{1}{\cos\theta_2} S|_{out}\right]} \qquad (6.27)$$

The most widely used backscattering arrangement employs a near-perpendicular angle of incidence for the probing ^{4}He beam and a detector placed at scattering angles near 180°. At usual energy resolution (15–18 keV) the resulting depth resolution ranges from 10 to 30 nm depending on the density of the material. Basically, there are three ways to improve the depth resolution of the backscattering measurement:

1. Maximizing the energy loss by using optimal beam parameters (energy and particle)
2. Improving the energy resolution of the measuring system
3. Optimizing the backscattering geometry

About 30% gain in depth resolution can be achieved by using ^{4}He ions with lower energies. Unfortunately, the use of low beam energies for backscattering has a number of practical and theoretical problems: deviation of the scattering cross section from the Rutherford cross section is one of them. The use of heavier ions, as mentioned above, in most cases does not necessarily lead to better resolution.

The energy resolution of the measuring system can be improved by more than an order of magnitude with the application of magnetic spectrometers and electrostatic analyzers. However, the increased depth resolution (2 nm for silicon) is achieved at the expense of the data collection time.

Optimization of the scattering geometry offers a relatively simple and powerful way to increase the depth resolution. The principle of increasing the depth resolution is simple: by tilting the sample with respect to the incident beam the depth scale of the backscattering spectrum scales with $1/\cos\theta$. Tilting the sample to $\theta > 85°$ increases the near-surface depth resolution to 2 nm. At the same time, the simplicity and advantages of the backscattering arrangement with a solid-state detector are preserved.

There are two major factors that should be taken into account when using the glancing angle arrangement for higher depth resolution. The sample tilting increases the effective area illuminated by the beam by a factor of $1/\cos\theta$. As a consequence, the usual effective beam diameter may increase to a size of several millimeters. Lateral inhomogeneity of thickness and/or composition of the measured structure may affect the depth resolution. Another problem may arise from the beam and detector collimation. For best results it is often necessary to use narrow beams and collimator slits in front of the particle detector. These optimization measures usually lead to an increase in the data collection time.

An unavoidable limitation of the glancing incidence method comes from the energy straggling. As a result of the energy straggling the depth resolution becomes increasingly worse at greater depths. For example, using a sample tilt of 85° for the analysis of an Al layer with a 2-MeV ^{4}He beam, the depth resolution is as good as 5 nm at the surface; on the other hand, at a depth of 100 nm it becomes 12–15 nm.

6.2.6.4. Sensitivity

The sensitivity of the backscattering measurement rapidly increases for heavier elements because of the Z^2 dependence of the scattering cross section. The detection limit of the backscattering method is set by sputtering. During ion sputtering, erosion of the target material takes place as a result of transfer of kinetic energy to the target nuclei. The useful backscattering yield for a thin surface layer containing N_S atoms/cm^2 is $Q_D = \sigma \Omega Q N_S$. The same number of ions produce a loss ΔN_S of the surface layer proportional to the sputtering yield Y: $\Delta N_S = YQ/a$, where a is the area irradiated by the beam. During measurement the loss relating to sputtering should be less than the original layer thickness, i.e., $\Delta N_S < N_S$, resulting in:

$$N_S > \left(\frac{Q_p Y}{\sigma a \Omega} \right) \tag{6.28}$$

Using 100 counts for the minimum number of counts necessary for a statistically significant measurement, we get about 10^{12} Au/cm^2 for the detectable limit. Naturally, the situation is much worse for lighter constituents for which high background counts from the heavier substrates cause an additional factor in the deterioration of the sensitivity.

6.3. Forward Recoil Elastic Scattering

A fundamental requirement for backscattering is that the target atoms be heavier than the bombarding particles. Laws of kinematics simply forbid scatterings to backward directions from lighter nuclei. In an experiment employing forward recoil elastic scattering instead of measuring the energy spectrum of the scattered ^{4}He particles, energies of the recoiled particles are measured. The detection of lighter atoms (hydrogen a and deuterium) is possible only if the background of the forward scattered ^{4}He particles is filtered out. For this purpose a thin foil (usually a 10-μm Mylar film) is placed in front of the detector. The penetration range of recoiled hydrogen or deuterium particles is much greater than that of ^{4}He particles. For example, a 1-MeV ^{1}H loses only 300 keV on its way through a 10-μm mylar foil while it completely stops ^{4}He particles with energies up to 3 MeV. The main advantage of using foils for filtering is its simplicity. The disadvantage is loss of depth resolution resulting from energy straggling introduced by the foil. The usual energy resolution at the surface of a sample is about 40 keV. Dunselman *et al.* (1987) have published a good summary on elastic recoil measurement.

To determine the depth scale of forward recoil scattering measurements, let us assume a symmetrical scattering geometry, where the sample is tilted to the beam at an angel α and the detector is placed at an angle 2α toward the forward direction. Under these conditions the particle's inward and outward paths are the same, $t/\sin\alpha$, where t is the layer thickness. Hydrogen (or deuterium) will recoil with an energy $E_2 = K'E_0$, where K' is the recoil kinematic factor given by Eq. (6.1a). The emerging energy of the recoiled particle is:

$$E_2 = K'E_0 - K'\Delta E_{He} - \Delta E_H \tag{6.29}$$

where ΔE_{He} is the energy loss of ^{4}He on the inward path and ΔE_H is the energy loss of hydrogen along the outward path. These quantities are:

$$\Delta E_{He} = \left.\frac{dE}{dx}\right|_{He} \cdot \frac{t}{\sin\alpha}$$

$$\Delta E_H = \left.\frac{dE}{dx}\right|_H \cdot \frac{t}{\sin\alpha}$$

The energy width ΔE is the energy difference between the recoils from the sample surface and depth t:

$$\Delta E = \frac{t}{\sin\alpha} \cdot \left.\frac{dE}{dx}\right|_{He} \left\{ K' + \frac{\left.\frac{dE}{dx}\right|_H}{\left.\frac{dE}{dx}\right|_{He}} \right\} \tag{6.30}$$

where the ratio of the stopping powers is about 1/6.

The cross section of a recoil at an angle of θ to the beam direction is given as:

$$\frac{d\sigma}{d\Omega} = [Z_1 Z_2 e^2 (M_1 + M_2)]^2 \frac{1}{4 M_2^2 E^2 \cos^3 \theta} \qquad (6.31)$$

6.4. Channeling

In our discussion we have viewed the target material as amorphous, where atoms are randomly positioned in three-dimensional space. In reality, most materials have some degree of crystallinity. The effect of channeling occurs because rows of atoms under certain geometrical conditions can "steer" incoming particles. When the direction of the incident collimated beam is nearly parallel to a crystalline axis, the particles will suffer mostly small-angle forward scatterings and they will "bounce" between the rows of atoms. This situation is shown in Fig. 6.3. The interaction of the impinging particle beam with a

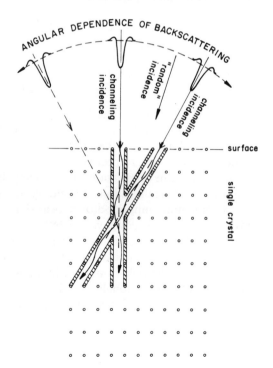

Figure 6.3. Schematic view of the channeling measurement. There are minima in the backscattering yield when the direction of the incoming beam is aligned parallel with one of the crystallographic axes.

crystal is monitored with the usual backscattering setup while the sample is oriented with one of its crystallographic axes parallel to the incident beam. Once the crystalline axis is aligned with the incoming beam, the number of backscattering events drops sharply. Depending on the beam parameters, this decrease in backscattering counts can be 100-fold for near-perfect crystals. The most common way to carry out this alignment is to measure the back-scattering yield at a fixed sample tilt angle while rotating the sample around the beam axis (azimuthal angular scan). The angular scan will then show the azimuthal coordinates of major planar channels of the crystal. The angular coordinates of the major channel can be derived from the angular scan using a simple graphical method (Chu *et al.*, 1976).

Channeling in a perfect crystal can be characterized by two basic quantities: the critical angle and the minimum yield. Barrett (1971) used the average interatomic potential V_{rs} to determine the critical angle of axial channeling as:

$$\Psi_{1/2} = 0.80[V_{rs}(1.2u_1)/E]^{1/2} \tag{6.32}$$

where u_1 is the one-dimensional rms thermal amplitude. The value for the minimal yield can be expressed as:

$$\chi_{min} = 18.8Ndu_1^2 \, (1 + \zeta^{-2})^{1/2}$$

where

$$\zeta \cong 126u_1/\Psi_{1/2}d \tag{6.33}$$

In Eq. (6.33), $\Psi_{1/2}$ is measured in degrees, N is the atomic concentration in atoms/cm^3, and d is the interatomic distance in angstroms. Typical values for the critical angle and the minimum yield for silicon at 2-MeV ^{4}He are about 0.6° and 2–3%, respectively.

An important application of the channeling measurement is determination of the lattice location of impurity atoms. In channeling of a perfect crystal containing impurity atoms located at lattice sites (substitutional impurities), no backscattering will be detected from these atoms because of the shadowing effect of the atoms of the host lattice. On the other hand, if all impurity atoms are located off lattice sites in a random location (interstitials), then the channeling yield from these atoms will show no angular dependence. The degree of substitutionality, R, is related to the ratio of the yield under channeling conditions versus random (nonchanneled) incidence:

$$R = \frac{1 - \chi_{impurity}}{1 - \chi_{host}} \tag{6.34}$$

The channeling measurement is usually carried out for impurities heavier than the atoms of the host lattice. The signals from lighter impurity atoms are usually swamped by the high background counts originating from the host lattice.

Channeling measurement has been widely used for studying different defects of the lattice. Csepregi *et al.* (1976) used channeling measurement to study the epitaxial regrowth of implanted amorphous silicon. The channeling spectrum of a single-crystalline substrate with a top amorphous layer contains two distinct regions. The box-shaped region corresponding to the amorphous layer has the same height as for random incidence. The backscattering counts drop significantly at lower energies because of the channeling in the crystalline substrate. The amorphous/crystalline interface is marked by a sharp drop in the backscattering yield. This technique gives an elegant way of determining the thickness of amorphous layers on the order of 0.6 μm or less. As the thickness of the amorphous layer decreases, so does the energy width of the box in the channeling spectrum. When this thickness becomes less than the energy resolution of the system (~ 30 nm), the box-shaped region corresponding to the amorphous layer shrinks to a surface peak. It is important to note that even in the case of a perfect crystal the channeling spectrum still shows a small characteristic surface peak. The presence of the surface peak in the channeling spectrum of perfect crystals can be explained simply. For the first few monolayers there is no channeling effect—the analyzing beam always "sees" these atoms. Therefore, the area of the surface peak corresponds to the number of atoms per unit area of the first monolayers of the crystal.

Mezey *et al.* (1978) used a collimated detector placed at grazing exit angle (97°) to improve the depth resolution of the channeling measurement. They reported increased depth resolution of 4–5 nm in silicon. In contrast to the grazing incidence measurement, the accuracy of channeling measurements is maintained but as a result of the scattering geometry the mass resolution somewhat suffers.

The useful information one can extract from a channeling measurement is actually based on the deviation of the channeled spectrum from an ideal crystal's channeled spectrum. Deterioration of channeling because of different defects in crystals is called dechanneling. Almost all defect types affect channeling, and channeling measurements can be a useful tool for studying defects.

Figure 6.4 summarizes the effect of basic defect types on the channeling spectrum. Dislocations are discontinuities of the crystal plane that cause distortion of the crystalline lattice. Picraux and Ray-Choudhury (1978) showed that in a distorted lattice the dechanneling is inversely proportional to the critical angle $\Psi_{1/2}$. As a consequence, energy dependence of dechanneling caused by dislocations exhibits a $\sigma_D \propto \sqrt{E}$ relationship.

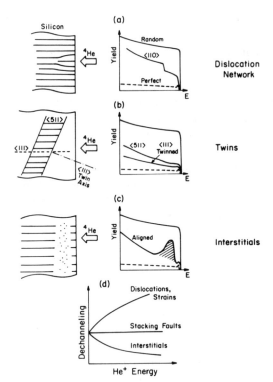

Figure 6.4. The effect of the basic defect types on the channeling spectrum: (a) dislocations, (b) twins, (c) interstitials. Panel (d) shows the energy dependence of dechanneling caused by different defect types.

Another important class of extended defect in crystals are twins. We talk about twins when a certain volume region of a crystal is rotated or inverted relative to the host lattice. Foti *et al.* (1977) showed that high density of twins creates significant dechanneling in a $\langle 111 \rangle$ which is weakly energy dependent.

Point defects and interstitials are most commonly created as a result of ion implantation of crystalline materials. The interaction of isolated scattering centers in the channels with the beam is described by Rutherford scattering cross section:

$$\frac{d\sigma}{d\Omega} \propto \frac{1}{E^2 \sin^4 \theta} \tag{6.35}$$

The dechanneled fraction of the scattering can be obtained by integrating the above expression over angles greater than $\Psi_{1/2}$. The resulting energy dependence of dechanneling by point defect is $\sigma_D \propto 1/E$.

In most cases it is very difficult to numerically evaluate channeling spectra. There are two basic reasons for this:

1. The dechanneling is characteristically a multiple-scattering phenomenon.

2. The physical model would require the use of a three-dimensional non-Coulomb interatomic potential.

Despite all of the difficulties of the theory of channeling, the application of channeling measurement to materials science and technology is most extensive and growing.

Feldman *et al.* (1982) have given the most detailed description on both theoretical and experimental aspects of channeling. The channeling effect is also well described in a number of review papers by Mayer *et al.* (1970), Morgan and Wood (1973), Picraux and Thomas (1973), and Boerma (1990).

6.5. Elastic Resonances

The Rutherford backscattering technique is very well suited for quantification of high-Z elements in low-Z substrates. Unfortunately, it suffers from low sensitivity for quantifying low-Z elements such as carbon, nitrogen, and oxygen. This low sensitivity is a result of the low scattering yield from light elements. The Rutherford cross section is proportional to the square of the target element atomic number Z. The poor sensitivity to light elements using backscattering makes quantification of the light elements in materials containing predominantly heavy elements even more difficult.

One approach to obtain enhanced cross sections for light elements is to employ incident beams for which the elastic scattering is resonant. From the point of view of nuclear reactions, charged particles such as ^{4}He cannot effectively react through nuclear forces unless they have an energy comparable to the Coulomb barrier of the target nucleus. If the α particles have sufficient energy to overcome the Coulomb barrier, they may be captured by the nucleus to form a compound nucleus, which is in a highly excited state. For light elements such as O, N, and C, their corresponding Coulomb barriers are relatively low, since the Coulomb barrier is proportional to atomic number. When the α particles have sufficient energy to overcome the Coulomb barrier for these light target elements, they can induce excitations in nuclear levels that yield nuclear reactions or resonance events. In this section we will discuss those channels of nuclear reactions when the deexcitation of the compound nucleus takes place by reemitting an α particle. Accordingly, the probability of the interaction between the α particle and the target nucleus can be significantly enhanced when the scattering is resonant. Since the difference in mass during this scattering is zero, this type of scattering can be treated from the point of view of kinematics just like an elastic scattering event.

The three common elastic scattering processes for oxygen, carbon, and nitrogen atoms are $^{16}O(\alpha,\alpha)^{16}O$, $^{12}C(\alpha,\alpha)^{12}C$, and $^{14}N(\alpha,\alpha)^{14}N$, respectively. For a long time some of these resonances were used only for accelerator calibration.

6.5.1. Cross Sections and Simulations

The α-O, α-N, and α-C cross sections for α energies in the region 2–5 MeV have been measured for backscattering analysis by a number of investigators (Leavitt et al., 1990; Herring, 1958). The cross section data for O, N, and C are shown in Fig. 6.5, which reveals the significant increase in cross sections at resonance energies. The cross sections for O, N, and C are 25, 2.5, and 150 times larger than their corresponding Rutherford cross sections at resonance energies of 3.045, 3.72, and 4.265 MeV, respectively. For energy calibration, which is crucial to the measurement, a special thin-film sample has been designed containing Au–Pd thin layer grown on thin silicon oxynitride film deposited on a carbon substrate. Because of ion beam contamination during measurement, a thin layer of carbon has been observed to form on the surface of this sample. By choosing resonance energies corresponding to oxygen, nitrogen, and carbon, the enhanced oxygen, nitrogen, and carbon peaks can be obtained.

The RUMP code by Doolittle (1985), which is generally used for conventional backscattering simulation, has been modified to handle resonance simulations. Since most ceramic materials contain light elements bonded with heavy elements, the influence of the background on the detection of a light element in a resonance spectrum must be considered during the quantification. The detection limit of a light element can be calculated assuming that the yield must be greater than the square root of the background yield (i.e., the statistical error) (Blanpain et al., 1988). Using this criterion, the detection limits for oxygen, nitrogen, and carbon in a matrix with a high atomic number have been plotted at their corresponding resonance energies (see Fig. 6.6). The highest sensitivity was obtained for carbon detection because of the significant increase of cross section at resonance energy, while nitrogen resonance measurements show a relatively low sensitivity for nitrogen in the heavy element matrix, since the non-Rutherford cross section at 3.65 MeV is only 2.5 times larger than the corresponding Rutherford cross section. Figure 6.5b shows three enhanced non-Rutherford cross section peaks corresponding to 3.60, 3.72, and 4.50 MeV for nitrogen. Since these peaks correspond to different enhanced cross sections at three energies, the corresponding detection limit for nitrogen is different in all three cases. The lowest detection limit (the highest sensitivity) for nitrogen is at 3.72 MeV.

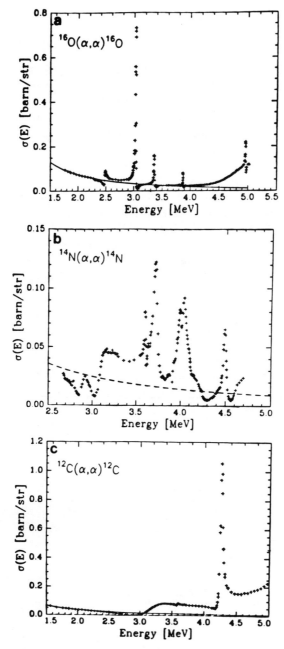

Figure 6.5. Measured elastic scattering cross sections (+) for ⁴He on (a) oxygen, (b) nitrogen, and (c) carbon for ⁴He laboratory energies. The solid and dashed lines represent the Rutherford cross sections for the three elements.

6.5.2. O, N, and C Quantification

6.5.2.1. Measurement of Oxygen Stoichiometry in YBCO Thin Films

The role of oxygen in superconducting oxides is of importance to the electrical properties of these materials. In the case of $YBa_2Cu_3O_{7-\delta}$, the material exhibits superconducting behavior for δ less than 0.3, but it is no longer superconducting for δ greater than 0.7. Methods for accurately determining oxygen stoichiometry are essential in gaining an understanding of the superconducting mechanism. Oxygen resonance measurement is applicable to direct quantification of oxygen concentration in the high-T_c superconducting oxides.

Figure 6.6 clearly shows that high-Z elements in the matrix strongly influence the sensitivity of oxygen detection. For example, the detection limit of oxygen in the bulk YBCO material is about 20 at. % because of the high background yield from Ba. However, for YBCO thin film (200 nm) on MgO substrate, the oxygen resonance peak is superimposed on the Mg shoulder, and the detection limit of oxygen is only about 2 at. %. By choosing an appropriate film thicknesses for YBCO or by tilting the sample to an angle β,

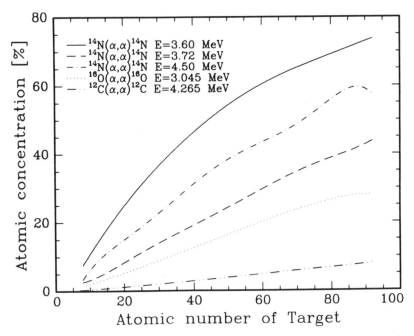

Figure 6.6. The detection limit for oxygen, nitrogen, and carbon in the high-Z metal oxides, nitrides, and carbides.

the backscattering energy of the Mg signal can be shifted to the low-energy direction. As a result, the oxygen resonance peak separates from the Mg background. Without the influence of the background yield, the accuracy of oxygen detection can be greatly enhanced. The basic requirement to separate the oxygen peak in a backscattering spectrum is: $E_{Mg} \leq K_O E_R \leq E_{Cu}$ where E_{Cu} and E_{Mg} correspond to the scattering energies of Cu and Mg. The range of the YBCO thickness t at $E_R = 3.045$ MeV can be determined as:

$$\frac{E_R(K_{Mg} - K_O)}{\dfrac{K_{Mg}N_tS(E_{in})}{\cos\beta} + \dfrac{N_tS(E_{out}^{Mg})}{\cos\theta}} \leq t \leq \frac{E_R(K_{Cu} - K_O)}{\dfrac{K_{Cu}N_tS(E_{in})}{\cos\beta} + \dfrac{N_tS(E_{out}^{Cu})}{\cos\theta}} \qquad (6.36)$$

where β is the tilt angle and θ is the angle which is determined by the scattering angle and sample stage geometry. K is the kinematic factor and N_t is the atomic density of the YBCO film. $S(E_{in})$ and $S(E_{out})$ are the stopping powers per atom corresponding to the average incoming energy E_{in} and average outgoing energy E_{out}. The values for $S(E_{in})$ and $S(E_{out})$ can be numerically calculated. Based on this calculation and RUMP simulation, we know that thicknesses in the range of 800 to 2000 nm of YBCO on MgO substrates are suitable for background peak separation without tilting and therefore sufficient to obtain more accurate measurement of oxygen stoichiometry. The oxygen resonance spectrum of YBCO film (600 nm) on MgO is shown in Fig. 6.7,

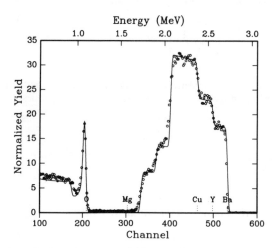

Figure 6.7. The determination of oxygen stoichiometry in $YBa_{1.9}Cu_{2.9}O_{6.8}$ film on MgO. 600 NM, 50° tilt.

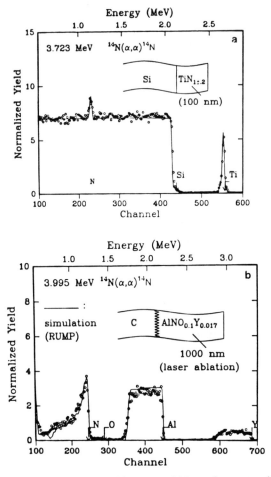

Figure 6.8. Experimental and simulated RBS spectra of (a) reactive-sputtered TiN films on Si substrate and (b) laser-ablated AlN film on carbon substrate. The impurities (O and Y) have been detected.

with RUMP simulation. At 50° tilting, the composition of the YBCO film is determined as $YBa_{1.9}Cu_{2.9}O_{6.8}$.

6.5.2.2. Nitrogen Resonance in TiN and AlN Films

TiN layers have been proven to be a strong candidate as diffusion barriers for metallization in Si devices. Quantification of nitrogen content in TiN films is important to understanding its properties and film fabrication. Olowolafe *et al.* (1991) employed nitrogen resonance to quantify the nitrogen content in TiN films made by reactive sputtering It in Ar/N_2 ambient on Si substrates. In Fig. 6.8a, RUMP simulation shows the best fit for an atomic ratio of It:N = 1:1. The error of this measurement is approximately 10 at % N in TiN with enhanced nitrogen peak superimposed on the Si background at resonance energy. The accuracy of the nitrogen content in TiN films by nitrogen resonance is also limited by the existence of the Si background.

AlN thin films have potential applications as passivation barriers and dielectric layers in integrated circuits. AlN thin films are commonly deposited by chemical vapor deposition (CVD). The fabrication of high-purity AlN films without residual carbon remains a challenge. AlN films approximately 1000 nm thick were made by laser ablation on carbon substrate. The nitrogen stoichoimetry measurement was performed by using $^{14}N(\alpha,\alpha)^{14}N$ resonance. As shown in Fig. 6.5b, the cross section for nitrogen is about 2.5 times larger than its corresponding Rutherford cross section at 3.70 MeV. In order to neglect the background influence on the accuracy, an AlN film grown on carbon substrate was used for resonance measurement. Three enhanced cross section peaks exist at energies near 3.70, 4.00, and 4.50 MeV. To characterize the N content in the AlN thin film, a 3.995-MeV He beam was employed. Based on the RUMP simulation of our results, the composition of this film is $AlNO_{0.1}Y_{0.017}$ (Fig. 6.8b).

6.5.2.3. Carbon Quantification

Figure 6.5c shows that a strong carbon resonance with FWHM = 70 keV occurs at 4.265 MeV. At this energy a significant increase in the carbon signal is obtained. This enhancement makes the carbon cross section similar to that of Er (Revesz, 1991).

The attainable sensitivity for measuring carbon in buried layers by way of resonance techniques is illustrated as follows. A Kapton-H sample was sequentially sputtered with a 20-nm layer of chromium and a 1200-nm layer of copper. A thin layer of an acrylate-based polymer film was spin-coated onto the copper film and this film was covered by a sputter-deposited layer of copper with a thickness of 120 nm. Figure 6.9a shows the backscattering spectrum (circles) taken with a beam energy of 4.312 MeV. The computer-simulated spectrum appears as the solid line. The wide step between channels 420 and 660 corresponds to the two copper layers since the organic film is too thin to produce any separation in the Cu signal. The carbon signal from the buried polymer layer appears at channel 170. The RUMP simulation gives 8×10^{15} C/cm^2 for the total number of carbon atoms in the buried layer. The smaller peak at channel 195 corresponds to a carbon impurity film on the surface of the copper film.

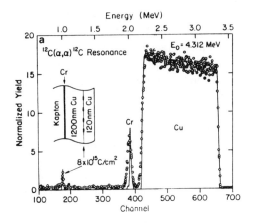

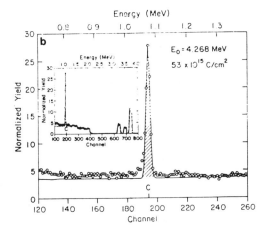

Figure 6.9. (a) 4.312-MeV measurement of thin buried carbon film inside of a copper layer deposited on Kapton substrate. (b) Carbon resonance measurement of the interaction of photoresist layer with high-T_c YBaCuO film.

An additional problem in the patterning of high-T_c film is the interaction of a photoresist layer with the oxide surface. Figure 6.9b shows the backscattering spectrum (circles) of a 250-nm-thick film of YBaCuO on a MgO substrate following a photopatterning process with a photoresist layer. The spectrum was measured at an energy of 4.268 MeV following photoresist removal. The computer-simulated spectrum is also given for comparison (solid line). The peak at channel 195 corresponds to the carbon resonance produced by residual photoresist on the high-T_c material. The estimated carbon level is 53 $\times 10^{15}$ C/cm^2. The insert shows the intensity of the carbon signal relative to the other components of the film.

In summary, the resonance scattering technique is a powerful tool to characterize light elements such as oxygen, nitrogen, and carbon in ceramic thin films because of its nondestructive nature and ease of quantification. Energy calibration and background subtraction are crucial to the accuracy of the measurement. The introduction of resonance capabilities in the standard computer simulation program (RUMP) allows for the estimation of extremely low levels of carbon in the sample.

6.6. Particle-Induced X-ray Emission (PIXE)

Elastic scattering is only one of the many ways an energetic ion interacts with the target material. As an ion passes by the atoms of the target, the time-dependent electric field causes inner-shell ionization. The excited state of the ionized atom may relax by emitting either an Auger electron or an x ray. The energy of the x ray is characteristic of the particular inner-shell transition that takes place during deexcitation. During the ionization process, vacancies of different inner shells may occur. The subsequent deexcitation results in emitting x rays with different energies corresponding to different transitions. The analytical edge of the PIXE method is a consequence of two basic factors: (1) the emitted characteristic x rays carry information about the kind and quantity of the irradiated atoms and (2) the use of proton (or heavier) ion beams gives the advantage of low bremsstrahlung background allowing trace element measurement. The reader can find a detailed description of the PIXE method in the book by Johansson and Campbell (1988).

The experimental setup of a basic PIXE measurement is relatively simple. The detection of the x rays in the kiloelectron-volt energy range is most commonly done with a lithium-drifted silicon (or germanium) detector. The Si(Li) and Ge(Li) detectors operate at liquid nitrogen temperature to preserve the integrity of the Li-drifted region. The surface of the detector in most cases is protected by a thin (5–10 μm) beryllium window. Measurements are usually performed with various filters placed in front of the detector. The purpose of the x-ray filter is to attenuate the intensity of the lower-energy x rays. The control of the intensity of the x rays is an important factor. Long pulse processing time ($\sim$60 μs) of the output signal from an x-ray detector (including the pulse-shaping electronics) seriously limits the allowable maximum x-ray count rates. Usual filter materials (foils of Mylar, Kapton, aluminum, etc.) do not alter the intensity of the higher-energy x rays but at lower energies ($<$5 keV) their attenuating effect is increasingly dominant. The most serious problem caused by the long pulse processing time is the pulse pileup. Pulse pileup occurs when during the pulse processing time two or more photons enter the detector and become recorded as a single event. This leads to false peaks and tails in the PIXE spectrum. The most effective method of reducing pulse pileup is the on-demand beam deflection as suggested by Jaklevic et al. (1972).

This technique uses a pair of electrostatic plates to deflect the beam from the specimen for the duration of 50–80 μs.

The use of high-energy proton beams (e > 2 MeV) offers the advantage of nonvacuum PIXE measurements. High-energy proton beams easily penetrate through a thin (5–20 μm) exit foil from vacuum into the laboratory atmosphere. The proton beam has a range of about 7 cm in air. The nonvacuum (or external) PIXE opens the door for measurements of vacuum-incompatible samples. Figure 6.10 shows the external PIXE spectrum of a red initial of a 12th century Spanish manuscript. The circles represent the measured data; the solid line is the result of computer simulation by a program created by Clayton (1983). Among the many x-ray lines originating from the target material, we can see the line belonging to Ar K-α emission. This argon signal is present in external PIXE spectra because the proton beam travels through a layer of air before it hits the target. This small amount of Ar in the air, therefore, gives its signature in the PIXE spectrum. The argon peak height is usually used for calibration purposes in external PIXE measurements.

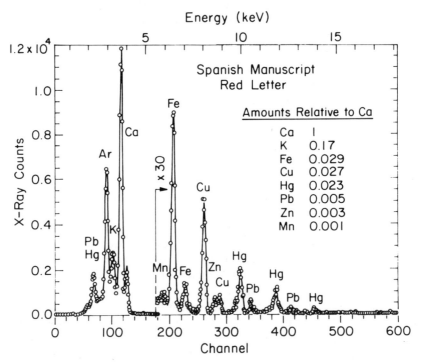

Figure 6.10. External PIXE spectrum of a red initial of a 12th century Spanish manuscript.

As mentioned above, the main benefit of the PIXE method is its ability to measure trace elements. The limit of detection is basically set by the criterion that the peak area should exceed the statistical uncertainty of the background count in the same spectral region. As shown by Folkmann (1976), the detection limit does not vary drastically with the trace element's atomic number. In the range of $Z = 10$–90 the detection limit is better than 1 ppm. For $Z > 50$ (Sb and heavier) it is usually preferable to detect L lines instead of K lines.

In the case of backscattering the depth information is quite simply coded in the spectrum: the deeper regions correspond to lower-energy regions in the spectrum. In PIXE the coding of the depth information in the spectrum is not so explicit. The energy of a particular x-ray line is independent of the location of the atoms. An element residing at different depths in the sample will produce an x-ray line with the same energy. The line intensity, on the other hand, depends on the depth because of two main factors: (1) the production of x rays changes as the incoming particle looses energy during penetration, (2) the outcoming x-ray is being attenuated by the target material. As a consequence, depth profiling by PIXE is not a competitive technique, except in certain well-defined situations. Information obtained from PIXE measurements characterizes the first few tens of micrometers of the sample.

The main advantage of the PIXE method remains its high sensitivity to trace elements with the added benefit of the possibility of measuring vacuum-incompatible specimens of arbitrary shape and size. The use of different beam-focusing techniques opens the way to analyze micron-size targets.

6.7. Conclusions

To solve a great number of problems in the field of materials science and technology, it is essential to have information about the surface micrometer of the material. As we have shown, methods employing megaelectron-volt ion beams can deliver important nonchemical information about the target material. Composition and composition depth dependence measurements can easily be carried out in a nondestructive way using the Rutherford backscattering technique. The sensitivity of this technique is sufficient for most studies in the fields of thin-film metallurgy and microelectronics technology. Backscattering technique is not sensitive to chemical characteristics, although with the addition of channeling measurement one can obtain important structural information as well.

For which applications would we recommend Rutherford backscattering as a main tool of investigation? Historically, studies involving thin films and ion implantation were the primary users of the backscattering technique. Within the last few years the scope of usage of backscattering has widened considerably. Today the technique is used extensively in the fields of ceramic and polymer research, to mention just two. The backscattering technique is quite effective when the main focus of study is a complicated multilayer structure with relative concentrations over 1 at. %. In the case of complicated structures the backscattering technique has a definite advantage. One spectrum contains depth and concentration information of all constituents. The use of computer programs makes the data reduction relatively simple and fast.

As we have pointed out, the sensitivity of the backscattering technique suffers regarding analysis of targets containing low-mass elements. The application of elastic resonances offers increased sensitivity for certain light elements.

Practically all laboratories that carry out backscattering are also equipped for channeling measurement. Channeling measurement has been an important tool for determining lattice damage, strain, and the location of different species. Although channeling is more difficult to carry out and harder to quantify than the usual backscattering, few techniques provide a comparable wealth of information.

A simple change in the geometry of the backscattering measurement enables us to determine light elements. The measurement of hydrogen and deuterium depth profiles by forward recoil elastic scattering has become a useful tool for polymer science.

With regard to the measurement of trace elements, the winner is PIXE. Its greatest advantages are short acquisition time, nondestructive probing, and it can be carried out in atmospheric ambient. This opens up a whole new class of (often quite exotic) materials for studies. PIXE measurement is easy to combine with the backscattering technique. Both use the same instrument. The two techniques also complement each other in depth and concentration sensitivity.

A number of laboratories use different attachments to their backscattering and PIXE apparatus to create small (micron) beam sizes (Cookson, 1991; Traxel, 1990). Although the addition of beam focusing by magnetic or electrostatic lenses does not add anything principally new to the analysis *per se,* still for a number of applications the possibility of studying extremely small areas with backscattering and PIXE is important. Combination of RBS and scanning focused ion beams has been employed for nondestructive two- or three-dimensional microanalysis (Doyle, 1986). From the analytical point of view, one cannot claim that a megaelectron-volt accelerator-based backscattering or PIXE ion microscope can be as easily and routinely operated as their electron-beam counterparts. PIXE and backscattering ion beam microscopy have been proven to be useful in a number of diverse areas ranging from microelectronics technology to biological research.

ACKNOWLEDGMENT. The authors thank Nicholas Szabo, Jr. for his contribution to this work.

References

Barrett, J. H. (1971). *Phys. Rev. B* **3**, 1527.

Blanpain, B., Revesz, P., Doolittle, L. R., Purser, K. H., and Mayer, J. W. (1988). *Nucl. Instrum. Methods* **B34**, 459.

Boerma, D. O. (1990). *Nucl. Instrum. Methods* **B50**, 77.

Bohr, N. (1915). *Philos. Mag.* **30**, 586.

Chu, W. K., Mayer, J. W., and Nicolet, M.-A. (1976). *Backscattering Spectrometry,* Academic Press, New York.

Clayton, E. (1983). *Nucl. Instrum. Methods Phys. Res.* **218**, 221.

Cookson, J. A. (1991). *Nucl. Instrum. Methods* **B54**, 433.

Csepregi, L., Mayer, J. W., and Sigmon, T. W. (1976). *Appl. Phys. Lett.* **29**, 92.

Doolittle, L. R. (1985). *Nucl. Instrum. Methods* **B9**, 344.

Doyle, B. L. (1986). *Nucl. Instrum. Methods* **B15**, 654.

Dunselman, C. M. P., Arnold, W. M., Habraken, F. H. P. M., and van der Veg, W. F. (1987). *MRS Bull.* Aug.–Sept., p. 35.

Everhart, E., Stone, G., and Carbone, R. J. (1955). *Phys. Rev.* **99**, 1287.

Feldman, L. C., and Mayer, J. W. (1986). *Fundamentals of Surface and Thin Film Analysis,* North-Holland, Amsterdam.

Feldman, L. C., Mayer, J. W., and Picraux, S. T. (1982). *Materials Analysis by Ion Channeling. Submicron Crystallography,* Academic Press, New York.

Folkmann, F. (1976). *Ion Beam Surface Layer Analysis.* Plenum Press, New York.

Foti, G., Csepregi, L., Kennedy, E. F., Pronko, P. P., and Mayer, J. W. (1977). *Phys. Lett.* **64A**, 265.

Herring, D. F. (1958). *Phys. Rev.* **112**, 1217.

Jaklevic, J. M., Goulding, F. S., and Landis, D. A. (1972). *IEEE Trans. Nucl. Sci.* **B9**, 71.

Johansson, S. A. E., and Campbell, J. L. (1988). *PIXE: A Novel Technique for Elemental Analysis,* Wiley, New York.

Leavitt, J. A., McIntyre, L. C., Jr., Ashbaugh, M. D., Oder, J. G., Lin, Z., and Dezfouly-Arjomandy, B. (1990). *Nucl. Instrum. Methods* **B44**, 260.

Olowolafe, J. O., Li, J., Mayer, J. M., and Colgau, E. G. (1991). *Appl. Phys. Lett.* **58**, 469.

Marion, J. B., and Young, F. C. (1968). *Nuclear Reaction Analysis, Graphs and Tables,* Wiley, New York.

Mayer, J. W., Eriksson, L., and Davies, J. A. (1970). *Ion Implantation in Semiconductors,* Academic Press, New York.

Mezey, G., Kotai, E., Lohner, T., Nagy, T., Gyulai, J., and Manuaba, A. (1978). *Nucl. Instrum. Methods* **149**, 235.

Morgan, V. D., and Wood, D. R. (1973). *Proc. R. Soc. London Ser. A* **335**, 509.

Picraux, S. T., and Ray-Choudhury, P. (1978). *Semiconductor Characterization Techniques,* Electrochemical Society, Princeton, N.J.

Picraux, S. T., and Thomas, G. J. (1973). *J. Appl. Phys.* **44**, 594.

Revesz, P., Li, J., Vizkelethy, G., Mayer, J. W., Matienzo, L. J., and Emmi, F. (1991). *Nucl. Instrum. Methods* **B58**, 132.

Rutherford, E. (1911). *Philos. Mag.* **21**, 669.

Traxel, K. (1990). *Nucl. Instrum. Methods* **B50**, 177.

Williams, J. S. (1978). *Nucl. Instrum. Methods* **149**, 207.

Ziegler, J. F., and Chu, W. K. (1974). *At. Data Tables* **13**, 483.

Part IV

Photon Beam Techniques

7

Confocal Microscopy

T. Wilson

7.1. Introduction

This chapter will review the physical principles of confocal scanning microscopy. The ability to modify and improve the optical imaging in these instruments results mainly from adoption of a scanning approach. In this way the
only requirement of the optical system is that it should be capable of imaging
one point of the object field at any time: the entire field is then built up by
scanning. This serial imaging has several advantages over the essentially parallel
processing of the conventional microscope. In particular, it provides the image
in an ideal form for subsequent image processing and display. A typical scanning system and computer interface is shown in Fig. 7.1.

The choice of how to scan in such systems is frequently, necessarily,
determined by the application. The simplest optical system, which results in
space-invariant imaging, involves mechanically scanning the object relative
to a stationary, finely focused, light beam (Minsky, 1961; Wilson and Sheppard, 1984; Brakenhoff et al., 1989; Wilson, 1990). However, such systems
are somewhat slow in their speed of image acquisition and in cases where
speed is important, or specimen environment prevents object scanning, beam
scanning systems may be constructed. The optical systems in these instruments
are a bit more complicated and care has to be taken in the optical design

T. Wilson • Department of Engineering Science, University of Oxford, Oxford OX1 3PJ,
England.

Microanalysis of Solids, edited by B. G. Yacobi *et al.* Plenum Press, New York, 1994.

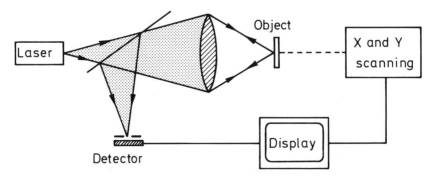

Figure 7.1. Schematic layout of a scanning optical microscope.

because the optics have to work equally at all angular positions of the scan. However, it is possible to obtain good images by this method. If the beam scanning is achieved by the use of vibrating galvanometer mirrors (Wilke, 1985), then the scanning speed is not usually dramatically faster than in specimen scanning schemes. This is important in the context of fluorescence imaging where too high a scanning speed is a disadvantage because it does not give time for the fluorescence to develop. However, acousto-optic beam deflectors permit essentially real-time imaging (Suzuki and Horikawa, 1986). An alternative approach is to use a rotating Nipkow disk to form a number of parallel confocal systems, each of which images only one portion of the specimen (Egger and Petran, 1967; Xiao and Kino, 1987). This is the tandem scanning microscope which permits direct observation of the image through a conventional microscope eyepiece. The imaging in these instruments is subtly different from that in the instruments we will now discuss in detail and so some of the following remarks will only apply qualitatively to these systems.

7.2. The Confocal Optical System

We now briefly review the imaging in conventional optical microscopes, introduce scanning, and present a plausible development of the conventional scanning microscope into a confocal system.

Figure 7.2a shows the optical system of a conventional microscope in its simplest form. Here the object is illuminated by a patch of light from an extended source through a condenser lens. The object is then imaged by the objective as shown, and the final image is viewed through an eyepiece. In this case the resolution is related primarily to the objective lens, while the aberrations of the condenser are unimportant. A scanning microscope using this arrangement could be realized by scanning a point detector through the image

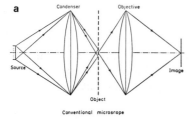

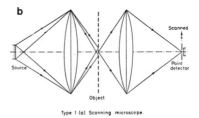

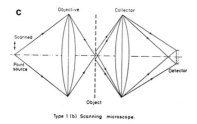

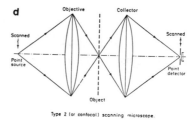

Figure 7.2. The optical arrangements of various forms of scanning optical microscopes: (a) conventional microscope; (b) type 1a scanning microscope; (c) type 1b scanning microscope; (d) confocal scanning microscope.

plane so that it detects light from one small region of the object at a time, thus building up a picture of the object point by point (Fig. 7.2b). The arrangement of Fig. 7.2c, using a second point source and an incoherent detector, has the same imaging properties as the microscope of Fig. 7.2b and the conventional microscope if the roles of the two lenses are exchanged.

This is the arrangement of the conventional scanning microscope. The point source illuminates one very small region of the object, while the large area detector measures the power transmitted by the collector lens. The arrangement shown in Fig. 7.2d is a combination of those in Fig. 7.2b and c. Here the point source illuminates one very small region of the object, and the point detector detects light only from the same area. An image is built up by scanning the source and detector in synchronism.

In this configuration we see that both lenses play equal parts in the imaging. We might expect that as two lenses are employed simultaneously to image the object, the resolution will be improved; and this prediction is borne out by both calculation and experiment. This arrangement has been named a confocal scanning microscope. The term *confocal* is used to indicate that both lenses are focused on the same point on the object. In practical arrangements, however, it is often more convenient to scan the object rather than the source and detector together.

This resolution improvement may at first seem to contravene the basic limits of optical resolution. It may be explained, however, by a principle, described by Lukosz (1966), which states that resolution may be improved at the expense of field of view. The field of view can be increased, however, by scanning.

As a particular example of the degree of improvement in the imaging we can consider the image of a single point object. This is given, in an aberration-free system with circular pupils, by

$$I(v) = \left(2 \frac{J_1(v)}{v}\right)^4 \tag{7.1}$$

where J_1 is a first-order Bessel function and v is a normalized optical coordinate which is related to the actual radial distance, r, by

$$v = 2\pi/\lambda r \sin\alpha \tag{7.2}$$

where λ is the wavelength and $\sin\alpha$ the numerical aperture.

This is the square of the image in a conventional microscope and, although the two images have the same zeroes, the squaring causes the half-width in the confocal case to be 1.4 times narrower than in the conventional case. The outer rings are also dramatically reduced. Although we will not discuss the imaging theory in great detail in this chapter, it is worth mentioning that if we consider the image of a thin object of transmissivity t, then we can write the intensity image in the form

$$I = |h_1 h_2 \otimes t|^2 \tag{7.3}$$

where the symbol $\otimes$ denotes the convolution operation and $h_1 h_2$ represents the product of the point spread functions of the two lenses. In essence, we have a simple coherent optical system.

This discussion allows us to expect that confocal images will be generally superior to conventional ones but this is not the whole story. The use of a point detector also serves to reduce the amount of scatter and flare light which is detected and hence present in the image in the form of noise which has the unwanted effect of reducing the quality and contrast of the image. The degree of signal-to-noise improvement is clearly a function of the size of the detector used. In practice it is impossible to use a point detector and so a pinhole of some finite diameter is placed in front of a photodetector. One immediate role of the finite-sized detector is in the reduction of the amount of flare and scattered light which is present in the image. It also affects the degree of confocality and it is found that to have true confocal operation (Wilson and Carlini, 1987) a pinhole radius of less than about 0.5 optical unit should be chosen.

However, it is perhaps in the imaging of structures with surface relief or specimens which vary with depth that the confocal microscope is particularly powerful. Again the effect is wholly the result of the presence of the point detector. The depth discrimination or optical sectioning property may be understood from Fig. 7.3: when the object is located in the focal plane and the reflected light is focused on the pinhole, then a large detected signal is produced; on the other hand, when the object is out of the focal plane, a defocused spot is formed at the pinhole, and the measured intensity is greatly reduced. If the pinhole were absent, as in a conventional microscope, all of the defocused light would be collected and no depth discrimination would be obtained.

In order to investigate the degree of sectioning that we can obtain, we consider the variation in detected signal as we scan a planar, perfectly reflecting, object axially. In the paraxial limit the result may be written (Wilson and Carlini, 1988).

$$I(u) = \left(\frac{\sin(u/2)}{u/2}\right)^2 \tag{7.4}$$

where u is a normalized axial coordinate given by

$$u = \frac{8\pi}{\lambda} z \sin^2(\alpha/2) \tag{7.5}$$

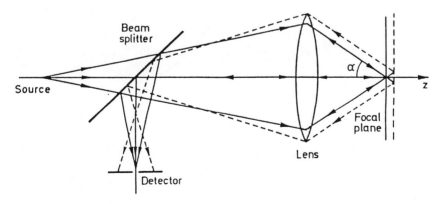

Figure 7.3. The origin of the optical sectioning property of confocal microscopy.

A more detailed theory, which is more appropriate at higher numerical apertures, is also available (Sheppard and Wilson, 1981). The major difference is that the side lobes no longer fall to zero at high apertures. We show in Fig. 7.4 the variation in the full-width-half-maximum of $I(u)$ as a function of numerical aperture based on this theory. This plot, which is arbitrarily drawn for red light, shows the degree of sectioning that might be obtained. It is clear that very sharp sectioning is possible even with red light. The sectioning would, of course, increase as the wavelength became shorter.

Examples of this sectioning have been presented in many other papers (Hamilton and Wilson, 1981; Wilson and Hamilton, 1982). The key property which this sectioning gives us is the ability to probe an object in three dimensions. In this way, we can simply regard the confocal microscope as a tool for imaging a thick specimen depth section by depth section. If we idealize the situation we can write the image, in general, as $I(x, y, z)$ where x and y are determined by the scanning and z is determined by the focal position.

It is clear that by axially scanning an object in a confocal microscope we can build up, in a computer, a three-dimensional data set. In essence, we have

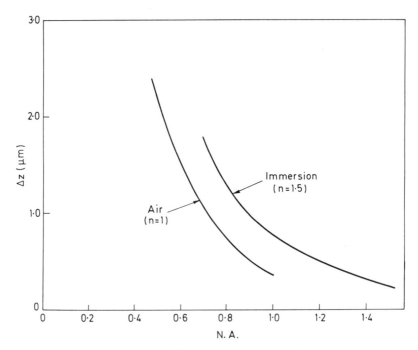

Figure 7.4. The optical sectioning width as a function of numerical aperture. The curves are for red light (0.6328-μm wavelength). Δz represents the FWHM of the curves like Eq. (7.4).

a series of x–y) images taken at different (z) focus settings. If we now choose to simply add together all of the (x–y) images, we can create an extended focus image (Wilson and Hamilton, 1982), Fig. 7.5. Mathematically, we can represent this as

$$I_{EF}(x, y) = \int I(x, y, z) \, dz \qquad (7.6)$$

As an alternative we can sift through the data set and instead of adding together the image values at a particular picture point we can simply choose to display the maximum image value at that particular picture point. In this way, we can create an autofocus image. Mathematically, we might write this as

$$I_{AF} = I(x, y, z_{max}) \qquad (7.7)$$

where z_{max} is the focus position corresponding to the maximum image value.

An inevitable by-product of the autofocus method is that we can obtain z_{max} and hence a surface profile or height profile of the specimen can be obtained (Hamilton and Wilson, 1982).

It is also clear that by suitable lateral displacement of the image slices on reconstruction we can arrange to view the object from a particular direction.

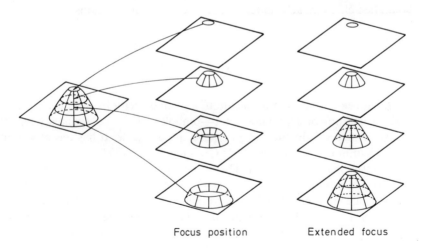

Focus position   Extended focus

Figure 7.5. An idealization of the optical sectioning property showing the ability to obtain a through-focus series of images, which may then be used to reconstruct the original thick, volume, object with high resolution.

This concept immediately lends itself to stereo viewing (Brakenhoff *et al.,* 1989). Mathematically, we might form two images from the data as

$$\int I(x \pm \gamma z, y, z) \, dz \qquad (7.8)$$

where the constant λ determines the angle of view. In some cases, it is not necessary to form two offset images and one offset and one extended or autofocus image will suffice.

An inevitable problem with this kind of imaging is how to display the information which the confocal microscope can give us. We essentially have reflectivity and height information at each sectional slice. In addition to the methods just described, we can also form an autofocus image where we scan axially to find the maximum image signal and use this to describe the image at that picture point. Instead of displaying the height information, we can alternatively convert these data to represent image brightness and hence produce a height image. A problem with this approach is that we have now lost the reflectivity information. However, if we are prepared to use pseudocolor we may, for example, code height by color and reflectivity by the brightness of that color. As an alternative we can produce an isometric display where reflectivity is coded as brightness and is superimposed on the height information. It is clear that we have a richness of choice and together with the possibility of presenting the data as an x–z scan rather than an x–y scan we should be able to make a suitable choice for most applications.

7.3. Alternative Forms of Detector

It has been usual to form the limiting point detector by placing a suitable aperture in front of a photodetector—pinholes, slits, and squares have all been used successfully and design criteria are available. It has recently been shown (Kimura and Wilson, 1991; Giniunas *et al.,* 1991) that single-mode optical fibers may also be used. The fiber replaces the pinhole and photodetector combination. If the optical field at the fiber face is arranged, via a suitable collection lens, to be large compared with the (single) mode profile of the fiber, then the fiber essentially detects the amplitude of the field. In this way confocal operation is achieved. An analysis of the design of single-mode fiber systems has been given by Kimura and Wilson (1991) using the full theoretical expressions for the mode profiles. However, a simpler model may be used if we merely assume that the mode profiles are Gaussian. Let us, therefore, consider a confocal microscope with a coherent source distribution given by

$$S \sim \exp - \frac{r^2}{\alpha^2} \tag{7.9}$$

and a coherent detector described by

$$D \sim \exp - \frac{r^2}{\beta^2} \tag{7.10}$$

The imaging of this coherent system is now given by (Wilson and Hewlett, 1989)

$$I \sim |h_{1\text{eff}} h_{2\text{eff}} \otimes t|^2 \tag{7.11}$$

where the effective point spread functions $h_{1\text{eff}}$ and $h_{2\text{eff}}$ are given by

$$h_{1\text{eff}} = h \otimes S \quad \text{and} \quad h_{2\text{eff}} = h \otimes D \tag{7.12}$$

where h is the point spread function of the lenses. Alternatively we can think of these effective point spread functions in terms of effective pupil functions given by the product of the actual pupil function with the Fourier transform of the source or detector distribution functions. Thus, we can write, introducing defocus via the optical coordinate u, where $u = 8\pi/\lambda \, z \, \sin^2(\alpha/2)$ where $\sin\alpha$ is the numerical aperture, λ the wavelength, and z the actual defocus distance,

$$h_{1\text{eff}} \sim \int_0^1 P(\rho) \exp - \frac{\alpha \rho^2}{4} \exp j \frac{u \rho^2}{2} \rho \, d\rho \tag{7.13}$$

or, introducing a complex variable $s = u + j\,\alpha/2$, we can write $h_{1\text{eff}}(u, \alpha, v) = h(s, v)$, that is, a point spread function with a complex value of u. This formal expression allows us to still use the standard expressions describing the behavior of coherent confocal microscopes. In particular, the response as a plane mirror is scanned axially through focus is given by

$$I(u) = \left| \frac{\sin([u + \delta/2]/2)}{u + \delta/2/2} \right|^2 \tag{7.14}$$

where $\delta = \alpha + \beta$. The expression reduces to the traditional confocal case when $\alpha = \beta = 0$. The case of one point and one fiber detector corresponds to $\alpha = 0$ and the reciprocal case to $\alpha = \beta$. It is clear that in this latter case it is even more important that the numerical apertures of the collimating lens be carefully chosen in order to achieve optimum results. Figure 7.6 shows the effects of a nonideal value of α in the reciprocal case.

In this case, as well as with more traditional confocal detectors, the imaging depends essentially on the amplitude reflection coefficient of the object. The microscope may be made much more versatile, however, if a two-mode fiber is used rather than a single-mode fiber. In this case we find that *differential* imaging is also possible (Juskaitis and Wilson, 1992). In the same way that the fundamental fiber mode detects the field, the second-order mode detects the differential of the field. These two modes then propagate along the optical fiber with different speeds. They, therefore, emerge from the fiber with a relative phase difference determined by the length of the fiber. This phase difference can then be tuned by varying the length of the fiber. In particular, we can tune the system to obtain either a confocal image, a differential amplitude image, or a differential phase image.

If we detect the signal in the far field of the end of the optical fiber with a detector split into two halves, we find (Juskaitis and Wilson, 1992) that

$$I \sim \mathrm{Re}\{a_1 a_2 * \mathrm{exp}j\delta\} \qquad (7.15)$$

where

$$a_1 = h^2 \otimes t \qquad \text{and} \qquad a_2 = h\frac{\partial h}{\partial x} \otimes t$$

where $\mathrm{Re}\{.\}$ denotes the real part and the asterisk complex conjugate. The phase factor δ is given by $\delta = (\beta_1 - \beta_2)l - \pi/2$ and so can be tuned by varying the length of the fiber.

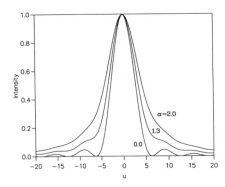

Figure 7.6. The variation in axial response of a reciprocal single-mode fiber confocal microscope for various values of α.

If we introduce the modulus b and phase φ of a_1 as

$$a_1 = b \, \text{exp} j\varphi \qquad (7.16)$$

and note that $a_2 \sim da_1/dx$, we can rewrite the image as

$$I = \frac{db^2}{dx} \cos\delta + 2b^2 \frac{d\varphi}{dx} \sin\delta \qquad (7.17)$$

Thus, by tuning the length of the fiber we can switch δ by $\pi/2$ to obtain either differential amplitude or differential phase imaging.

Let us now consider the differential phase imaging case and ask how the system responds as a perfect reflector is moved through focus, Fig. 7.7. We show, for each of explanation, the analogous case where we choose to move the fiber tip axially through the focal region of the collection lens. If the fiber is properly centered on the optic axis and moved axially, the wave-front gradient (differential phase) it experiences is always zero and hence no signal will be detected.

If, however, the fiber is slightly *offset,* the behavior is completely different. As the fiber moves along the line 1–2–3 it experiences regions of negative

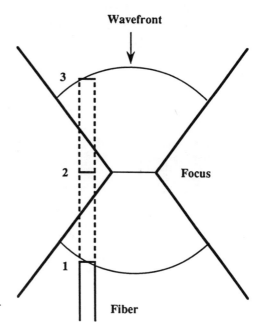

Figure 7.7. Schematic diagram to explain the origin of the height contrast.

gradient (1), zero gradient in the focal plane (2), and positive gradient (3). Thus, a signal is obtained which is zero in the focal plane and rises or falls as we move away from the focal plane. It is clear that this system could also act as a noncontacting surface profilometer simply by scanning the specimen axially at each picture point and noting the axial scan distance required to produce a null signal.

Theoretical calculations show that the axial response in this differential case is given by the derivative of the traditional confocal axial response provided that the offset is not too large. Uncalibrated height profiles may be obtained for surface height steps within the linear region of this curve. Figure 7.8 shows several predictions of images. The system is seen to work well until the step height becomes too high. In all of the cases modeled here, the microscope was assumed to be focused on the left-hand side of the edge.

7.4. Conclusions

We have discussed the operation of the confocal scanning microscope. This is an optical system which is most easily implemented in a scanning system and results in a purely coherent imaging system with improved lateral resolution. A necessary consequence is that the system possesses a unique optical sectioning property which gives images very low depth of field. The introduction of axial scanning allows images of arbitrary depth of focus to be

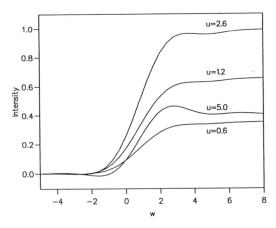

Figure 7.8. Height images corresponding to various step heights for the case of small lateral offset. The vertical scale is in arbitrary units.

obtained as well as permitting the instrument to serve as a noncontacting surface profilometer.

If we place limiting apertures of various shapes in front of a photodetector, the imaging will be confocal in the sense that depth discrimination will usually result—the better sectioning arises, of course, if the aperture is as small as possible. In all cases, however, the imaging will be nondifferential in the sense that image contrast will depend mainly on the reflectivity of the object under study. In order to obtain differential contrast, we may either incorporate a Nomarski-type system into the confocal arrangement or, more simply, use a two-mode optical fiber as a detector. In this way the system becomes vastly more versatile and permits, for example, confocal, differential amplitude and differential phase imaging to be performed simultaneously. The system may also be used as a surface profilometer by introducing a small lateral offset. If a reciprocal fiber microscope is considered where the same fiber is used both to launch the light into the microscope as well as to detect it, the alignment tolerances are *greatly* reduced and the imaging possibilities extended. In this case we can choose to launch either the fundamental or second-order mode, or both. We can then detect either the fundamental or second order mode, or both. An analysis of these possibilities reveals that surface profilometry can be obtained in this case without the need to laterally offset the fiber.

We have limited our discussion to bright-field reflection imaging as this is likely to be of most application in semiconductor applications. However, we should also mention that confocal fluorescence imaging is also available and is finding great application in biology and the life sciences (Pawley, 1989; Wilson, 1990). Another important application of scanning optical microscopy concerns using the laser beam to optically excite carriers in semiconductors. This gives rise, among other things, to a nondestructive optical analogue of the EBIC of scanning electron microscopy (see, e.g., Werner and Grasserbauer, 1991).

References

Brakenhoff, G. J., van der Voort, H. T. M., van Spronsen, E. A., and Nanninga, N. (1989). *J. Microsc. (Oxford)* **153**, 151.

Egger, M. D., and Petran, M. (1967). *Science* **157**, 305.

Giniunas, L., Juskaitis, R., and Shatalin, S. V. (1991). *Electron. Lett.* **27**, 724.

Hamilton, D. K., and Wilson, T. (1981). *Opt. Lett.* **6**, 625.

Hamilton, D. K., and Wilson, T. (1982). *Appl. Phys.* **B27**, 211.

Juskaitis, R., and Wilson, T. (1991). *Appl. Opt.* **31**, 898.

Kimura, S., and Wilson, T. (1991). *Appl. Opt.* **30**, 2143.

Lukosz, W. (1966). *J. Opt. Soc. Am.* **56**, 1463.

Minsky, M. (1961). U.S. Patent No. 3,013,467.

Pawley, J. (1989). *Handbook of Biological Confocal Microscopy,* I.M.R. Press, University of Wisconsin.

Sheppard, C. J. R., and Wilson, T. (1981). *Appl. Phys. Lett.* **38,** 858.

Suziki, T., and Horikawa, Y. (1986). *Appl. Opt.* **25,** 4115.

Wener, H., and Grasserbauer, M., (1991). *Analysis of Microelectronic Materials and Devices,* Wiley, New York.

Wilke, W. (1985). *Scanning* **7,** 88.

Wilson, T. (ed.) (1990). *Confocal Microscopy,* Academic Press, New York.

Wilson, T., and Carlini, A. R. (1987). *Opt. Lett.* **12,** 227.

Wilson, T., and Carlini, A. R. (1988). *J. Microsc. (Oxford)* **149,** 51.

Wilson, T., and Hamilton, D. K. (1982). *J. Microsc. (Oxford)* **128,** 139.

Wilson, T., and Hewlett, S. J. (1989). *Int. Phys. Conf. Ser.* No. 98, 629.

Wilson, T., and Sheppard, C. J. R. (1984). *Theory and Practice of Scanning Optical Microscopy,* Academic Press, New York.

Xiao, G. Q., and Kino, G. S. (1987). *Proc. SPIE* **809,** 107.

8

X-ray Microscopy

A. G. Michette and A. W. Potts

8.1. Introduction

Ever since Röntgen discovered x rays almost 100 years ago, the possibility of using them for high-resolution imaging of both biological and nonbiological materials has been investigated. The early work reached a peak in the 1950s and 1960s through the work of Cosslett and Nixon (1960) on the development of the projection x-ray microscope, but interest then waned for several reasons. Principal among these were the rapid development of electron microscopy and the lack of sufficiently good x-ray sources and optics to allow suboptical resolutions to be routinely obtained. It was not until the late 1970s, when synchrotron radiation sources and microfabrication techniques for x-ray optics began to become available, that serious consideration was once more given to x-ray microscopy. By then electron microscopy was such a well-developed technique that it has proved difficult to find a place for x-ray microscopy, but this is now changing.

A comparison of x-ray microscopy with other forms of imaging has appeared recently (Duke and Michette, 1990). Here the main advantages of x-ray microscopy over conventional forms of optical and electron microscopy are only briefly discussed. Because of their shorter wavelengths, x rays can be used to give much higher resolutions than is possible with visible light; cur-

A. G. Michette and A. W. Potts • Department of Physics, King's College, London WC2R 2LS, England.

Microanalysis of Solids, edited by B. G. Yacobi *et al.* Plenum Press, New York, 1994.

rently, resolutions are limited by the x-ray optics to something in the region of 50 nm in practice, but ~10 nm should be possible in the not too distant future. Although this is much worse than the resolution possible with electron microscopy, the ways in which x rays interact with material mean that specimens can be examined in their natural environments (e.g., aqueous medium at atmospheric pressure) and so the potentially artifact-inducing and lengthy preparation procedures used in electron microscopy are not necessary. Natural contrast arises in x-ray microscopy through differences in absorption [mainly in the wavelength range between the oxygen K (2.3 nm) and carbon K (4.4 nm) absorption edges, the water window] or phase modulation (over a less restricted wavelength range) between water and carbon-containing material such as protein. It is also possible to enhance the contrast by making use of sharp changes in absorption across x-ray absorption edges, and comparing images taken at wavelengths on either side of the edge.

In the last few years x-ray microscopy has developed rapidly both in instrumentation and in the range of applications. This rapid development has been described in a recent review article (Michette, 1988) and can be followed in the proceedings of several conferences dedicated to x-ray microscopy (Schmahl and Rudolph, 1984; Cheng and Jan, 1987; Sayre *et al.*, 1988; Shinohara *et al.*, 1990; Michette *et al.*, 1992). Several x-ray imaging techniques are now sufficiently advanced to allow serious applications to be considered. These techniques include projection microscopy, contact microscopy (or, more correctly, contact microradiography), microtomography, and x-ray holography. None of these involve focusing of the x-ray beam and so, in a sense, are not truly microscopy. Techniques which involve focusing of the x rays are transmission x-ray microscopy (TXM; the direct analogue of transmission electron microscopy) and scanning transmission x-ray microscopy (STXM; the direct analogue of scanning transmission electron microscopy). Of particular interest, for the current chapter, are the techniques in which x rays are used to excite photoelectron emission from the sample.

In order to examine the information available from the various forms of x-ray photoemission microscopy, a brief consideration of the possible x-ray interactions must first be given. The primary interaction occurring in the energy range currently used for x-ray microscopy (~100–800 eV) is that of photoionization. Secondary processes that can be of use for imaging are Auger electron emission and x-ray fluorescence. Inelastic scattering of primary photoelectrons, leading to the production of secondary electrons, contributes to an enhanced total electron yield. Although x-ray fluorescence imaging is important for the electron beam microprobe, it has not so far been extensively used as a way of obtaining information in x-ray microscopy. This is presumably because of the limited energy range of the current x-ray microprobes and also signal considerations. The short escape depths which apply for low-energy electrons (Seah and Dench, 1979) mean that photoemission microscopes are essentially surface science instruments and that useful imaging must be carried out in a UHV environment with clean surfaces.

Imaging with energy-selected primary photoelectrons from a particular atomic core level can lead to elemental surface maps and, where chemical shifts in core ionization energies can be resolved, to maps indicating the distribution of a particular chemical state of an atom. The technique of core-level photoelectron microscopy also provides the possibility of small-spot ESCA studies of samples (Coxon *et al.*, 1990). Use of Auger electrons for imaging should lead to information of the type already obtainable from Auger imaging with an electron-beam microprobe. It is anticipated, however, that an x-ray microprobe should be considerably less damaging than the electron-beam microprobe for sensitive surface processes. Using the total electron yield for imaging provides a means of obtaining rapid surface images. It has been shown that the total electron yield can be proportional to the x-ray absorption coefficient of surface layers (Harp *et al.*, 1990). The total yield intensity plotted as a function of x-ray wavelength can therefore be used to provide small-spot x-ray absorption near edge structure (XANES). Such structure, which is the result of multiple scattering effects for the outgoing photoelectron, is not as simply related to atomic and chemical information as is the structure in an x-ray photoelectron spectrum, but contains information which can frequently be used in a fingerprint fashion to establish the chemical states of surface species (Norman, 1986).

It should be mentioned at this point that, largely as a result of signal considerations, the resolution currently obtained in the type of photoelectron microscopes to be discussed is limited to 200 nm–10 μm, although the resolution limits of 10–30 nm associated with zone-plate optics should be achievable with sufficiently intense sources.

8.2. X-ray Sources

Until recently, all high-resolution x-ray microscopes had to rely on synchrotron sources in order to allow imaging in reasonably short times. This is because such sources are several orders of magnitude brighter than conventional ones. Current synchrotron sources, such as the National Synchrotron Light Source (NSLS) at Brookhaven in the United States, Bessy in Berlin, and the Synchrotron Radiation Source (SRS) at Daresbury near Liverpool, allow images to be obtained on the order of minutes or less. Sources which are now under construction, such as the Advanced Light Source (ALS) at Berkeley in California and Elettra at Trieste in Italy, as well as several more around the world, will be brighter still and will allow (almost) real-time images to be obtained when they are available for use in the mid-1990s. However, even the so-called "table-top" synchrotrons will never be cheap enough to be affordable to individual research laboratories, and so if x-ray microscopy is ever to become a routine analytical tool like optical and electron microscopies it is vital that smaller, cheaper sources be developed.

In the past few years, major advances toward a laboratory-sized source have been made. These sources use the intense x-ray emission from plasmas, which can be generated in various ways. For TXM a good possibility is to use a plasma focus device, and such a source has been built by the Fraunhofer Institute for Laser Technology in Aachen, Germany (Lebert *et al.*, 1992), for use with the Göttingen TXM. Presently, several hundred pulses are needed to form an image, but the source/microscope combination is by no means optimized, and there is a very good possibility that single pulse imaging will be achieved—which means that the whole image can be formed in a few nanoseconds. For STXM, plasmas generated by high-repetition-rate lasers are most suitable; with correct optimization of the plasma generation, enough x rays are emitted per laser pulse to form one image pixel (Turcu *et al.*, 1992) and so the imaging time is limited only by the laser repetition rate which can be several hundred hertz with suitable modern lasers.

8.3. X-ray Optics

As far as soft x-ray optics is concerned, the best resolutions have still been obtained using zone plates (Michette, 1986), which are circular diffraction gratings with radially increasing line densities. The spatial resolution in an image is essentially equal to the outermost zone width. Advances in micro-fabrication techniques mean that outer zone widths of 30 nm or less can now routinely be made, either by interference (holographic) techniques (which, however, are limited to ~50 nm) (Rudolph, 1987) or by various forms of electron-beam lithography (Bögli *et al.*, 1988; Anderson and Kern, 1992; Charalambous and Morris, 1992).

A problem with zone plate optics in the past has been that they have operated principally by modulating the amplitude of the x-ray beam, leading to low diffraction efficiencies so that only a few percent of the incident radiation could be brought to the focus. Again, however, advances in technology can allow this to be considerably improved by making use of the phase-changing properties of the zone-plate material. This should allow first-order diffraction efficiencies of up to about 25% to be compared with the maximum of 10% for amplitude zone plates. These values are for zone plates in which the individual zones are correctly positioned, such that the areas of all of the zones are equal, and are reduced if this is not the case. Typically, zone boundaries should be placed in their correct positions to within about a quarter to a third of the outer zone width for good imaging qualities.

Other possible x-ray optical components, namely grazing incidence reflectors (Dhez, 1992) and multilayer mirrors (Chauvineau, 1989), have to date not been so successfully used as have zone plates in terms of the attainable resolutions. This is also now changing and some very sophisticated optics [e.g., multilayer zone plates, also known as Bragg–Fresnel lenses (Aristov *et al.*, 1992) which can give good resolution coupled with high efficiency] are under serious consideration.

8.4. X-ray Microscopes

There are four operating x-ray microscopes that have shown the capability of high resolution using transmitted x rays. In addition, several more are being constructed or are at various stages of planning and microscopes which form the image with photoelectrons emitted when x rays are absorbed in the specimen are now starting to operate.

8.4.1. Transmission X-ray Microscopes

The Göttingen TXM, which is operated at the Bessy synchrotron in Berlin, uses zone plates for both the condenser and the objective lenses (there are many papers describing all aspects of this microscope in the following references: (Schmahl and Rudolph, 1984; Cheng and Jan, 1987; Sayre *et al.*, 1988; Shinohara *et al.*, 1990; Michette *et al.*, 1992). High resolution (<100 nm) images can be obtained in a few seconds using either photographic film or a CCD (charge-coupled device) camera, and thus it is beginning to become possible to consider imaging of dynamic processes with this microscope. The major drawback (apart from the technical one of making condenser zone plates of sufficient quality, which has been overcome in Göttingen) concerns radiation damage to the specimen. This is increased by the use of the postspecimen objective zone plate which is typically about 5% efficient so that only 1 in 20 of the x-ray photons transmitted by the specimen can contribute to the image. Recent images taken with this microscope have been obtained at the short-wavelength end of the water window (2.4 nm) where more x rays are transmitted and the contrast between water and (e.g.) protein is still high, and some attempts have been made to employ phase-contrast imaging by use of a postobjective phase plate. Both of these can reduce the radiation damage to a certain extent.

This problem of radiation damage with respect to x-ray imaging has not been considered much in the past, although it is arguably the most important factor affecting the eventual usefulness of x-ray microscopy. Serious efforts are now being made to understand how soft x rays cause damage as a first step toward limiting the problem, but it is too early for any significant conclusions to be drawn. A major reason for using scanning transmission x-ray microscopes is that they have no postspecimen optics and therefore the radiation damage is less than in TXM. The three operating STXMs are at the NSLS, Bessy, and the SRS; these have been developed by the Stony Brook, Göttingen, and King's College groups, respectively (see references in preceding paragraph). They are all similar in principle, using zone plates to form the probe across which the specimen is mechanically scanned, with the transmitted x rays detected by single-wire flow proportional counters. The image is stored in a computer and displayed as it is formed, and is directly amenable to later image processing. The main difference between the three STXMs is in the construction of the scanning stages, which are capable of steps in the range 10–100 nm over fields of several millimeters. The King's College and Stony Brook microscopes use custom-built piezoelectric stages, while the Göttingen system is based on long-arm lever reduction.

8.4.2 Photoelectron Microscopes

The obvious possibilities of the photoelectron microscope have led to the development of a variety of microscopes in various laboratories around the world. These may loosely be divided into three groups.

For full-field microscopes the specimen is flooded with monochromatic x rays and the specimen is imaged in terms of the total or energy-selected electron yield. The image is magnified using either an electrostatic lens or a magnetic lens system (Fig. 8.1). Since the whole image is recorded simultaneously, this method can provide particularly rapid imaging. The electrostatic

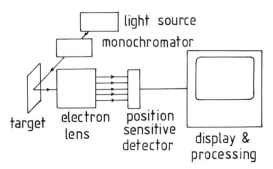

Figure 8.1. Schematic diagram of a full-field photoelectron microscope.

system employs an essentially simple immersion lens, either directly or via further lens elements, to project a magnified image onto a detector (Tonner and Harp, 1989). The main resolution limitation with this system is from the chromatic aberration which arises when it is used for total-yield imaging. This type of system has been used with considerable success for small-spot XANES spectroscopy using a tunable synchrotron source. With the magnetic system the target is positioned within a strong magnetic field, normally produced by a superconducting magnet. Photoelectrons spiral along the field lines from the high-field target region to the low-field detector region where they are detected, producing a magnified image (Turner *et al.*, 1986). Energy resolution is possible with this device using either a retarding field analyzer or a bandpass filter (Turner *et al.*, 1988).

In the second type of photoelectron microscope a focused x-ray microprobe is formed using either zone plate (Ade *et al.*, 1990a) or reflection optics (Underwood *et al.*, 1992). The sample is scanned through the x-ray microprobe and an image is built up in terms of either the total electron yield or the energy-resolved signal (Fig. 8.2). All reported microscopes of this type have, or intend to incorporate, a deflection electron-energy analyzer. The cylindrical mirror analyzer (CMA) appears to be most popular for this at present, although the hemispherical deflection analyzer is also used. This type of microscope takes longer to produce a complete image than the full-field microscope does, but the enhanced signal from the focused microprobe should mean superior performance for small-spot spectroscopy and hence microanalysis.

Finally, systems have been reported that make use of the focusing properties of the hemispherical electron-energy analyzer and combine this with the magnification properties of an electrostatic lens system to provide limited imaging and small-spot ESCA (Coxon *et al.*, 1990). Rather than discuss in-

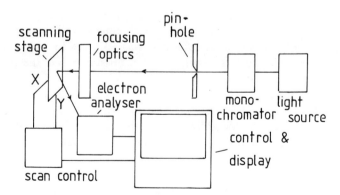

Figure 8.2. Schematic diagram of a scanning photoelectron microscope.

dividual systems in detail, the main performance parameters and reported characteristics of the instruments are collected together in Table 8.1. It should be noted that the term *photoemission microscope* is frequently used for microscopes imaging in terms of threshold electrons produced by UV irradiation (3–5 eV) (Griffith, 1986; Telieps and Bauer, 1985). Such instruments, which have high resolution capability (20–30 nm), have not been considered here since they do not have the possibility of providing microanalytical information.

8.5. Applications

A comprehensive coverage of the most recent applications of the various forms of x-ray microscopy is given in Michette *et al.* (1992). These include imaging of hydrated biological material such as chromosomes, fibroblasts, and zymogen granules, and a range of nonbiological test specimens (mostly drawn from solid-state physics and imaged by photoelectron microscopy). A study which will be considered in more detail here is of cement, imaged by the King's College STXM.

The study of cement is a good example of the use of imaging across an absorption edge to enhance the contrast. As a result of its atomic structure, the variation in transmission of calcium in the neighborhood of its L absorption edge is very much more marked than that across a normal edge. By taking images at two energies where there is a large difference in absorption it is possible, by subtracting or taking the ratio of the two images, to map the concentration of calcium. This can also be done for other elements with absorption edges in the wavelength range used.

Finely ground particles of ordinary portland cement, with the approximate stoichiometry Ca_3SiO_5, were hydrated in a very dilute water–cement mixture and then examined in the x-ray microscope at regular intervals over a period of about 25 h. Figure 8.3 shows three pairs of images that indicate the slow change in the structure of a single cement grain that took place during this time. One image in each pair was taken at an x-ray energy of 349.3 eV where there is a minimum in transmission, while the other was taken at the transmission maximum of 350.6 eV.

The sequence in Fig. 8.4 shows another cement sample after 36 h of hydration. Images (a) and (b) were taken at 350.6 and 349.3 eV, respectively, and (c) shows the natural logarithm of the ratio of (a) to (b). The differences in images (a) and (b) are highlighted in this way since the x-ray transmission through the specimen follows an exponential law, and the only major change in transmission over the energy range used is due to calcium. The light areas in Fig. 8.4c show the regions where the concentration of calcium is the greatest. Figure 8.4d is the same as (c) except that the gray levels have been contoured

Table 8.1. Performance Parameters for Existing Photoelectron Microscopes

Name/ref.	Microscope type	Radiation source	Reported spatial resolution	Energy analyzer	Imaging time
X1 SPEM (Ade et al., 1990a,b)	Scanning; zone plate	Synchrotron (NSLS)	200 nm	CMA	~60 min
MAXIMUM (Underwood et al., 1991)	Scanning; multilayer Schwarzschild	Synchrotron (Aladdin)	500 nm	CMA	
Tonner and Harp (1989), Harp et al. (1990)	Full-field; electrostatic focusing	Synchrotron (Aladdin)	2 μm		30 ms
Pianetta et al. (1990)	Full-field; magnetic focusing	Synchrotron (SSRL)	5 μm	Retarding field	
ESCASCOPE (Coxon et al., 1990)	Hemispherical analyzer; electron lens	Al K_α, Mg K_α	<10 μm	Hemispherical	20–40 min
PESM (Turner et al., 1988, 1986)	Full-field; magnetic focusing	He I, Al K_α, etc.	1–20 μm	Retarding field; bandpass filter	
ESCA 300 (Gelius et al., 1990)	Electron lens; hemispherical analyzer	Al K_α, Mg K_α	20 μm	Hemispherical	
Anjum et al. (1992)	Scanning; zone-plate focusing	He I	5 μm	CMA	~30 min

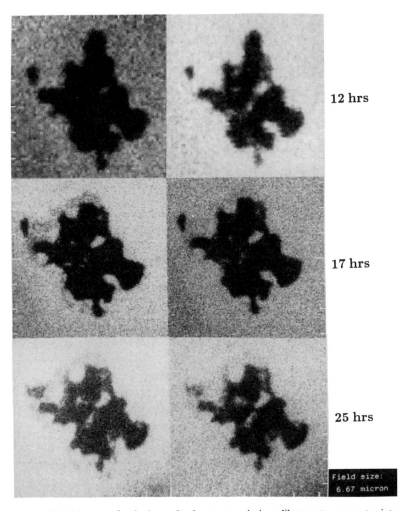

Figure 8.3. STXM images of a single portland cement grain in a dilute water–cement mixture. By comparing left–right pairs of images the calcium concentration can be inferred (areas bright in the left-hand but dark in the right-hand image), while from top to bottom the time evolution of the grain can be seen. The field size of each image is 6.67 μm.

at equal linear intervals to make changes in the calcium concentration easier to identify.

In terms of applications, the field of photoemission microscopy is still in its infancy and much of the published work illustrates potential rather than definite applications. Ade *et al.* (1990b) have reported imaging results using

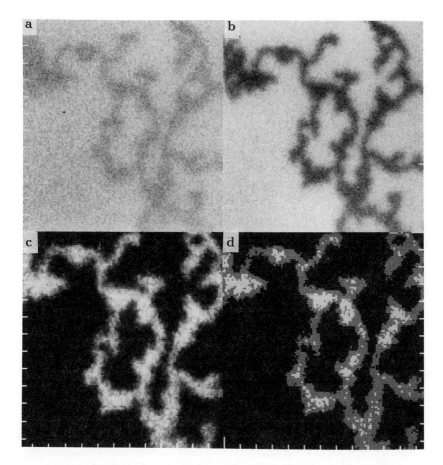

Figure 8.4. STXM images of portland cement after 36 h of hydration. (a) was taken at the transmission maximum of calcium at an x-ray energy of 350.6 eV and (b) at a transmission minimum of 349.3 eV. (c) is the natural logarithm of (a)/(b) and shows the calcium distribution in the sample (bright areas corresponding to high concentrations of calcium). (d) is the same as (c) except that the gray levels have been contoured at equal linear intervals to highlight the concentration changes. The field size of each image is 7.69 μm.

their scanning photoemission microscope at the NSLS with an artificial structure consisting of alternate bars of SiO_2 and Al on a Si substrate. Imaging with 300-nm resolution was possible in terms of Al $2p$ electrons, O KVV Auger electrons and with electrons from both the Si $2p$ and chemically shifted Si $2p$ levels. Perhaps a more genuine application of photoelectron microscopy was reported by Tonner and colleagues (1990), who used their full-field electrostatic microscope to perform small-spot (5 μm) XANES studies on SiO_2 layers during thermal desorption. From the characteristic XANES spectra they were able to infer the presence of both Si^{2+} and Si^{3+} at the Si–SiO_2 interface and on the surface of the Si substrate. These results had not been observed using other surface microscopies and perhaps indicate the damage limitation with x-ray excitation. Other indications of the use of photoemission microscopy are provided by the demonstration that the full-field magnetic-focusing microscope has been used both for small-spot ESCA of Al/Si targets and for XANES studies of the O 1s edge in high-temperature superconductors (Pianetta et al., 1990). In both of these cases, however, the spectroscopy was carried out on areas with dimensions of several tens of micrometers. The possibility of imaging with electrons from a variety of core levels (W 4f, Si 2p, Al 2p, O 1s, and C 1s) has been demonstrated with the Vacuum Generators ESCASCOPE using standard Al K_α and Mg K_α laboratory x-ray sources (Coxon et al., 1990). Small-spot (10 μm) ESCA (Coxon et al., 1990) studies of polymer surfaces have demonstrated the possibility of working with insulating materials.

8.6. The Future

There has been a steady growth in interest in x-ray microscopy over the past few years, while at the same time related advances in many aspects of x-ray science and technology have been gathering pace, and the number of potential applications of the various forms of x-ray imaging have increased.

For photoelectron microscopy future developments are likely to arise from the increased signal to be expected with the next generation of synchrotron sources. These sources should enable the full potential resolution of zone plates to be achieved together with much shorter imaging times. Current resolutions, ranging from a fraction of a micrometer to several micrometers, limit the scope for applications, but it seems likely that even with these modest resolutions the technique will be able to contribute much to the field of metallurgy and to the understanding of surface reactions at polycrystalline surfaces. Study of the segregation of atoms both at surfaces and at grain boundaries should be possible, and the effect of an inhomogeneous metal surface on gas phase–surface reactions will be an obvious area for study. Applications of the technique will increase as resolutions get better; for example, if a resolution of 10 nm could be achieved, then the study of dispersed catalysts will become possible. It is also likely that, once rapid imaging at submicrometer resolution is achieved, the technique of photoelectron microscopy will find applications in microelectronics.

ACKNOWLEDGMENTS. The King's College STXM is funded by the SERC, and has been developed over several years through the hard work and dedication of our colleagues at King's College, at Daresbury, and elsewhere. It is impossible to name everyone who has contributed to the project, but at present the principal workers to whom the authors extend thanks are Professor Ron Burge, Drs. Graeme Morrison, Chris Buckley, Mike Browne, Pambos Charalambous, Andrew Yacoot, and Pauline Bennett, and Pete Anastasi, Guy Foster, Dean Morris, Jake Palmer, Saqib Anjum, Mike Schulz, and Bill Luckhurst (all King's), and Professor Phil Duke, Dr. Howard Padmore, and Ciaran Mythen (all Daresbury).

References

Ade, H., Kirz, J., Hulbert, S., Johnson, E., Anderson, E., and Kern, D. (1990a). *Nucl. Instrum. Methods* **A291**, 126–131.

Ade, H., Kirz, J., Hulbert, S. L., Johnson, E. D., Anderson, E., and Kern, D. (1990b). *Appl. Phys. Lett.* **56**, 1841–1843.

Anderson, E. H., and Kern, D. (1992). In *X-Ray Microscopy III, Springer Series in Optical Sciences* (A. G. Michette, G. R. Morrison, and C. J. Buckley, eds.), Springer-Verlag, Berlin, pp. 75–78.

Anjum, S., Burge, R. E., Potts, A. W., and Yacoot, A. (1992). In *X-Ray Microscopy III, Springer Series in Optical Sciences* (A. G. Michette, G. R. Morrison, and C. J. Buckley, eds.), Springer-Verlag, Berlin, pp. 241–3.

Aristov, V. V., Basov, Y. A., Snigirev, A. A., Belakhovsky, M., Dhez, P., and Freund, A. (1992). In *X-Ray Microscopy III, Springer Series in Optical Sciences* (A. G. Michette, G. R. Morrison, and C. J. Buckley, eds.), Springer-Verlag, Berlin, pp. 70–74.

Bögli, V., Unger, P., Beneking, H., Greinke, B., Guttmann, P., Niemann, B., Rudolph, D., and Schmahl, G. (1988). In *X-Ray Microscopy II, Springer Series in Optical Sciences* Vol. **56** (D. Sayre, M. Howells, J. Kirz, and H. Rarback, eds.), Springer-Verlag, Berlin, pp. 80–87.

Charalambous, P., and Morris, D. (1992). In *X-Ray Microscopy III, Springer Series in Optical Sciences* (A. G. Michette, G. R. Morrison, and C. J. Buckley, eds.), Springer-Verlag, Berlin, pp. 79–82.

Chauvineau, J. P. (1989). In *X-Ray Instrumentation in Medicine and Biology, Plasma Physics, Astrophysics, and Synchrotron Radiation* (R. Benattar, ed.), Proc. SPIE **1140**, 440–447.

Cheng, P. C., and Jan, G. J. (eds.) (1987). *X-Ray Microscopy: Instrumentation and Biological Applications*, Springer-Verlag, Berlin.

Cosslett, V. E., and Nixon, W. C. (eds.) (1960). *X-Ray Microscopy*, Cambridge University Press, London.

Coxon, P., Kirzek, J., Humpherson, M., and Wardell, I. R. M. (1990). *J. Electron Spectrosc. Relat. Phenom.* **52**, 821–836.

Dhez, P. (1992). In *X-Ray Microscopy III, Springer Series in Optical Sciences* (A. G. Michette, G. R. Morrison, and C. J. Buckley, eds.), Springer-Verlag, Berlin, pp. 107–113.

Duke, P. J., and Michette, A. G. (eds.) (1990). *Modern Microscopies: Techniques and Applications*, Plenum Press, New York.

Gelius, U., Wannberg, B., Baltzer, P., Fellner-Feldegg, H., Carlsson, G., Johansson, C. G., Larsson, J., Munger, P., and Vegerfors, G. (1990). *J. Electron Spectrosc.* **52**, 747–785.

Griffith, O. H. (1986). *Appl. Surf. Sci.* **26**, 265–279.

Harp, G. R., Han, Z. L., and Tonner, B. P. (1990). *J. Vac. Sci. Technol. A* **8**, 2566–2569.

Lebert, R., Holz, R., Rothweiler, D., Richter, F., and Neff, W. (1992). In *X-Ray Microscopy III, Springer Series in Optical Sciences* (A. G. Michette, G. R. Morrison, and C. J. Buckley, eds.), Springer-Verlag, Berlin, pp. 62–65.

Michette, A. G. (1986). *Optical Systems for Soft X Rays,* Plenum Press, New York.

Michette, A. G. (1988). *Rep. Prog. Phys.* **51,** 1525–1606.

Michette, A. G., Morrison, G. R., and Buckley, C. J. (eds.) (1992). *X-Ray Microscopy III, Springer Series in Optical Sciences,* Springer-Verlag, Berlin,

Norman, D. (1986). *J. Phys. C* **19,** 3273–3311.

Pianetta, P., King, P. L., Borg, A., Kim, C., Landau, I., Knapp, G., Keenlyside, M., and Browning, R. (1990). *J. Electron Spectrosc.* **52,** 797–810.

Rudolph, D. (1987). In *Soft X-Ray Optics and Technology* (E.-E. Koch and G. Schmahl, eds.), Proc. SPIE **733,** 294–300.

Sayre, D., Howells, M., Kirz, J., and Rarback, H. (eds.) (1988). *X-Ray Microscopy II, Springer Series in Optical Sciences* Vol. 56, Springer-Verlag, Berlin.

Schmahl, G., and Rudolph, D. (eds.) (1984). *X-Ray Microscopy, Springer Series in Optical Sciences* Vol. 43, Springer-Verlag, Berlin.

Seah, M. P., and Dench, W. A. (1979). *Surf. Interface Anal.* **1,** 2–11.

Shinohara, K., Yada, K., Kihara, H., and Saito, T. (eds.) (1990). *X-ray Microscopy in Biology and Medicine,* Japan Scientific Societies Press, Tokyo, and Springer-Verlag, Berlin.

Telieps, W., and Bauer, E. (1985). *Ultramicroscopy* **17,** 57–66.

Tonner, B. P., and Harp, G. R. (1989). *Rev. Sci. Instrum.* **59,** 853–858.

Turcu, I. C. E., Tallents, G. J., Schulz, M. S., and Michette, A. G. (1992). In *X-Ray Microscopy III, Springer Series in Optical Sciences* (A. G. Michette, G. R. Morrison, and C. J. Buckley, eds.), Springer-Verlag, Berlin, pp. 54–57.

Turner, D. W., Plummer, I. R., and Porter, H. Q. (1988). *Rev. Sci. Instrum.* **59,** 45–48.

Turner, D. W., Plummer, I. R., and Porter, H. Q. (1986). *Philos. Trans. R. Soc. London Ser. A* **318,** 219–241.

Underwood, J. H., Perera, R. C. C., Kortright, J. B., Batson, P. J., Capasso, C., Liang, S. H., Ng, W., Ray-Chaudhuri, A. K., Cole, R. K., Chen, G., Guo, Z. Y., Wallace, J., Welnak, J., Margaritondo, G., Cerrina, F., De Stasio, G., Mercanti, D., and Ciotti, M. T. (1992). In *X-Ray Microscopy III, Springer Series in Optical Sciences* (A. G. Michette, G. R. Morrison, and C. J. Buckley, eds.), Springer-Verlag, Berlin, pp. 220–225.

9

X-ray Photoemission Spectroscopy

A. J. Nelson

9.1. Introduction and Background

The development and implementation of advanced technologies requires a thorough understanding of the effects of fabrication and processing on the design and performance of a specific product. Measurements of the surface composition, elemental distributions, segregated impurity species, and interfacial chemical reactions are of fundamental importance in determining and understanding performance and possible failure mechanisms. X-ray photoelectron spectroscopy (XPS) has demonstrated its effectiveness in the solution of many of these technological problems.

XPS [also known as electron spectroscopy for chemical analysis (ESCA)] is a widely accepted and powerful technique for studying solids, surfaces, and interfaces. The discovery of the photoemission phenomena by Hertz in 1887 and its interpretation by Einstein in 1905 provided the basis for the development of this technique. Initial development of this analytical technique began with the early work of Robinson and Rawlinson in 1914 and of de Broglie in 1921. However, the utilization of the technique was hampered by the unavailability of analyzers with sufficient resolution to distinguish between the various peaks in the resulting XPS energy spectra. The progression of the photoemission practice to its present state of the art has occurred within the last 40 years beginning with the pioneering work by Siegbahn and co-workers

A. J. Nelson • National Renewable Energy Laboratory, Golden, Colorado 80401.

Microanalysis of Solids, edited by B. G. Yacobi *et al.* Plenum Press, New York, 1994.

at Uppsala University in the early 1950s. This group introduced a high-resolution spectrometer for the detection of low-energy electrons produced by soft x-ray radiation and recognized that the precision in determining the energy of photoelectron peaks was accurate enough to study the mechanics of atomic orbitals. Additionally, Siegbahn's group determined that the technique was surface sensitive. Siegbahn's work laid the foundations for XPS and culminated in a landmark publication (Siegbahn *et al.*, 1967). These theoretical and experimental investigations led to the commercialization of XPS as well as to the application of the methodology in the understanding of many advanced technological developments.

9.2. Fundamental Principles

This section describes various aspects of the photoemission process to provide a sound theoretical basis for subsequent discussions of specific experimental observations and applications.

9.2.1. Photoemission Process

XPS involves the energy analysis of photoemitted electrons from a sample with energies characteristic of the target elemental composition and its chemical state. Photoelectrons are emitted from the material (solid, liquid, or gas) as a result of absorption of a photon with energy $E = h\nu$. The less tightly bound valence electrons and shallow core levels may be photoionized with UV photons (UPS), while deeper core-level electrons may be photoionized using x-ray photons (XPS). Both types of photoelectrons are ejected with discrete kinetic energy E_k given by

$$E_k = h\nu - E_b - \phi_s \qquad (9.1)$$

(see Fig. 9.1) where $h\nu$ is the energy of the incident photon, E_b is the binding energy of the photoelectron relative to the Fermi level, and ϕ_s is the work function of the spectrometer. The Fermi level is used as the zero binding energy level by definition. Each photoelectron has a discrete energy representative of the element from which it was emitted, thus allowing one to identify the atomic species present. Furthermore, the binding energy of an electron is influenced by its chemical environment and thus can be used to identify the chemical state of a given atom in the sample. In addition to core-level ionization, x-ray-induced Auger electrons are emitted because of relaxation of the energetic ions left after photoemission. Each of these processes has a different probability or cross section as described in the next section. Therefore,

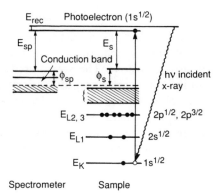

Figure 9.1. Schematic representation of the electronic levels involved in the photoemission process from a solid and the measurement of the electron energy by an analyzer. (Adapted from Katz, 1981.)

an XPS spectrum contains core-level emission and Auger emission spectral lines and provides information on both the elemental composition and the chemical state of atoms in a given sample.

9.2.2. Photoionization Cross Sections

The absorption of a photon by atoms or molecules in a solid, liquid, or gas and the subsequent emission of a photoelectron defines the photoionization process. The probability that photon absorption with electron emission will occur is called the photoionization cross section, σ, i.e., the transition probability per unit time of a core-level electron with a unit incident photon flux of $1 \text{ cm}^{-2}\text{s}^{-1}$. Photoionization cross sections are a function of photon energy $h\nu$, core-level binding energy E_b, atomic number Z, and the relative directions of photon incidence and electron emission (ϕ). These cross sections allow one to predict the intensities of photoelectron peaks and are absolutely necessary in quantitative XPS compositional analysis.

Relativistic and nonrelativistic calculations of total atomic subshell cross sections for relevant XPS photon energies are discussed in detail in the literature. Briefly, Scofield used a relativistic analogue of the single-potential Hartree–Slater atomic model to calculate total subshell cross sections for Z = 1–96 for Mg $K\alpha$ and Al $K\alpha$ x-ray photon energies (Scofield, 1976). These calculated cross sections are widely used for quantitative XPS using Mg and Al x-ray sources. More recently, Yeh and Lindau calculated atomic subshell photoionization cross sections and asymmetry parameters with the Hartree–Fock–Slater one-electron central potential model for all elements Z = 1–103 (Yeh and Lindau, 1985). These tabulated calculations allow convenient access to cross sections in the energy range 0–1500 eV (UPS and XPS photon energies) as well as providing a comprehensive reference for photoemission researchers.

9.2.3. Photoelectron Escape Depth

Photoelectrons ejected from an atomic or molecular energy level must escape from the material into the vacuum for detection without experiencing inelastic collisions. The x-rays used for excitation typically penetrate 1–10 μm into the sample and thus it is essential to know the mean escape depth of the photoemitted electrons. The escape depth can be understood by defining the mean free path (MFP) λ, which is the characteristic length across which the peak intensity decreases exponentially because of elastic and inelastic scattering according to the standard expression for exponential decay

$$I(z) = I_0 \exp(-z/\lambda) \tag{9.2}$$

The inelastic mean free path (IMFP) is the equivalent length related only to inelastic scattering and is of primary importance in XPS analysis. The relation between MFP and IMFP depends on the cross sections for elastic and inelastic interactions and varies by only 3% in the energy range $3 \le E \le 3000$ eV (Seah and Dench, 1979). The probability or cross section for inelastic scattering events is determined by the kinetic energy of the electron and material through which it is traveling. Material properties which influence scattering cross sections include the electronic structure, surface composition, lattice geometry, and surface roughness.

Seah and Dench compiled experimental IMFP data on electrons with energies between 0 and 10 keV for elements, organic and inorganic compounds, and adsorbed gases, and suggest empirical formulas for each. Basically, they found that the path lengths are very high at low energies, decrease to between 0.1 and 0.8 nm for energies of 30–100 eV, and then increase approximately as $E^{1/2}$. Additionally, low-Z materials are more permeable than high-Z materials. From these measurements, a universal escape depth curve (Fig. 9.2) was generated which is a standard reference for photoemission researchers.

The previous discussion details the specific reasons why XPS is a surface-sensitive technique. However, in practical applications, one may be required to provide even more surface sensitivity. This may be accomplished either by varying the incident radiation to get the photoemitted electron's kinetic energy at the minimum of the escape depth curve, or by varying the emission angle. Either of these methods can be used to measure an overlayer thickness by comparing the relative intensities of the substrate signal and the overlayer signal.

9.2.4. Chemical Shifts

The binding energy E_b of an electron is influenced by its chemical environment and thus the measured energy [Eq. (9.1)] can be used to probe the

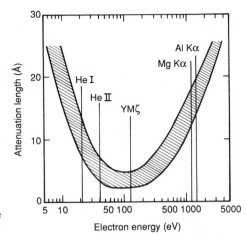

Figure 9.2. Universal escape depth curve for photoemitted electrons.

chemical environment. When comparing XPS lines from the same element in different chemical environments, shifts in the peak position are observed (Fig. 9.3). This fact was first realized by Hagström *et al.* (1964) in the identification of chemically different species involving the same atom. Further work by Siegbahn *et al.* (1969) fully exploited this phenomenon for the investigation of small gaseous molecules using electron spectroscopy.

Chemical shifts of core levels can be understood by examining the basic physics involved with the change in binding energy. Specifically, the attractive potential of a nucleus and the repulsive Coulomb interaction with other electrons determines the energy of an electron in a tightly bound core state. When the chemical environment of the atom changes, a spatial rearrangement of the average charge distribution occurs because of the creation of a different potential by the nuclear and electronic charges of the other atoms in the compound. The magnitude of the shift is determined by the type and strength of the bond, i.e., covalent or ionic, and can be interpreted qualitatively or semiquantitatively by applying the concept of electronegativity (Pauling, 1960). Identification of the chemical environment of an atom can also be accomplished by comparing the binding energies of the same atom in various reference compounds. The National Institute of Standards and Technology (NIST) has recently published a comprehensive tabulation of core-level binding energies for a multitude of organic and inorganic compounds, thus providing researchers with a useful aid in the identification of XPS chemical shifts.

9.2.5. Additional Spectral Features

Major peaks in the XPS spectra are associated with core-level emission and Auger electron emission from atoms near the surface. These spectral

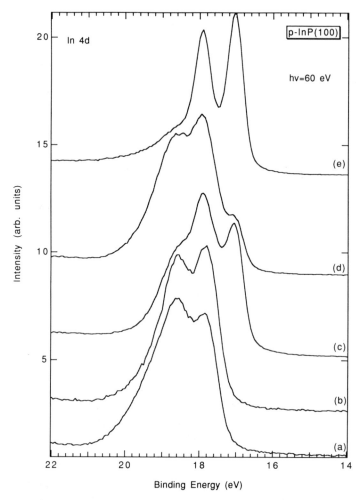

Figure 9.3. X-ray photoemission measurements on H_2 plasma processed InP, specifically, In $4d$ spectra of InP(100) for (a) Br_2 etch, (b) Br_2 etch + 250°C, (c) 1-kV Ar sputter, (d) Ar sputter + 280°C, and (e) H_2 plasma etch showing an example of the chemical shift.

features are usually dominant, but the spectra may contain other features that can confuse the interpretation. The additional peaks that are sometimes observed, and are not fundamental to the XPS process, include:

- *Satellite peaks.* The emission spectrum of the x-ray probe (e.g., Mg $K\alpha_{1,2}$) contains not only the characteristic x ray, but also some minor components at higher photon energies (e.g., Mg $K\alpha_3$, $K\alpha_4$, $K\alpha_5$, $K\alpha_6$, and $K\beta$). Thus, each of these can give rise to a family of peaks in the XPS spectrum at binding energies below the expected transitions.

- *Ghost peaks.* It is possible that another element rather than the expected source material may inadvertently give rise to unwanted x-radiation. This contaminating radiation may come from an impurity in the anode, from material near the anode, or from x rays generated within the sample itself. This contaminating radiation causes minor lines called ghost lines. Ghost lines are rare and do not usually cause difficulties.

- *Shake-up lines.* There is a finite probability that an ion can be left in an excited state several electron volts above the ground state after the photoelectric process. When this occurs, the kinetic energy of the emitted photoelectron is reduced by the corresponding energy difference between the ground state and the excited state. This results in the formation of a satellite peak at a position a few electron volts higher in binding energy (lower in kinetic energy) than the major transition peak. These satellite lines can cause problems in interpretation since their intensities can approach those of the major line. Positive identification using the Auger line in the XPS spectrum can aid in the identification of the actual chemical state.

- *Energy loss lines.* Photoelectrons can lose specific amounts of energy because of interactions with other electrons in the surface region of the sample. Energy loss lines usually occur at binding energies higher than the major transition and are located at periodic intervals. The energy between the primary peak and the loss peak is called the plasmon energy. Plasmon or energy loss lines are usually more prominent in conductors.

- *Valence lines.* Photoelectron emission from the valence band can provide low-intensity lines in the XPS spectrum in the low-binding-energy region, i.e., within 10 eV of the Fermi energy reference. UV photoemission is the preferred technique for studying valence-band electronic structure because of the higher photoemission cross sections at UV photon energies.

9.3. Experimental Considerations

The components necessary for performing an XPS experiment are a vacuum chamber, a radiation source, an electron energy analyzer, electron detector and control electronics, and a specimen of interest. In addition to these basic components, one may choose to incorporate an ion sputtering source for *in situ* sample cleaning, quadrupole mass analyzer for residual gas analysis, a sample preparation chamber with interlock, and so forth. The essential aspects of each of the basic components are considered in detail below.

9.3.1. Radiation Sources

The standard x-ray source consists of a heated filament (cathode) and a target anode with a potential (5–20 kV) applied between these elements to accelerate electrons from the filament toward the target (Fig. 9.4). Electron bombardment of the anode creates holes in the inner shells of the atoms which are then filled by electrons from higher-lying orbitals resulting in x-ray emission. An additional source of radiation is a continuous background of bremsstrahlung caused by the interaction of the rapidly decelerating electrons with the nuclei and electrons from the target.

The most commonly used anode materials are Mg and Al; Au, Ag, Cu, Si, Zr, and Ti anodes are also commercially available. The Mg, Al, Si, and Ti anode materials produce an x-ray spectrum that is dominated by a very intense, unresolved, $K\alpha_1$–$K\alpha_2$ doublet resulting from transitions of the type $2p_{3/2} \rightarrow 1s$ and $2p_{1/2} \rightarrow 1s$, respectively. $L\alpha$ radiation results from transitions of the type $3p \rightarrow 2s$. The x-ray energies produced in such sources are: Mg $K\alpha_{1,2}$ 1253.6 eV, Al $K\alpha_{1,2}$ 1486.6 eV, Ag $L\alpha_1$ 2984.3 eV, Cu $L\alpha_{1,2}$ 929.7 eV, Si $K\alpha_{1,2}$ 1739.5 eV, Zr $L\alpha_1$ 2042.4 eV, and Ti $K\alpha_1$ 4510.9 eV. Higher-energy $M\alpha$ radiation from transitions of the type $4p_{3/2} \rightarrow 3d_{5/2}$ include Au $M\alpha_1$ at 2122.9 eV. For these non-monochromatized x-ray sources, the primary factor limiting energy resolution is the natural linewidth of the x-ray line. The resolution limit is thus determined by the full-width-at-half-maximum intensity

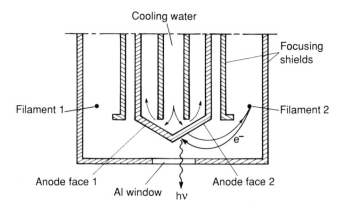

Figure 9.4. Commercial dual-anode x-ray source allowing for the operation of either Al or Mg $K\alpha$ radiation. The thick Al and Mg films are deposited on a copper block which is water-cooled. Separate filaments are located near their respective anode.

(FWHM) which, for the more commonly used anodes, is approximately 0.6 eV for Mg $K\alpha_{1,2}$ and 0.8 eV for Al $K\alpha_{1,2}$. For the higher-energy sources, the FWHM linewidths are 1.0 eV for Si $K\alpha_{1,2}$, 1.65 eV for Zr $L\alpha_1$, 2.4 eV for Au $M\alpha_1$, 2.8 eV for Ag $L\alpha_1$, and 2.1 eV for Ti $K\alpha_1$. Obviously, from this discussion, selection of the most suitable x-ray anode requires the consideration of the range of transitions that can be excited as well as the analytical resolution and sensitivity of those transitions for the materials of interest.

The standard UV light source for UPS is commonly a helium discharge lamp which produces the two resonance radiations He[I] at 21.22 eV and He[II] at 40.8 eV. The gas discharge is operated at pressures between 0.1 and 10 torr and is separated from the high-vacuum analysis chamber by a series of low-conductance capillaries which are differentially pumped. The operating pressure determines the resonance radiation which is produced. These sources are attractive for studying valence-band structure since valence levels have a higher photoelectric cross section at the lower photon energies. Synchrotron radiation light sources are the ultimate light source and provide radiation which can be tuned over the entire energy region from UV to x-rays.

The monochromatization of the aforementioned x-ray lines significantly improves spectral quality by eliminating satellite lines and bremsstrahlung. Monochromatization and focusing of x-ray lines is achieved by Bragg reflection from a single spherically bent quartz crystal (Fig. 9.5). In order to efficiently collect x-ray intensity from the source, the ratio of the crystal diameter to the radius of curvature must be large within the limitations of geometrical aberrations. The concave surface of the spherically bent crystal on which the x-rays impinge must be of optical quality and have a precise radius of curvature equal to the diameter of the appropriate Rowland circle. Additionally, focusing of the x-ray source for small-spot XPS analysis is achieved by utilizing x-ray back-diffraction from a suitable quartz crystal, i.e., toroidially bent and mechanically polished. This geometry produces an x-ray-irradiated area on the order of ≤1 mm. Further microfocusing of x-rays to ≤200 μm requires re-

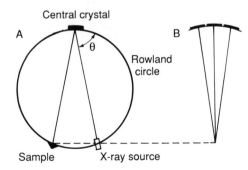

Figure 9.5. Basic elements of a typical x-ray monochromator.

duction of the power dissipation in the anode, thus reducing the x-ray flux as well as the photoelectron signal intensity. For this reason, most instrument manufacturers prefer electro-optic aperturing over the microfocus mono-chromator for state-of-the-art spatially resolved XPS. Commercially available x-ray monochromator systems based on the aforementioned design criteria will provide higher energy resolution and low noise figures, but at higher cost. Furthermore, as previously mentioned, monochromatized and focused sources provide a significantly lower x-ray flux at the specimen when compared with conventional sources. For comparison, Fig. 9.6 displays XPS spectra for clean $CuInSe_2$ acquired with a conventional Al source and with a monochromatized Al source. One can see that the background noise intensity is significantly reduced with the monochromatized x-ray source.

9.3.2. Electron Energy Analyzers

Advances in electron spectrometer design coupled with the aforemen-tioned improvements in photon sources have yielded significant improvements

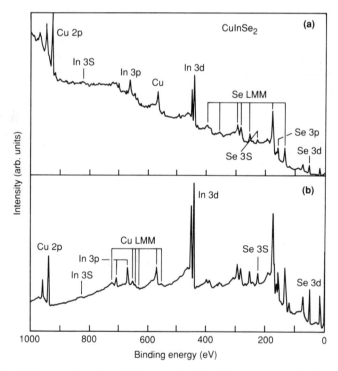

Figure 9.6. XPS spectra for clean $CuInSe_2$ acquired (a) with a conventional Al source and (b) with a monochromatized Al source.

in energy resolution as well as spatial resolution. The three major analyzer configurations commercially available are: 1) spherical sector capacitor, 2) cylindrical mirror, and 3) nondispersive energy filter. All three analyzers require a suitable vacuum chamber with pressures $<10^{-5}$ torr in order to minimize electron scattering through collisions with residual gas molecules. The specific details on the design and operation of each analyzer are presented in the following discussions.

9.3.2.1. Spherical Sector Capacitor Analyzer

The spherical sector analyzer is a dispersive electrostatic energy analyzer of the deflector or "prism" type (Purcell, 1938), i.e., the main electron trajectory follows an equipotential surface (Fig. 9.7a). Specifically, electrons are dispersed along a radial coordinate, based on their kinetic energies, and are brought to a focus at the exit slit of the instrument. The central trajectory lies in the symmetry plane with a radius of curvature equivalent to the radius of the corresponding equipotential surface in that plane. Ignoring aberrations, a point source on a specimen is reimaged as a point, thus demonstrating space focusing for this analyzer (Wannberg et al., 1974). XPS spectra are produced by applying a sweep voltage to the spherical deflector plates of the analyzer which allows electrons of successive kinetic energies to pass through the exit slit to the detector. This mode of operation produces constant energy resolution over the entire spectrum. Alternatively, the voltage applied to the deflector plates is fixed and the retarding field is ramped. The photoelectrons are slowed by the retarding field while the fixed sector voltage determines the kinetic energy of the electrons which pass through to the detector. This mode of operation provides better sensitivity to electrons with low initial kinetic energy.

In order to achieve spatial resolution $\leq 100\ \mu$m with good photoelectron signal intensity for state-of-the-art small-spot XPS analysis, apertures are placed in the analyzer. Specifically, photoelectrons are imaged through the primary analyzer lens onto an area-defined aperture. This aperture defines the actual analysis area on the sample from which photoelectrons are detected. The spatial resolution achieved by this methodology is only limited by the spherical aberration of the primary lens. However, in practical applications, the photoelectron brightness will be the factor limiting the ultimate spatial resolution (≈ 5–$20\ \mu$m).

Multidetector techniques can be readily employed with this electrostatic deflection instrument since the analyzer has a well-defined focal plane. Photoelectrons transmitted by the analyzer impinge on an array of detectors located at the focal plane. This detection method will be describe in detail in a subsequent section.

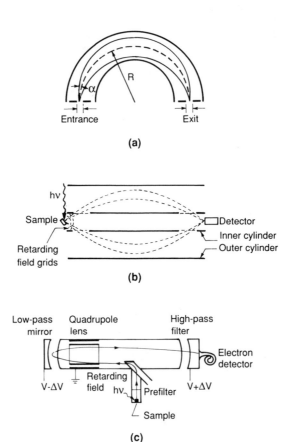

Figure 9.7. Schematic representations of electron energy analyzers: (a) spherical sector analyzer; (b) cylindrical mirror analyzer; (c) nondispersive energy analyzer.

9.3.2.2. Cylindrical Mirror Analyzer

The cylindrical mirror analyzer (CMA) with a coaxial electron gun is a popular configuration which has the advantage of higher transmission (Fig. 9.7b). The theoretical design criteria as well as practical construction and operation of this analyzer are discussed in detail by Palmberg (1975). Basically, the CMA consists of two coaxial cylinders. The analyzer operates by applying a sweep voltage to the outer cylinder of the analyzer which determines the electron energies permitted to pass through to the detector. The energy resolution of the analyzer, R, is defined in terms of the pass energy, ΔE, and the range of electron energies transmitted, E,

$$R = \Delta E / E \qquad (9.3)$$

For XPS analysis, the analyzer is usually operated in retarding mode which gives an energy distribution function with intensity proportional to E^{-1}. This is accomplished by using the retarding grids at the front end on the analyzer to scan the spectrum while the CMA is operated at constant pass energy, thus resulting in constant energy resolution across the entire energy spectrum. The size of the analysis area is determined by circular apertures within the CMA and is typically several millimeters. Thus, the energy distribution, energy resolution, and analysis area are all strictly a function of the analyzer.

9.3.2.3. Nondispersive Energy Filter Analyzer

Nondispersive instruments are quite different from the previously described analyzers since they are not imaging and consequently are used only to detect one electron energy at a time. This limitation is offset by the very large luminosity (defined as the product of the entrance aperture, entrance solid angle, and transmission) which can be attained.

The nondispersive energy analyzer applies two spherically symmetric retarding fields in the energy analysis of photoemitted electrons (Lee, 1973). Specifically, the first retarding field is operated in a reflection mode and acts as a low-energy pass filter, while the second retarding field is operated in a transmission mode and acts as a high-energy pass filter (Fig. 9.7c). There exists a small energy overlap between these two filters and thus the analyzer transmits electrons in a narrow energy range. The low- and high-energy filters select electron energies nondispersively and, since the fields are spherically symmetric, require no spherical aberration compensation. The product of these filter functions yields the analyzer response function. Photoelectrons from the sample are first prefiltered by a retarding field to match their energy to that of the analyzer, thus preventing the production of a high spectral background. Following this operation, the electrons are deflected through an entrance aperture toward the low-pass energy filter. Electrons with the proper kinetic energy are reflected and focused back through a central aperture and aligned with a quadrupole lens for transmission through the aperture. These transmitted electrons are then screened by the high-pass energy filter which allows electrons with the proper kinetic energy to be transmitted to the detector. A spectrum is produced by scanning the retarding field while maintaining the analyzer at a fixed pass energy.

Table 9.1 summarizes the types and characteristics of the above-mentioned electron energy analyzers.

Table 9.1. Summary of Electron Energy Analyzer Characteristics

Type	Resolution range (%)	Transmission (%)	Comments
Spherical sector analyzer (SSA)	<0.2	—	Highest resolution for XPS
Cylindrical mirror analyzer (CMA)	0.2–5.0	6–9	High resolution for XPS
Nondispersive energy analyzer	0.5–1.0	10	High luminosity, high resolution for XPS

9.3.3. Detection and Signal Processing

Commercial instrumentation employs either a single-channel electron multiplier (channeltron) or a large-area, position-sensitive multichannel detector.

Channeltrons are electrostatic devices which consist of a small lead-doped glass tube with a thin semiconducting inner surface having a high secondary electron emissivity. High voltage is applied between the ends of the tube and secondary electron multiplication (10^6–10^8) is achieved by multiple collisions on the inner wall. The output of the multiplier consists of a series of pulses which are fed into an amplifier/discriminator and then into a computer for data acquisition.

Multichannel detection can be employed for analyzers which have a well-defined focal plane, e.g., the spherical sector analyzer. This detection methodology provides superb sensitivity, excellent resolution, and high dynamic range for x-ray photoemission. The detector's sensitivity can be optimized by utilizing the entire focal plane of the analyzer while high resolution is achieved by dividing the focal plane into multiple channels. The high dynamic range is achieved by providing discrete amplifiers, discriminators, and counters for each channel. A multichannel detector consists of either an array of channeltrons or a set of microchannel plates located at the focal plane of the specific analyzer (Fig. 9.8). Photoelectrons which are transmitted by the analyzer to the exit slit impinge on the detector array and generate signal pulses which are fed into an amplifier/discriminator, a counter, and then processed by a data acquisition computer. If the single exit slit is replaced by N parallel exit slits, a range of electron energies can be detected simultaneously since, in this configuration, the detector yields a factor N multiplex enhancement. Additionally, the use of multichannel detection in conjunction with a monochromatized x-ray source partially compensates for the lower x-ray flux incident on the sample because of the monochromator. The combination of

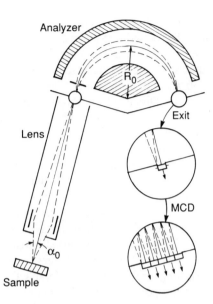

Figure 9.8. Schematic representation of a spherical sector analyzer with a multichannel detector at the exit focal plane.

a monochromatized x-ray source, an apertured spherical sector analyzer, and multichannel detection provides a powerful surface analytical tool with high spatial resolution and superior energy resolution. This instrumental configuration has become the industry standard for state-of-the-art small-spot XPS analysis.

9.3.4. Specimen Preparation

X-ray photoemission spectra have been recorded for solid, liquid, and gas specimens. Sample preparation and handling procedures require specific attention to the preservation of a clean surface. Contamination of the surface can severely limit the intensity of the substrate photoemission lines, because of the short MFP of the photoelectrons, as well as influence the interpretation of the experimental results. Specific preparation and handling procedures for the different types of samples are outlined below.

9.3.4.1. Solid Specimens

Solid samples, consisting of conductors, semiconductors, or insulators, require various preparation methods to ensure a clean surface for accurate XPS analysis. Removal of volatile residues prior to introduction into the vacuum chamber may simply consist of ultrasonic cleaning in solvents (e.g., trichloroethylene, acetone, and methanol). Additionally, powdered samples may be pressed into pellets, mounted on double-sided adhesive tape, or pressed into indium foil. The latter two mounting techniques for powders have the undesirable characteristics of limiting temperature excursions (In melts at 150°C) and, in the case of tape, providing a source of carbonaceous contamination. Dissolved materials or colloidal suspensions can be deposited from solution onto a substrate by evaporating off the solvent.

Once the sample has been mounted and placed in the analysis chamber, the surface may be prepared for analysis by cleaving, scraping, or ion milling. Cleaving a sample in vacuum results in a new surface free from atmospheric contamination. A single crystal can be cleaved along one of its low-index lattice planes with a sharp edge while polycrystalline samples will preferentially cleave along grain boundaries. In general, cleaved single-crystal surfaces are usually less reactive than the corresponding polycrystalline surfaces because of differences in the potential barrier at these interfaces.

Removal of surface layers by brute force mechanical means, i.e., filing, brushing, or scraping, is preferable for compounds and alloys with high-vapor-pressure components. The method does not change the surface composition of interest and introduces less damage on an atomic scale than ion beam sputtering.

Ion beam sputter etching is probably the most generally accepted method for removing contaminated or nonintrinsic surface layers. Argon and xenon gases are commonly used because of their inert nature at pressures on the order of 10^{-5}–10^{-7} torr. The gas is ionized by electron bombardment and the ions are accelerated through an electro-optic column toward the sample. Sputter etching physically removes material and results in a highly damaged surface region. Additionally, the surface region may be nonstoichiometric because of preferential sputtering effects. Annealing of the sample at elevated temperatures following sputter etching may be necessary to reestablish surface composition and structure.

In addition to the aforementioned surface cleaning techniques, photoemission measurements on solid samples require a precise energy reference. The Fermi energy (E_F) is the preferred binding energy reference and is defined as the highest occupied energy level in a metal at absolute zero; this is also more or less true for metals at normal experimental temperatures. Experimentally, the Fermi edge is obtained by acquiring a high-resolution photoemission spectrum from a clean metal surface (e.g., Au, Ni, Mo, Ta). The Fermi energy is deduced from measurement of the midpoint on the leading edge in the valence-band spectra (Fig. 9.9). The position of E_F as determined from the reference metal specimen can be applied in the accurate analysis of valence and core-level binding energies for semiconducting and insulating specimens since states at E_F are not occupied for these materials. This tech-

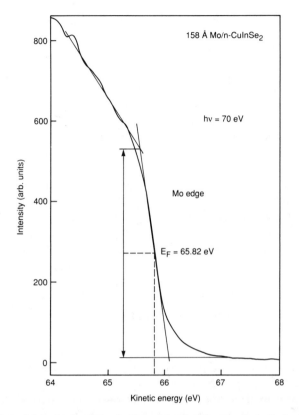

158 Å Mo/n-CuInSe$_2$

hv = 70 eV

Mo edge

E$_F$ = 65.82 eV

Figure 9.9. Methodology for the determination of the Fermi energy from the leading edge of a valence-band spectrum for a clean metal surface.

nique assumes that electronic equilibrium has been established between the sample, the reference, and the spectrometer but does not account for possible charging effects.

9.3.4.2. Liquid Specimens

Photoemission measurements on liquid samples with relatively high vapor pressures can be quite difficult because of the requirement that pressures near the analyzer be maintained at $\leq 10^{-4}$ torr. Siegbahn's group developed techniques involving the continuous replenishment of liquid on a wire in conjunction with high-speed differential pumping between the specimen chamber and the analyzer for the analysis of such samples. Alternatively, liquid metals have been examined with relatively little special equipment (Eastman, 1972).

Liquid specimens are not ordinarily encountered (or desired) and various other analytical techniques are usually applied.

9.3.4.3. Gaseous Specimens

Gas phase studies require a chamber fitted with an x-ray transparent window which separates the gas from the x-ray source and a small capillary or slit to allow the photoelectrons to exit into the energy analyzer. Differential pumping is also required between the exit slit and the analyzer in order to minimize gas-phase inelastic scattering effects. Typical gas chamber pressures range between 10^{-2} and 1 torr with specimen volumes on the order of 1 cm^3. The gas specimen can be furnished by a room-temperature gas-phase source or can be generated by heating liquid- or solid-phase reservoirs. These latter devices allow unique photoemission investigations of vaporizable solids in the gas phase. In general, for gas-phase spectra, the vacuum level is the energy reference level.

9.4. Quantitative XPS

The two major schemes used for quantifying XPS results utilize *peak area sensitivity factors* and *peak height sensitivity factors.* The peak heights method parallels the procedure used for AES with the best accuracy gained by comparison to structurally similar standards. The method based on the measurement of peak areas is the most accurate since this area exactly represents the photoelectrons emitted from a given transition (Fig. 9.10). However, interferences from other peaks or other chemical species of a specific transition can alter the shape of the peak and cause significant error. Therefore, substantial care must be exercised to gain the required accuracy in the utilization of the peak area method.

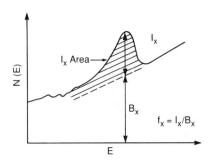

Figure 9.10. Definition of peak height and peak area determination for quantitative XPS analysis.

Quantitative XPS can be understood in terms of the basic process. As photons penetrate the surface, they are either absorbed or transmitted through the bulk. The probability that photon absorption and the consequent electron emission will occur is called the photoionization cross section, σ. The variation of the photoionization cross section as a function of element, core level, and photon energy were discussed in detail in a previous section. Reiterating, the cross section is defined as:

$$\sigma = \left(\frac{N_{abs}}{N_{inc}}\right) d \qquad (9.4)$$

where N_{abs} is the number of photons absorbed/unit area, N_{inc} is the number of incident photons/unit area, and d is the atomic density of material. As previously mentioned, core-level photoionization cross sections for the elements as a function of photon energy are tabulated in a comprehensive reference by Yeh and Lindau (1985).

Since it is known that the integrated intensity of a photoelectron signal I_A is directly proportional to the fractional atomic concentration C_A of species A, one may develop an expression for the atomic concentration ratio of two species (for a CMA):

$$\frac{C_1}{C_2} = \frac{I_1}{I_2} \frac{E_{K1}}{E_{K2}} \frac{\sigma_2}{\sigma_1} \frac{\lambda_2}{\lambda_1} \frac{\phi_2}{\phi_1} \frac{y_2}{y_1} \qquad (9.5)$$

where C is the atomic concentration, I is the area under a photoemission line, σ is the photoionization cross section, E_K is the kinetic energy of the electron, λ is the escape depth, ϕ is the angular intensity factor, and y is the fraction of ground-state electrons. The sensitivity factor can be defined from Eq. (9.5) as

$$S = \frac{\sigma \lambda \phi y}{E_K} \qquad (9.6)$$

and the atomic concentration ratio for two species can be written

$$\frac{C_1}{C_2} = \frac{I_1}{I_2} \frac{S_2}{S_1} \qquad (9.7)$$

Or, analogous to the determination of the atomic fraction of any constituent in a sample previously discussed for AES,

$$[A] = \frac{C_A}{\sum_j C_j} \tag{9.8}$$

The values of the sensitivity factors based on peak area measurements have been tabulated in the literature. They apply to a specific analyzer type, e.g., the double-pass CMA. The use of atomic sensitivity factors in quantitative XPS analysis normally provides results within ±0.5 at. %. Further deviations from accuracy result when examining nonhomogeneous samples, samples in which peak interferences occur, including oxide and satellite peaks, and samples containing transition metals.

9.5. Summary

This chapter has presented an overview of XPS and has discussed both fundamental and applied aspects of the technique. The fundamental principles governing the photoemission process and how it relates to excitation energy and specific atomic properties of a material have been described. These principles have been related to observed spectral features one might encounter in an XPS spectrum and will aid in the interpretation of these measurements.

The practical aspects of instrument design governing specific analytical applications of the XPS technique have also been communicated. Photon sources, monochromators, energy analyzers, and detectors have been described and their relationship to energy and spatial resolution has been discussed.

It has been shown that XPS is a nondestructive analytical technique which can provide quantitative compositional analysis of technological materials with an accuracy of ±0.5 at. %. Additionally, XPS can also provide information on elemental distribution, segregated impurity species, and interfacial chemical reactions. XPS is thus a very effective technique for the microcharacterization of solids.

References

de Broglie, M. (1921). *Compt. Rend.* **172**, 274.
Eastman, D. E. (1972). In *Electron Spectroscopy* (D. A. Shirley, ed.), North-Holland, Amsterdam, p. 487.
Einstein, A. (1905). *Ann. Phys.* **17**, 132.
Hagström, S., Nordling, C., and Siegbahn, K. (1964). *Z. Phys.* **178**, 439.
Hertz, H. (1887). *Ann. Phys.* **31**, 983.
Katz, W. (1981). *Microbeam Analysis—1981. Proceedings of the 16th Annual Conference on Microbeam Analysis* (R. H. Geiss, ed.), San Francisco Press, San Francisco, pp. 287–295.

Lee, J. D. (1973). *Rev. Sci. Instrum.* **38,** 1210.

Palmberg, P. W. (1975). *J. Vac. Sci. Technol.* **12,** 379–384.

Pauling, L. (1960). *The Nature of the Chemical Bond,* Cornell University Press, New York.

Purcell, E. M. (1938). *Phys. Rev.* **54,** 818.

Robinson, H., and Rawlinson, W. F. (1914). *Philos. Mag.* **28,** 277.

Scofield, J. H. (1976). *J. Electron Spectrosc. Relat. Phenom.* **8,** 129–137.

Seah, M. P., and Dench, W. A. (1979). *Surf. Interface Anal.* **1,** 1–11.

Siegbahn, K., Nordling, C., Fahlman, A., Nordberg, R., Hamrin, K., Hedman, J., Johansson, G., Bergmark, T., Karlsson, S.-E., Lindgren, I., and Lindgren, B. (1967). *Nova Acta Regiae Soc. Sci. Ups.* Ser. IV, Vol. 20.

Siegbahn, K., Nordling, C., Johansson, G., Hedman, J., Hedén, P.-F., Hamrin, K., Gelius, U., Bergmark, T., Werme, L. O., Manne, R., and Baer, Y. (1969). *ESCA Applied to Free Molecules,* North-Holland, Amsterdam.

Wannberg, B., Gelius, U., and Siegbahn, K. (1974). *J. Phys. E* **7,** 149.

Yeh, J. J., and Lindau, I. (1985). *At. Data Nucl. Data Tables* **32,** 1–155.

10

Laser Ionization Mass Spectrometry

R. W. Odom and F. Radicati di Brozolo

10.1. Introduction

Laser ionization mass spectrometry technique (LIMS) is a microanalytical materials analysis technique which employs a finely focused UV laser pulse (5–10 ns) to vaporize and ionize a microvolume of material.* The ions produced by the laser pulse are accelerated into a time-of-flight mass spectrometer where they are analyzed according to mass and signal intensity. Each laser shot produces a range of differing mass ions and a complete mass spectrum, typically covering the range 0 to 250 atomic mass units (amu), can be acquired with each laser pulse. LIMS is a microanalytical technique since typical laser spot sizes are 1 to 5 μm in diameter and typical sampling depths are 0.01 to 0.5 μm. LIMS analyses generally involve some form of survey characterization of the inorganic or organic constituents on or in a solid surface and the technique is often used in failure microanalysis. Elemental detection sensitivities with this technique are typically 1 to 100 part per million atomic, and it detects all elements in the periodic table. LIMS can also provide molecular information on a microanalytical scale and this capability is becoming more

---

* Excellent papers on early laser microprobe applications are contained in *Fresenius Z. Anal. Chem.* **308** (1981).

R. W. Odom and F. Radicati di Brozolo • Charles Evans & Associates, Redwood City, California 94063.

Microanalysis of Solids, edited by B. G. Yacobi *et al.* Plenum Press, New York, 1994.

important as the understanding of laser desorption and ionization mechanisms increases (Odom and Schueler, 1990).

The energy transfer from laser irradiation of a solid sample and the formation of ions in this laser/solid interaction are complex phenomena. When the absorbed energy exceeds the solid's heat of vaporization, neutrals are emitted from the solid surface. At sufficiently high laser irradiance values, ions are produced through a number of mechanisms including plasma (ionized vapor) formation, ion/atom or ion/molecule reactions, and direct laser photoionization (Ready, 1971). This latter ionization occurs primarily via nonresonant multiphoton ionization (NRMPI) (Morellec et al., 1982). Ion intensities are a function of laser power density, optical properties, and chemical state of the material. Typically, the ion species observed in LIMS analysis include singly charged elemental ions, elemental cluster ions, and molecular fragment ions. Multiply charged ions are rarely observed, which sets an approximate upper limit on the energy that is effectively transferred to the material.

The material evaporated by the laser pulse is representative of the composition of the solid; however, the ion signals that are actually measured by the mass spectrometer must be interpreted with regard to the ionization efficiencies of the different species. A comprehensive model for ion formation from solids under typical LIMS conditions does not exist. However, conditions of high laser irradiance ($\geq 10^{10}$ W/cm^2) produce an abundance of elemental ions and the detection limits vary from approximately 1 ppm atomic for easily ionized elements (such as the alkalies in positive ion spectroscopy or the halogens in negative ion spectroscopy) to 100 to 200 ppm atomic for elements with high ionization potentials (e.g., Zn or As). The observation of poor ion yields in conventional LIMS for some elemental and molecular species has motivated the development of alternative techniques which ionize the much more abundant neutral species ablated from the sample surface. These techniques are referred to as postionization processes and the most common form of LIMS postionization utilizes a second laser pulse to photoionize neutrals ablated by the primary laser pulse. Laser postionization can be achieved using either resonant or nonresonant transitions in the ablated atoms, atomic clusters, or molecular species ablated by the primary laser pulse (Schueler and Odom, 1987). Thus, NRMPI, resonant multiphoton ionization (REMPI), and single-photon near-resonant photoionization processes have all provided useful enhancements in ion yields in LIMS analyses.

10.2. Instrumentation

A typical LIMS instrument consists of the following components illustrated schematically in Fig. 10.1:

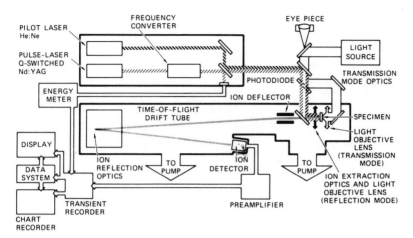

Figure 10.1. Schematic diagram of LIMA 2A.

- A Q-switched UV laser (typically a frequency-quadrupled Nd:YAG laser operating at a wavelength of 266 nm) and appropriate optical components that focus the laser pulse onto the sample surface. The typical laser spot size in these instruments is approximately 2 μm. A He:Ne pilot laser, coaxial with the UV laser, assists in locating the analytical area. A calibrated photodiode is used to monitor the laser energy.
- A time-of-flight mass spectrometer (TOF-MS) which includes a sample stage equipped with x–y–z motion, the ion extraction region, and the ion flight tube (approximately 2 m in length) with energy-focusing capabilities. The mass spectrometer, sample chamber, and ion detector are under high to ultrahigh vacuum. The mass spectrometer is normally equipped with an ion reflector located at one end of the flight path. This ion reflector reverses the direction of the ion path and provides kinetic energy filtering of the ions. This filtering improves the ultimate mass resolution of the TOF-MS (Mamyrin *et al.*, 1973).
- An ion detection system which consists of a high gain electron multiplier and the signal digitizing system, along with the computer for data acquisition and manipulation.

Figure 10.2 presents a schematic view of the ion source region of a LIMS instrument configured for laser postionization. In this example, a second, high-irradiance, frequency-quadrupled pulsed Nd:YAG laser is focused parallel to and above the sample surface, where it intercepts the plume of neutral species that are produced by the ablating laser. Appropriate focusing optics and pulse time delay circuitry are used in this configuration.

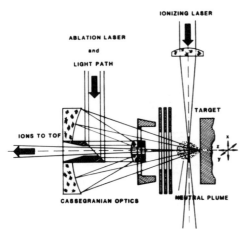

Figure 10.2. View of the ion source region of the Charles Evans & Associates' instrument in postionization (PI) configuration.

A typical single laser analysis is performed by positioning the region of interest using the sample manipulator and He:Ne laser beam and firing the UV laser. The laser pulse creates a burst of ions of different masses from the analytical crater produced by the laser pulse. These ions are accelerated to a nominally fixed kinetic energy and are injected into the spectrometer flight tube. As the ions travel through the flight tube and through the energy-focusing region, small differences in kinetic energy of ions having the same mass are compensated. Discrete packets of ions arrive at the detector and give rise to amplified voltage signals that are input to the transient recorder. The function of the transient recorder is to digitize the analog signal from the electron multiplier, providing a digitized record of both the arrival time and the intensity of the signals associated with each mass. The data are then transferred to the computer for further manipulation, the transient recorder is cleared and rearmed, and the instrument is ready for the acquisition of another spectrum.

Typical instrumental conditions include a UV laser pulse duration of 5 to 10 ns, an ion flight path approximately 2 m long, and an ion accelerating potential of 3 kV. Under these conditions, an H^+ ion arrives at the detector in approximately 3 μs, and a U^+ ion arrives at the detector in approximately 40 μs. Since the time width of an individual signal can be as short as several tens of nanoseconds, a high-speed detection and digitizing system must be employed.

Typical mass resolution values range from 250 to 750 at m/z 100. The parameter that appears to have the most influence on the measured mass resolving power is the duration of the ionization event, which may be longer than the duration of the laser pulse. Ion signals are also broadened in time by the time constants associated with the detection electronics.

The intensity of an ion signal recorded by the transient recorder is proportional to the number of detected ions. This proportionality is limited by nonlinear output of the electron multiplier detector at high input ion signals and by the dynamic range of the digitizer. For example, the dynamic range of typical venetian-blind-type electron multipliers to fast signals is less than 10,000:1. Electron multipliers characterized by other geometries (mesh type) are currently being evaluated and may provide a larger inherent dynamic range.

The second limiting factor in the quantitative measurement of ion signal intensity is associated with the digitization of the electron multiplier output signal by the transient recorder. For example, a Sony-Tektronix 390 AD transient recorder on the LIMA 2A (manufactured by Cambridge Mass Spectrometry, Ltd.) instrument is a 10-bit digitizer with an effective dynamic range of 6.5 bits for 10-MHz signals. This device provides approximately 90 discrete voltage output levels at input frequencies of typical ion signals. The limitations in digitizer dynamic range can be overcome with the use of multiple transient recorders operating at different sensitivities, or with the addition of logarithmic preamplifiers in the detection system (Simons, 1983).

Instrument operation in the postionization mode differs only slightly from the above analytical protocol. In this mode of operation, the ionizing laser fires after a suitable delay from the ablator laser trigger pulse. This delay is necessary to ensure that the neutral species emanating from the sample surface have traveled to the region where they are intercepted by the ionizing laser. Any ions formed directly by the ablating laser at the sample surface can be filtered out with a suitable electrical bias of the energy filter in the flight tube. A time-resolved study over a range of delays supplies information on the velocity distribution and mass of the neutral species originating from the sample surface, as is described in more detail below.

10.3. Applications

LIMS applications span a large number of scientific disciplines and the LIMS technique has been applied to the characterization of biological materials (Seydel and Lindner, 1987), polymers (Wilk and Hercules, 1987), proteins and peptides (Karas and Hillenkamp, 1988), optical materials (Odom and Schueler, 1987), and semiconductor devices and materials (Odom and Schueler, 1990). With regard to the latter applications, LIMS is often utilized for various failure analyses. A microanalytical failure analysis problem may involve, for example, the determination of the cause of corrosion in a metallization line of an integrated circuit. LIMS mass spectra produced from survey analysis of the corroded and uncorroded regions are compared, and this comparison of mass spectra may reveal the presence of additional elemental or molecular species in the defective region. These extraneous species are often the cause or by-products of the corrosion. In this type of analysis, the selection of a relevant control sample is very important.

LIMS analytical applications are classified as elemental or molecular survey analyses. The former can be further subdivided into surface or bulk analyses while molecular analyses are generally applicable only to surface contamination. The applications described below are representative of the range of utility of the LIMS technique. All of these applications were performed in our laboratory.

10.4. Bulk Analysis

One example of the application of LIMS to bulk contamination micro-analysis is the analysis of low-level contamination in GaP light-emitting diodes (LED). The light emission characteristics of GaP LED can be severely affected by the presence of relatively low levels of transition elements. Although the nature of the poisoning species may be suspected or inferred from intentional contamination experiments, the determination of elemental contaminants in actual failures is a difficult analytical problem, in particular because of the

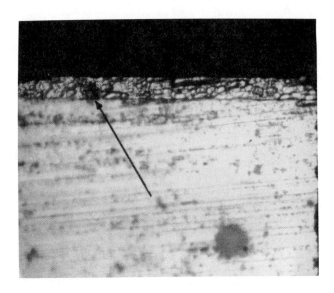

Figure 10.3. Photomicrograph of LED cross section (400× magnification). The arrow indicates one of the sampling locations at the *p–n* boundary.

small size and complex geometry of the parts. Figure 10.3 is an optical micrograph of the cross section of a GaP LED which was thought to be contaminated. Figures 10.4 and 10.5 show two positive ion mass spectra that were acquired from cross sections of a defective and a nondefective GaP LED, respectively. The laser power density employed in this analysis was high ($\sim 10^{10}$ W/cm^2) in order to maximize the detection of low-level contaminants. The depth of sampling is estimated to be 1000 to 1500 Å. The two mass spectra exhibit intense signals of Ga$^+$, along with moderately intense signals of P$^+$. The defective LED also exhibits readily recognizable signals at m/z 63 and 65, matching in relative intensity the two Cu isotopes. The presence of Cu in the defective LED can explain its anomalous optical behavior. The unique advantages of LIMS over other analytical techniques in this example include speed of analysis, the ability to analyze a small sample of nonplanar geometry without time-consuming sample preparation, and its sensitivity which is superior to that of most electron beam techniques.

10.5. Surface Analysis

An example of elemental contamination surface microanalysis is shown in Fig. 10.6. This is a negative ion mass spectrum acquired from a small (~ 4

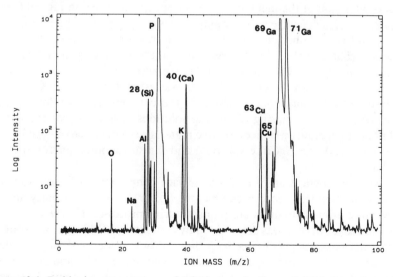

Figure 10.4. Positive ion mass spectrum of defective sample. Copper ion signals are observed at m/z 63 and 65.

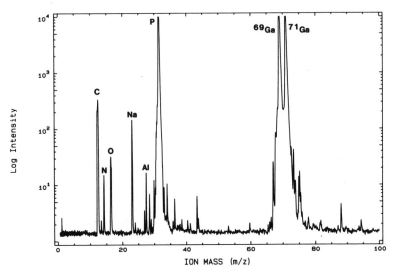

Figure 10.5. Positive ion mass spectrum of control sample. Copper ion signals are absent.

μm) window etched through a photoresist layer deposited onto a HgCdTe substrate. An Al film is then deposited in these windows to provide electrical contact with the substrate. These windows were found to be defective because of poor adhesion of the metallic layer. The spectrum shown in Fig. 10.6 was acquired from a defective window and reveals the presence of intense signals of Cl⁻ and Br⁻, neither of which is observed in similar regions having good adhesion characteristics. In this case, the photoresist had been etched with solutions containing Cl and Br. The laser power density employed in this analysis was low and the sampling depth was estimated to be ≤500 Å. This analysis indicates that poor adhesion of the contaminated windows is the result of incomplete rinsing of etching solutions. The ability of the LIMS technique to operate on nonconductive materials is a major advantage in this case, since both the HgCdTe substrate and the surrounding photoresist are insulating.

Another application of LIMS to surface contamination is the study of solder dewetting from copper conductors. Solder dewetting seriously affects the yield of printed circuit board manufacturing processes (Hirt *et al.*, 1989). The cause of dewetting was determined by comparing LIMS spectra acquired from a defective copper surface with those of pure soldermask on a printed circuit board. These mass spectra are illustrated in Figs. 10.7 and 10.8. Both spectra exhibit numerous signals formed from suspected aromatic fragment ions (m/z 77, 91, 165, 213, and 215). These peaks confirm the suspicion that

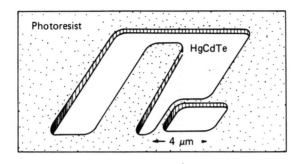

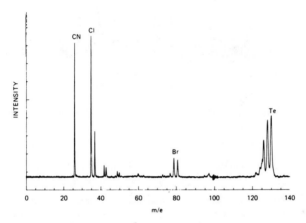

Figure 10.6. Windows cut in photoresist on HgCdTe substrate (top). Negative ion spectrum from defective window shows intense signals of Cl⁻ and Br⁻ (bottom).

soldermask residues are present on the copper surface since the soldermask formulation consisted of an epoxy resin containing bisphenol A which contains two aromatic rings.

10.6. Laser Postionization

A ZnSe on GaAs epitaxial layer required sensitive survey of near-surface contamination. Laser postionization analysis was selected for ZnSe analysis because the major constituents and many of the expected impurities are elements that have poor ion yields in conventional LIMS. Figures 10.9 and 10.10 are two mass spectra acquired from the ZnSe epitaxial layer, using

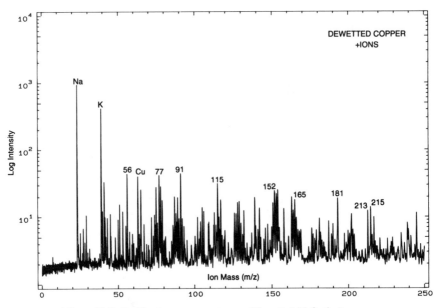

Figure 10.7. Positive ion mass spectrum of dewetted (defective) copper.

conventional single-laser LIMS and the postionization technique, respectively. The single-laser spectrum in Fig. 10.9 exhibits primarily the Zn^+ and Se^+ signals along with weak signals of Cr^+ and Fe^+. The high background signal level following the intense Se signal is related to detector saturation. The ablator laser irradiance for this spectrum was estimated to be $\geq 10^{10}$ W/cm^2, hence the high background signal.

In contrast, the mass spectrum illustrated in Fig. 10.10 was produced from a laser postionization analysis and exhibits readily observable signals of Cd^+ and Te^+ in addition to the Zn and Se signals. The Cd and Te in this near-surface region are contaminants in the ZnSe layer and these contaminants could degrade the optical performance of this material. Note also the low background in the region that follows the Se signal. The ablator laser irradiance in the postionization spectrum was approximately 10^9 W/cm^2, a factor of ten lower than in the single-laser analysis. The lower ablator laser irradiance samples the top 100 Å of the sample as compared with an estimated 1000-Å sampling depth for the single-laser analysis. Thus, laser postionization provides better surface sensitivity in this analysis. In conclusion, the laser postionization form of LIMS is especially useful when the elements present have high ionization potentials which preclude efficient ion detection via conventional LIMS analysis, and in those cases when a higher surface sensitivity is desired.

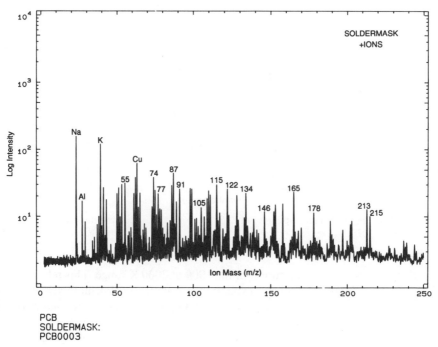

PCB
SOLDERMASK:
PCB0003

Figure 10.8. Positive ion mass spectrum of soldermask, consisting of epoxy resin.

Laser postionization was also employed to measure the kinetic energy of neutral species emitted from the surface of a high-T_c superconductor exposed to high laser irradiances (Venkatesan *et al.*, 1988). The goal of this study was to attempt to better understand pulsed laser sputtering of high-T_c target materials used in the fabrication of high-T_c thin films on low-temperature substrates. These films had a high degree of crystallinity and were produced with very reproducible stoichiometries (Dijkkamp *et al.*, 1987). This observation is somewhat surprising given the random nature of laser vaporization of solids. The specific objective of this study was the determination of the kinetic energy of laser-ablated neutrals produced by laser sputtering of target materials. Figure 10.11 illustrates the velocity distributions of neutral species produced from UV (266 nm) laser ablation of a $YBa_2Cu_3O_{7-x}$ target. These velocity distributions were determined by varying the delay time between the ablation and ionizing laser pulses and measuring the ion intensities for the respective neutrals. Each plot in Fig. 10.11 represents the different positive ions formed from these respective neutral species. The dashed curve is the velocity distribution of a BaO molecule having 1.35 eV of kinetic energy. Most of the neutrals produced by UV laser ablation have relatively high kinetic energies

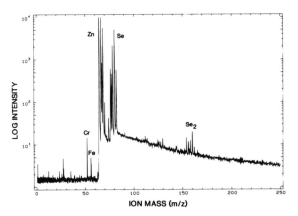

Figure 10.9. Positive ion mass spectrum of ZnSe epitaxial layer, using conventional single-laser technique.

corresponding to "temperatures" in the 1000 to 2000 K range. This relatively high neutral kinetic energy helps explain the formation of stoichiometric high-T_c films from laser ablation of the target material. This observation supports the hypothesis that the high kinetic energy is converted to thermal energy on the surface facilitating growth of crystalline films at low substrate temperatures.

10.7. Polymer Analysis

The LIMS technique can be very useful for the characterization of different polymeric materials (Mattern and Hercules, 1985). Although this tech-

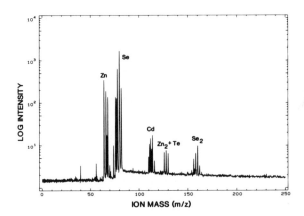

Figure 10.10. Positive ion mass spectrum of ZnSe epitaxial layer, using two-laser (PI) configuration.

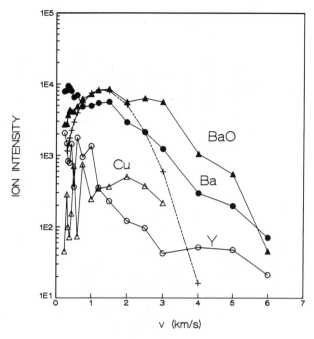

Figure 10.11. Velocity distribution of neutral Cu, Y, Ba, and BaO formed at ablator laser irradiance of 7×10^9 W/cm². The dashed line is the Maxwell–Boltzmann velocity distribution for the BaO molecule with average thermal velocity of 1.35 eV.

nique can potentially produce high-mass ions from polymer oligomers, the low laser irradiance conditions required for these chemical structure analyses are often not sufficiently sensitive for useful failure analysis applications. An alternative approach to polymer analysis in LIMS is the generation of high laser irradiance (high sensitivity) data and performing data analysis using pattern recognition techniques (Radicati di Brozolo *et al.,* 1990). The basic premise in this approach is that although the high irradiance conditions will significantly fragment the polymeric structures, the ion intensities and masses produced from different polymers will reflect their basic chemical structure. Examples of this pattern recognition approach to polymer analysis are illustrated in Figs. 10.12 and 10.13. Figure 10.12 illustrates high-irradiance ($\sim 10^{10}$ W/cm²) mass spectra produced from thin films of commercial photoresist and polyimide on silicon substrates. These spectra contain intense low-mass ion signals indicating that the polymers have been significantly fragmented by the laser irradiation. These data as such do not provide unique information on the chemical structure of the polymers. However, by accumulating a num-

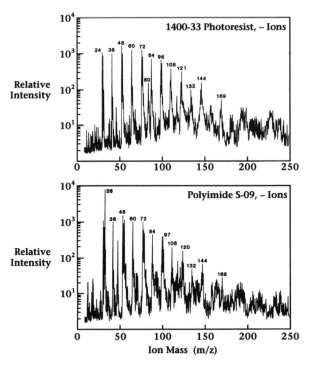

Figure 10.12. High-irradiance negative ion LIMS spectra of photoresist and polyimide samples.

ber of spectra from a group of polymers, it is possible using pattern recognition techniques to determine if the mass spectral signatures or fingerprints of each polymer type are unique. This type of pattern recognition analysis is illustrated in Fig. 10.13 in which the principal component scores or score plots are plotted for a group of polymers consisting of five different commercially available polyimides and one photoresist formulation. The component scores are formed from principal component analysis of the mass spectra in a data set. These principal components are linear combinations of the most significant mass values which best describe the data set (Massart *et al.,* 1988). For the data illustrated in Fig. 10.13, 15 spectra were acquired from each polymer at 15 different locations using high laser irradiance. Component score plots provide a qualitative measure of the uniqueness of each group of spectra in the data set as well as the reproducibility of the analysis. For example, in Fig. 10.13, the individual numbers on the plot illustrate individual spectra and the tight clustering of each group of numbers indicates that the reproducibility for each group of polymers was high. The separation of one group of spectra from another is a measure of the uniqueness of the different sample types.

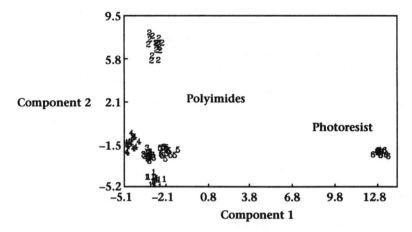

Figure 10.13. Component score plot for the first two linear discriminants for a data set consisting of high-irradiance negative ion LIMS mass spectra of five polyimide and one photoresist sample. Numbers 1 through 5 identify polyimides, 6's identify photoresist.

Thus, in Fig. 10.13 it is apparent that the photoresist spectra are quite unique compared with the polyimide spectra. The results of this principal component analysis can also be used to measure the confidence of uniqueness of the different groups in the data base. The data in Fig. 10.13 had a 75% cross validation accuracy, indicating that one could predict with 75% accuracy a specific polymer type from these data.

Principal component analyses such as these are most useful in failure analysis applications if a data set can be built of the various types of polymers. For example, a data base consisting of polyimide or photoresist spectra could be applied to the analysis of residues of these compounds on device materials at various phases of processing. This pattern analysis approach to LIMS data analysis can also be applied to correlating physical properties of polymers to LIMS mass spectra (Harrington *et al.,* 1992).

References

Dijkkamp, D., Venkatesan, T., Wu, X. D., Shaheen, S. A., Jisrawi, N., Min-Lee, Y. H., McLean, W. L., and Croft, M. (1987). *Appl. Phys. Lett.* **51,** 619.

Harrington, P. B., Voorhees, K. J., Radicati di Brozolo, F., and Odom, R. W. (1992), unpublished results.

Hirt, A. M., Artaki, I., Radicati, F., Odom, R. W., Jabbar, A. H., and Morton, K. L. (1989). *Printed Circuit Fabrication* **12,** No. 2.

Karas, M., and Hillenkamp, F. (1988). *Anal. Chem.* **60,** 229.

Mamyrin, B. A., Karatev, V. I., Shmikk, D. V., and Zagulin, V. A. (1973). *Sov. Phys. JETP* **37**, 45.

Massart, D. L., Vandeginste, B. G. M., Deming, S. N., Michotte, Y., and Kaufman, L. (1988). *Chemometrics: A Textbook,* Elsevier, Amsterdam.

Mattern, D. E., and Hercules, D. M. (1985). *Anal. Chem.* **57**, 2041.

Morellec, J., Normand, D., and Petite, G. (1982). In *Advances in Atomic and Molecular Physics,* Academic Press, New York, p. 97.

Odom, R. W., and Schueler, B. (1987). *Thin Solid Films* **154**, 1.

Odom, R. W., and Schueler, B. (1990). In *Lasers and Mass Spectrometry* (D. M. Lubman, ed.), Oxford University Press, London.

Radicati di Brozolo, F., Odom, R. W., Harrington, P. B., and Voorhees, K. J. (1990). *J. Appl. Polym. Sci.* **41**, 1737.

Ready, J. F. (1971). *Effects of High-Power Laser Irradiation,* Academic Press, New York.

Schueler, B., and Odom, R. W. (1987). *J. Appl. Phys.* **61**, 4652.

Seydel, U., and Lindner, B. (1987). In *Microbeam Analysis—1987* (R. H. Geiss, ed.), San Francisco Press, San Francisco, p. 353.

Simons, D. S. (1983). *Int. J. Mass Spectrom. Ion Process* **55**, 15.

Venkatesan, T., Wu, X. D., Inam, A., Jeon, Y., Croft, M., Chase, E. W., Chang, C. C., Wachtman, J. B., Odom, R. W., Radicati di Brozolo, F., and Magee, C. E. (1988). *Appl. Phys. Lett.* **53**, 1431.

Wilk, Z., and Hercules, D. M. (1987). *Anal. Chem.* **59**, 1819.

11

Microellipsometry

R. F. Cohn

11.1. Introduction

A polarized electromagnetic plane wave will undergo a relative state change upon reflection or transmission at an interface or surface. The field of ellipsometry comprises the theory, instrumentation, measurement, and analysis of this relative polarization state change. In the case of a single uncoated surface, this polarization change is solely a function of the refractive indices of the surface and the surrounding medium. For multilayer surfaces and interfaces, whenever the layers are at least partially transmissive, the polarization change is dependent on the layer refractive indices and thicknesses. Therefore, single layer thicknesses and refractive indices are directly derivable from the measurement, when the properties of the surrounding medium are known. Multilayer films require more complex comparisons between the assumed theoretical model and multiple independent ellipsometric measurements. The major advantage of ellipsometry is that extremely thin films, down to average thicknesses of less than a single atomic monolayer, may be detected and measured. Furthermore, these measurements are nondestructive and can be performed *in situ,* permitting direct, real-time measurements of film change, growth, or dissolution processes. The primary disadvantage inherent in ellipsometric measurements is the difficulty encountered in analyzing rough or

R. F. Cohn • 51–5 Jacqueline Rd., Waltham, Massachusetts 02154.

Microanalysis of Solids, edited by B. G. Yacobi *et al.* Plenum Press, New York, 1994.

contaminatcd interfaces, or surfaces with more than a single layer film coating. The modeling and interpretation of multilayer film measurements remains a primary direction of current research.

Conventional ellipsometry involves the measurement of the polarization change at the interface using a manually adjusted instrument. The measurement records average values over surface areas, typically ranging from 1 to 100 mm^2 requiring at least 1 min for completion, making repetitive measurements a very tedious procedure. Modern instrumentation developments have followed several different interdependent directions. High-speed automatic ellipsometers, with reported measurement times as fast as 10 μs, follow rapid *in situ* experiments (kinetic ellipsometry). Multiple wavelength (spectroscopic ellipsometry) and multiple angle of incidence (AOI) measurements are required for complete characterization of systems of two or more film layers, and to obtain the optimum sensitivity for monochromatic studies. Imaging ellipsometry (or spatially resolved ellipsometry) involves the measurement of full or partial ellipsometric parameters over a large number of densely spaced points on the surface or interface under examination. Microellipsometry commences when the resolvable feature size of the imaged interface becomes sufficiently small.

This section focuses on microellipsometry (the measurement of ellipsometric parameters with a high spatial resolution), which permits both the imaging of spatial variations in interface or film parameters and the detailed study of the properties at a selected localized site. For the purposes of this discussion, a system will be included if it can resolve features of 100-μm diameter or less. Instruments that measure single microscopic points or produce high-resolution one- or two-dimensional surface images are described. First, an overview of the general principles of ellipsometry will be given. Predominant ellipsometer designs will then be described, followed by a brief overview of the major applications of ellipsometry found in the literature.

Specific instruments which fall under the classification of microellipsometry will then be described. These instruments may be generally classified as mechanically scanned, microspot ellipsometers and full-field imaging ellipsometers. Included is a table compiling capabilities of the microellipsometers for which sufficient details were available, is included. Published applications of microellipsometry in various research fields are then briefly described. Finally, conclusions and some suggestions of possible future directions for microellipsometer development are discussed.

11.2. Principles of Ellipsometry

The theory, techniques, and applications of ellipsometry have been thoroughly discussed in the literature. The indispensable monograph by Azzam and Bashara (1987) provides a comprehensive source of information on polarization and ellipsometric theory, as well as instrument designs, detailed error analysis, and applications. The paper by Muller (1973) provides a good concise introduction to the basic concepts and principles of ellipsometry and is particularly recommended for the novice to the field. The SPIE Milestone series of selected reprints on ellipsometry, edited by Azzam (1990), contains a selection of papers covering "the history, definitions and conventions, theoretical foundation, modeling and data interpretation, instrumentation and novel techniques, precision and accuracy considerations, and recent developments that broaden the range and scope of this method." A number of the papers referenced in this chapter are included in this volume. Greef (1984) provides a very useful overview of the field which includes an introduction to Jones and Mueller calculus. Also, Aspnes (1983) provides a useful summary of spectroscopic ellipsometry and a recent review by Collins (1990) covers operation and calibration of rotating element ellipsometers. Additional valuable resources for information are the papers collected in the proceedings of the five international conferences on ellipsometry [Passaglia *et al.*, 1963; Bashara *et al.*, 1969; Bashara and Azzam, 1976; Muller *et al.*, 1980; *Journal de Physique* **44** (Colloque C10, supplement to No. 12) (1983)].

11.2.1. Basic Theory

11.2.1.1. Optical Polarization

The most general state of a completely polarized (monochromatic) electromagnetic plane wave is that of elliptical polarization. Here the electric field vector rotates about the direction of travel, tracing out an ellipse as shown in Fig. 11.1. The electric field vector for a wave incident upon an interface is generally represented by the vector components parallel to the plane of incidence (p) and perpendicular to the plane of incidence (s). For complete description of the polarization state of the plane wave, four quantities must be specified (Azzam and Bashara, 1987):

1. The azimuth angle (θ) of the major axis of the ellipse with the x axis
2. The ratio of the minor axis length (b) relative to the major axis length (a) known as the ellipticity (ϵ), wherein the sign determines the handedness (direction of electric field rotation)
3. The amplitude defined by $(a^2 + b^2)^{1/2}$
4. The absolute phase (δ) of the electric field vector relative to a specified time t_0

Ellipsometry is only concerned with the change in the relative state of polarization, and therefore the amplitude and absolute phase are ignored. The

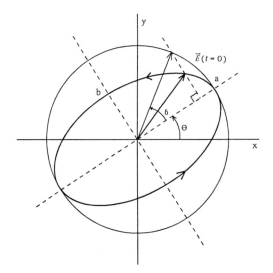

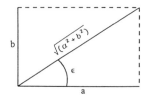

Figure 11.1. Ellipse of polarization showing the path traced by the electric field vector. Definitions of the azimuth angle θ, ellipticity, ϵ, initial phase δ, and electric field magnitude, all relative to the major (a) and minor (b) axes are defined.

relative polarization state can be expressed by a single complex polarization variable specified by the ratio of the electric field components where

$$\chi = \frac{E_p}{E_s} = \frac{|E_p|\,e^{j(\omega t + \delta_p)}}{|E_s|\,e^{j(\omega t + \delta_s)}} \tag{11.1}$$

To simplify Eq. (11.1) the parameters Δ and ψ are defined as

$$\Delta = \delta_p - \delta_s \tag{11.2}$$

$$\tan\psi = \frac{|E_p|}{|E_s|} \tag{11.3}$$

Δ and ψ specify the same information about the elliptical state as do θ and ϵ, as may be shown by observing that the equation of the complex polarization variable may also be expressed as (Azzam and Bashara, 1987)

$$\chi = \tan\psi\, e^{j\Delta} = \frac{\tan\theta + j\,\tan\epsilon}{1 - j\,\tan\theta\,\tan\epsilon} \tag{11.4}$$

11.2.1.2. Reflection of Polarized Light

Upon reflection from a surface (assuming that any existing films are not birefringent) the change in magnitude and phase of the electric field is modeled using two complex reflection coefficients as follows:

$$r_p = |r_p| \exp(j\delta_p) \tag{11.5}$$

$$r_s = |r_s| \exp(j\delta_s) \tag{11.6}$$

Again by using the ratio of these two components to represent the relative polarization change on reflection (ρ), the equation of ellipsometry is derived as

$$\rho = \frac{|r_p| e^{j\delta_p}}{|r_s| e^{j\delta_s}} = \tan\psi e^{j\Delta} \tag{11.7}$$

Analogous results are obtained when considering optical transmission. For the sake of brevity and because reflection ellipsometry is more commonly applied than transmission ellipsometry, this discussion will be directed to the reflection case. However, virtually all concepts discussed are equally applicable to the transmission case as well.

Δ and ψ (or $\cos\Delta$ and $\tan\psi$) are the values most frequently measured in ellipsometric applications. To understand the usefulness of these values, first we consider the case of reflection at an interface between two different media. With the use of Snell's law combined with the Fresnel equations (Hecht, 1987) for reflection it may be shown that (Azzam and Bashara, 1987)

$$n_1 = n_0 \sin\phi_0 \left[1 + \left(\frac{1 - \rho}{1 + \rho} \right)^2 \tan^2\phi_0 \right]^{1/2} \tag{11.8}$$

where n_0 is the refractive index of the surrounding medium, n_1 is the refractive index of the reflecting surface, and ϕ_0 is the angle of incidence. From Eq. (11.8) the complex refractive index of the reflection medium may be directly determined when n_0 and ϕ_0 are known, and ρ is determined by measuring Δ and ψ. The reflecting medium is not required to be transparent. Note that only two independent parameters Δ and ψ can be measured at a single optical wavelength and angle of incidence. This allows the direct solution for at most two unknowns, the real and imaginary parts of the complex index of refraction for this case.

Next, we consider the case of a transparent film on a substrate which is not required to be transparent. All media are assumed to be homogeneous and isotropic. As Fig. 11.2 shows, the optical plane wave will undergo multiple reflections within the film and the resulting reflection coefficients, first derived by Drude (1889), when combined with the equation of ellipsometry yield (Azzam and Bashara, 1987)

$$\rho = \frac{r_{01p} + r_{12p}e^{-j2\beta}}{1 + r_{01p}r_{12p}e^{-j2\beta}} \times \frac{1 + r_{01s}r_{12s}e^{-j2\beta}}{r_{01s} + r_{12s}e^{-j2\beta}} \tag{11.9}$$

where r_{01} is the reflection at the first interface, r_{12} at the second, and β is the optical phase change resulting from the wave traveling from one interface to the other within the film. If the properties of the surrounding media 0 and 2 are known, together with angle of incidence, the thickness and refractive index of a nonabsorbing film layer (real n) may be directly derived. The complex exponential terms in Eq. (11.9) are periodic functions, which thereby cause the polarization states to cycle, i.e., repeating a previous value when increasing film thickness causes an optical path length change of one wavelength. This is referred to as the film thickness period or order. To resolve the ambiguity in thickness, either additional physical information, or multiple measurements at different wavelengths or angles of incidence are required. It should be noted that Eq. (11.9) does not depend on optical interference as do many techniques for thin film measurement. Rather, the phase and amplitude shifts of the two orthogonal components specifying polarization are measured. Therefore, this technique is not limited to measuring films with minimum thicknesses of the same order of magnitude as the wavelength of the incident radiation, thus allowing extremely thin films to be measured. Detection of partial surface coverage by a single atomic layer has been shown (Azzam and Bashara, 1987; Archer and Gobeli, 1965).

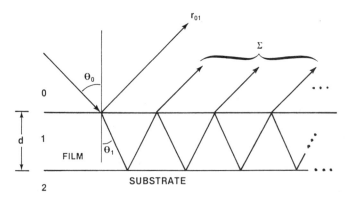

Figure 11.2. Optical reflection from a single-layer film of thickness d on a substrate.

Exact solutions cannot be generally found for more complicated cases, involving films with a complex refractive index n, additional unknown quantities, or multiple film layers. Numerical solutions are available for specific cases (Collins, 1990; Aspnes et al., 1979; McMarr et al., 1986), and this remains an active area of research. Solving for a larger number of unknowns requires additional input parameters beyond the single Δ, ψ pair considered to this point. Additional independent parameters may be obtained through spectroscopic ellipsometry (Azzam and Bashara, 1987; Aspnes, 1983), where Δ and ψ are measured at various optical wavelengths, by performing measurements at multiple angles of index, or by their combination, variable angle spectroscopic ellipsometry (VASE) (Alterovitz et al., 1988).

11.2.2. Ellipsometer Designs

The majority of ellipsometers are based on similar optical designs. The most concise notation (and therefore the most intuitive approach) is to use Jones calculus (Azzam and Bashara, 1987) to analyze ellipsometric instruments. In Jones calculus, polarized electromagnetic waves are modeled as two-element vectors and optical components which alter the polarization state are represented by 2×2 matrices. The polarization transfer function for multielement optical systems may be directly computed by multiplying a series of appropriate matrices. Using Jones calculus for ellipsometer analysis requires the assumption that the system contains no depolarizing elements, which is generally reasonable in a high-quality ellipsometer. Some authors (Hauge, 1980; Lauer and Marxer, 1986b) prefer the use of the four-element Stokes vectors and 4×4 Mueller matrices (Azzam and Bashara, 1987; Hauge et al., 1980). The larger number of terms results in equations which are less straightforward to intuitively understand, but maintains the capability of modeling depolarizing optics and specimens.

Probably the most common fundamental ellipsometric arrangement is the polarizer, specimen, compensator, and analyzer (PSCA), schematically shown in Fig. 11.3. A nonnormal angle of incidence is shown because most reflection ellipsometry is performed near the Brewster angle (Hecht, 1987) as this achieves the maximum sensitivity (Azzam and Bashara, 1987). The optical irradiance measured at the detector may be shown to be (Azzam and Bashara, 1987; Cohn et al., 1988b)

$$I(A) = \frac{I_0}{4} K_{PAC} |r_p| \{ [1 + \cos2C \cos2(P - C)] \cos^2A \, \tan^2\psi$$

$$+ [1 - \cos2C \cos2(P - C)] \sin^2A + [\sin2C \cos2(P - C) \cos\Delta$$

$$- \sin2(P - C) \sin\Delta] \sin2A \, \tan\psi \} \quad (11.10)$$

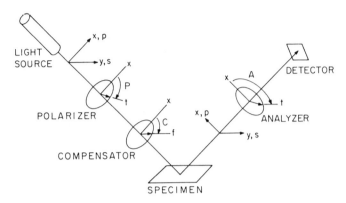

Figure 11.3. Schematic of a basic PCSA (polarizer, compensator, specimen, and analyzer) ellipsometer. Definitions of component angles *P, C,* and *A* relative to the p (parallel) and s (perpendicular) directions.

where P, A, and C represent the angles of the respective components, the compensator is assumed to be an ideal quarter-wave plate, K_{PAC} represents the lumped attenuation constants of the components, and the optical irradiance incident on the polarizer (I_0) is assumed to be completely depolarized. Equation (11.10) will be used to explain the basic concepts of null and radiometric ellipsometer operation. There are a variety of alternative component

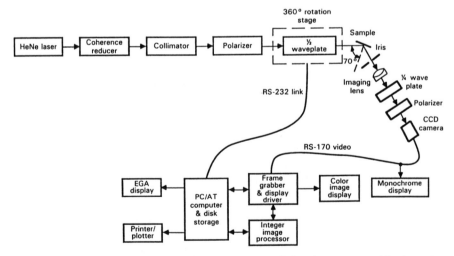

Figure 11.4. Schematic diagram of full optical and electronic imaging system used for dynamic imaging microellipsometry (Cohn and Wagner, 1989a).

arrangements employed, with PSA and PSCA systems finding the most common usage (Hauge, 1980). Figure 11.4 shows a schematic diagram of a complete system for an imaging microellipsometer which is typical of the types of system variations which are found. This microellipsometer employs a modified PSCA system using a fixed polarizer, rotating half-wave plate combination (PHSCA) to replace the rotating polarizer of the standard PSCA. Note that the incorporation of a compensator either restricts ellipsometer operation to a single wavelength or requires that wavelength error corrections be made. Therefore, the PSA system and its variants are commonly employed in spectroscopic ellipsometry.

11.2.2.1. Null Ellipsometers

Null ellipsometry is the basic approach commonly used in manual instruments. The measurements are performed by extinguishing or "nulling" the optical irradiance at the detector. This approach avoids errors induced by noise in the direct measurement of irradiance. The component settings at which the null occurs are then used to determine Δ and ψ (Azzam and Bashara, 1987; Muller, 1973). Accurate results can be obtained, but when manually performed the process is very tedious and slow, particularly whenever multiple measurements are required. Automated null designs have been developed which rectify these problems.

The basis for the null technique may be found by setting the electric field amplitude at the detector of a PCSA system to zero, which results in (Azzam and Bashara, 1987)

$$\tan\psi e^{j\Delta} = -\tan A \left[\frac{\tan C + \rho_C \tan(P - C)}{1 - \rho_C \tan(P - C)} \right] \qquad (11.11)$$

where ρ_c is the retardation of the compensator. The most general approach to null ellipsometry is to use a quarter-wave plate compensator fixed at an azimuth angle (C) of either $+45°$ or $-45°$, and then adjust the angles of the polarizer and analyzer to extinguish the optical irradiance at the detector. This reduces Eq. (11.11) to

$$\tan\psi e^{j\Delta} = \mp\tan A e^{\mp j2(P \mp \pi/4)} \qquad (11.12)$$

where the sign of the compensator angle determines the correct sign choice. There are two pairs of P, A values that satisfy Eq. (11.12) at each of the compensator settings, resulting in four distinct sets of pairs (ignoring those values resulting from a rotation of P or A by $180°$). These pairs are defined to each be located in one of four different *zones* or regions of the Poincaré sphere (Azzam and Bashara, 1987; Muller, 1973). Averaging results from different zones (2, 4, and 16 zone averaging) is a standard technique for reducing systematic error.

Automation of the null ellipsometer can be accomplished by rotating the compensator or polarizers using either servo or stepping motors under computer control. An additional modulator is frequently included to impart a small ac variation to the polarization state ($\sim 1°$) leading to an ac variation in the measured irradiance. This enables a more accurate determination of the null angles, free from low-frequency noise and systematic drifts found in dc irradiance measurements. For faster measurements, electro-optical rotation has replaced mechanical means. Ellipsometers using either Faraday rotators or electro-optic crystals such as KDP have been developed (Azzam and Bashara, 1987; Lauer and Marxer, 1986b).

11.2.2.2. Radiometric Ellipsometers

Radiometric ellipsometers (also termed photometric ellipsometers) use multiple measurements of irradiance to compute the ellipsometric parameters Δ and ψ. Consider Eq. (11.10), specifying the irradiance at the detector in a PCSA system. Assuming that the component settings are known, and normalizing the measured irradiance by the incident irradiance, three unknowns remain: Δ, ψ, and $K_{PAC}|r_p|$. Setting a system component, for example the angle A, to three or more values, then measuring the irradiance, produces three independent equations from which the three unknowns may be determined (Azzam and Bashara, 1987; Cohn et al., 1988b; Budde, 1962). This method and its variants are referred to as static radiometric ellipsometry. In practice only Δ and ψ are of interest; therefore, only their results are actually computed. The simplest equations are obtained for the case of four analyzer settings (Cohn et al., 1988b) of $A = -45, 0, 45,$ and $90°$ as follows:

$$\psi = \tan^{-1}\left[\sqrt{\frac{1 - \cos 2C \cos 2(P - C)}{1 + \cos 2C \cos 2(P - C)}} \sqrt{\frac{I(0°)}{I(90°)}}\right] \quad (11.13)$$

$$\Delta = \cos^{-1}\left(\frac{I(45°) - I(-45°)}{2\sqrt{I(0°)I(90°)}}\right) - \tan^{-1}\left(\frac{\tan 2(P - C)}{\sin 2C}\right) \quad (11.14)$$

Note that these equations are free from a gain calibration requirement when a single detector is used (with constant characteristics during the measurement period). This is an important consideration in imaging ellipsometry where a solid-state imaging sensor may contain over 2.5×10^5 individual detectors.

The dark signal level must be removed, but this is a relatively simple operation compared with a gain calibration. In dynamic radiometric ellipsometry, one or more optical elements are continuously rotated or modulated. The signal received by the detector is Fourier analyzed to determine the ellipsometric parameters (Azzam and Bashara, 1987; Collins, 1990; Hauge, 1980; Azzam, 1976, 1977).

Radiometric measurements are susceptible to systematic error (offset errors in measured ellipsometric parameters) caused by imperfect components and misalignment, and noise sources (random errors in irradiance determination) such as shot noise, random error in the modulation component's setting, electronics noise, and digital quantization (Hauge and Dill, 1973; Cohn and Wagner, 1989a; Cohn, 1990; Azzam and Bashara, 1974). Noise reduction may be accomplished by averaging multiple measurements of Δ and ψ, or by including a larger number of measurements of $I(A)$ in their determination. When more than four irradiance measurements are to be used, a common approach (Azzam and Bashara, 1987; Hauge and Dill, 1973; Cohn, 1990) is to observe that Eq. (11.10) may be rearranged into

$$I(A) = \frac{I_0}{4} K_{PAC} |r_{\mathrm{p}}| \{u + v \cos 2A + w \sin 2A\} \qquad (11.15)$$

where u, v, and w are functions of Δ, ψ, C, and P. Equation (11.15) represents a simple harmonic series and the Fourier coefficients (u, v, and w) can be found from a set of irradiance measurements at uniform steps of angle A. The ellipsometric parameters Δ and ψ are then directly computed from u, v, and w. This allows large numbers of measurements to be included, reducing the overall measurement uncertainty. When the number of irradiance measurements required becomes so large as to cause an excessive increase in the data acquisition period, multiple measurements of Δ and ψ from data subsets should be averaged to minimize errors caused by systematic drifts.

A wide range of radiometric instrument designs exist. Hauge (1980) provides a summary of different design options for polarimetry (determination of the complete or partial Stokes vector) and ellipsometry. Systems are broken into polarization state generators (PSG) and detectors (PSD). Many variations are described, listed under major classifications including: rotating element polarimeters, oscillating element polarimeters, phase modulation polarimeters, instruments with rotating elements or modulation elements in both the PSD and the PSG arms, pulsed ellipsometry, and return path ellipsometry. Collins (1990) provides detailed analysis for several rotating element systems. To achieve higher speeds than are possible with a mechanically rotating analyzer or polarizer, polarization modulation is often accomplished through the use of electro-optic crystals (Pockels cells) or photoelastic modulators.

As previously discussed, PSA, PSCA, and more complex designs are commonly used in radiometric ellipsometers. PSA systems have obvious advantages in terms of simplicity of calculations, spectroscopic capability, and freedom from systematic errors caused by compensator imperfections. However, the random error level is a function of both Δ and ψ values, rendering a significant portion of the ellipsometric range unusable. When a compensator is present, it can be used in conjunction with the fixed polarizer to tune the system, so that the desired range of ellipsometric values may be covered (Cohn et al., 1988b; Hauge and Dill, 1973). Applications, such as the measurement of passive film growth and breakdown on metallic surfaces, exhibit limited Δ and ψ ranges, falling within the useful range of a PSA system (McBee and Kruger, 1974). In spectroscopic ellipsometry, portions of the spectrum will fall within the usable range upon which the subsequent analysis can then concentrate.

11.2.3. Applications

A brief summary of general applications of ellipsometry follows. A more thorough summary is found in Azzam and Bashara (1987). Theeten and Aspnes (1981) provide a summary of both real-time applications using the three-phase thin film model in kinetic ellipsometry, and of multilayer films and interfaces. In reflection ellipsometry, applications involve the surface characterization of bulk material properties, single and multilayer thin film properties and thicknesses, and surface adsorption. Transmission ellipsometry applications involve characterization of bulk material properties and thin films on transparent substrates. Holzapfel and Riss (1987) include a discussion of high-accuracy transmission ellipsometry.

A fundamental concern in ellipsometric applications is the accuracy of the measurement. An extensive body of work has dealt with the improvement of ellipsometer accuracy (Azzam and Bashara, 1974, 1987; Muller, 1973; Azzam, 1990; Hauge and Dill, 1973; Cohn and Wagner, 1989a; Cohn, 1990; Aspnes, 1971) leading to the minimization of systematic and random errors in current instruments. As discussed by Aspnes (1983), the major current limitation to accuracy is sample preparation rather than instrumental error. Inadequate cleaning and polishing, surface contamination, and environmental variables (e.g., convection, diffusion layers) are serious limitations that all ellipsometric applications must address.

11.2.3.1. Bulk Material Properties

Ellipsometry can determine the complex refractive index (Azzam and Bashara, 1987; Aspnes, 1985) which is alternatively stated as a complex dielectric constant. It is a particularly valuable tool for studies on nontransmissive materials. Spectroscopic ellipsometry is frequently used to characterize material properties over wavelengths ranging from the IR to the UV. For accurate determination, surfaces must be free of contamination, and the cleaning and measurement process should be maintained in an inert environment or vacuum. Furthermore, consideration must be given to the microstructural properties of the material. Significant differences may be found between studies performed on aggregate properties of polycrystalline samples when compared with those on single crystals.

Thin film studies, not directly concerned with the determination of bulk material properties, are often required to ascertain the refractive index of their particular substrate before proceeding to analyze film formation. The accuracy of the determination of the thin film properties will depend in part on the accuracy to which substrate uniformity and properties are known.

Transmission ellipsometry can be applied to the evaluation of optically transparent materials, although few applications are found in the literature. Nonuniformities in optical elements caused by material anisotropies or stress birefringence can be quantified. As noted by Holzapfel and Riss (1987), "there is a major need of transmission ellipsometric investigations and quality control, e.g. in the field of substrates of laser mirrors, Brewster windows, laser windows, complete gas laser amplifiers, electrooptical or photoelastic modulators."

11.2.3.2. Thin Film Properties

The complex refractive index and thickness of films adsorbed, condensed, or deposited on substrates have been determined by ellipsometry. Frequently, spectroscopic ellipsometry is used to measure the film parameters over the optical spectrum. Applications have included (Azzam and Bashara, 1987): adsorption of water vapor into silicon surfaces, kinetic ellipsometry study of the chemical adsorption of a monolayer of oxygen onto a cleaved silicon surface (Archer and Gobeli, 1965), refractive index measurement of thin films of condensed gases on gold and copper substrates at 4.2 K.

11.2.3.3. Oxidation of Semiconductors

The largest application for commercial ellipsometers is found in the semiconductor industry. The thickness and uniformity of oxide layers on semiconductor wafers can be characterized. Ion implantation may be measured from the change in the optical absorption coefficient (k). To improve quality control, measurements and wafer mapping may be performed on-line, at stages in production, to quickly eliminate defective wafers and identify processes which require adjustment (Hauge and Dill, 1973; Bloem *et al.*,

1980). Research applications have included the studies of oxide growth rates (Azzam and Bashara, 1987) and the quality of the surface after chemical treatments (Erman *et al.*, 1987). *In situ* studies by Collins (1987, 1988) and by Collins and Cavese (1987) provide previously unobtainable information on material and interface problems, and microstructural evolution.

11.2.3.4. Electrochemistry and Corrosion

Ellipsometry is widely used as a probe of processes occurring at electrochemical electrodes. Ellipsometric measurements which are sensitive to very small surface changes may be performed *in situ* within electrochemical cells. Studies have included (Kruger, 1973) ion, gas, and organic molecule adsorption at surfaces, studies of double-layer effects at the interface between the electrode and the electrolyte, and the growth/dissolution of passive oxide films. The corrosion properties of many metals depend on very thin (<5 nm) passive surface films, making their *in situ* study of fundamental importance (Kruger, 1988).

11.2.3.5. Surface Roughness Measurements

Ellipsometric measurements can be used to estimate surface roughness, texture, and to measure the thickness and void fraction of rough films (Nee, 1988). Wherever applicable, this provides a noncontact alternative to mechanical stylus measurements, permitting rapid scanning and preventing surface scratches. Development of models for the accurate determination of film properties on rough surfaces will widen the range of applications of ellipsometry.

11.2.3.6. Biology and Medicine

Applications in biology and medicine have included (Azzam and Bashara, 1987) blood interactions with foreign surfaces (coagulation), immunological reactions, and measurements of cell surface coatings. When blood contacts a foreign surface, coagulation begins through the transformation of the protein fibrinogen into filaments of fibrin. Ellipsometry has been used to study the kinetics of the protein adsorption (Azzam and Bashara, 1987).

Measurements of surface antigen–antibody reactions have been performed by coating a surface with a thin layer of an organic substance. When this surface is brought into contact with an antiserum solution, an additional surface layer will be adsorbed which may be measured using ellipsometry. Because these layers may be extremely thin (5 nm or less), ellipsometry is a very powerful tool for their measurement which also allows *in situ* observation. Care must be taken in evaluating the thickness contribution of any nonspecific adsorption to the overall film thickness (Azzam and Bashara, 1987; Place *et al.*, 1985).

Studies of the proteins coating cell surfaces have been performed by culturing a layer of cells which attach to a substrate's surface. These cells are then mechanically sheared off the surface, leaving a residual coating. Ellipsometric studies have measured the growth of the coating thickness under a variety of conditions. Cell coating thickness from virus-infected cells have been measured, showing both increased and decreased thicknesses, dependent on the specific cell–virus combination (Azzam and Bashara, 1987).

11.3. Microellipsometry

In order to obtain the high spatial resolution required in microellipsometry, two fundamental approaches have been followed, denoted here as microspot and full-field imaging ellipsometry. The microspot approach incorporates focusing optics with a standard ellipsometer (usually an automatic ellipsometer), and then measures the ellipsometric parameters over a small region of a sample's surface, typically 10 to 200 μm in diameter. When more than a single spot must be measured, the sample is typically mechanically scanned, either in a single line or surface mapping mode. This technique maintains the precision of an automatic ellipsometer, but may require long time periods to scan the required number of surface locations.

The full-field imaging approach replaces the standard detector in an ellipsometer with an imaging detector such as a solid-state CCD (charge-coupled device) camera and imaging optics. A collimated beam illuminates a broad region of the sample and magnification is subsequently applied by the imaging optics and the spatial resolution (pixel dimensions) of the image detector. Measurement uncertainty levels (random error or noise) typically increase in this approach, but ellipsometric images may now be acquired rapidly, enabling kinetic ellipsometry studies to be performed.

11.3.1. Microspot Scanning Ellipsometers

Focusing and light-collecting optics may be incorporated in most conventional ellipsometers. Many researchers (Lauer and Marxer, 1986b; Holzapfel and Riss, 1987; Erman and Theeten, 1986; Sugimoto and Matsuda, 1983; Dunlavy *et al.*, 1981; Erman *et al.*) have pursued this approach and several commercial ellipsometers provide this option.* Details of the individual

---

* Rudolph Research Technical Bulletins 613 and SAL.PUB.616 through 618, Rudolph Research, Flanders, N.J.; "Auto Gain Ellipsometers," Bulletin EE, p. 13, Gaertner Scientific Corp., Chicago, Ill.; MOSS Multilayer Optical Spectrometric Scanner, SOPRA, Bois-Colombes, France (1989); Tri-Beam, Photo Acoustic Technology Inc., Westlake Village, Calif. (1990).

instruments will not be independently discussed. Most microspot instruments are automatic ellipsometers because line or surface scans are commonly performed, requiring far too many individual measurements to be practical with a manual instrument. Table 11.1 presents the most pertinent available specifications for several microspot instruments.

Several important points pertaining to both microspot ellipsometers and general microellipsometry must be considered. First, the theory of ellipsometry is based on the interaction of electromagnetic plane waves at one or more interfaces. Clearly, when lenses are utilized to focus the incident light, the plane wave case no longer directly applies. Erman and Theeten (1986) have considered this issue in an important paper which analyzes errors caused by both focusing and nonuniform samples. They refer to their microellipsometric imaging technique as "spatially resolved ellipsometry" (SRE). A Fourier optics (Hecht, 1987) approach is employed to model the range of incidence angles resulting from finite aperture focusing. The component of angle orthogonal to the plane of incidence is shown to have a second-order effect on the results and is therefore ignored. The authors then derive equations for irradiance, $\tan\psi$, and $\cos\Delta$ for both coherent and incoherent optical sources. Take for example the results for $\tan\psi$, wherein the modulation transfer functions (Hecht, 1987) $g_x(\theta)$ and $g'_x(\theta)$ are for the focusing and detection optics, respectively, and θ is the angle relative to the average angle of incidence θ_0. The conventional case of plane wave illumination leads to the usual result

$$\tan\psi = \frac{\|R_p\|}{\|R_s\|} \tag{11.16}$$

However, for focused coherent illumination, where amplitude functions are convolved, this changes to

$$\tan\psi = \frac{\|[R_p(\theta) * g_x g'_x(\theta)](\theta_0)\|}{\|[R_s(\theta) * g_x g'_x(\theta)](\theta_0)\|} \tag{11.17}$$

and finally, for focused incoherent illumination, where irradiance functions are convolved, they find

$$\tan\psi = \left(\frac{\|R_p(\theta)\|^2 * \|g_x g'_x(\theta)\|^2(\theta_0)}{\|R_s(\theta)\|^2 * \|g_x g'_x(\theta)\|^2(\theta_0)}\right)^{1/2} \tag{11.18}$$

Similar results are also derived for the case of a nonuniform sample.

The authors then analyze the error found in measured complex dielectric functions from a uniform sample, assuming different monochromator aper-

Table 11.1. Table of Reported Instrument Capabilities

Source	Type	Spatial resolution	Speed	Accuracy	Uncertainty	Wavelength	Comments
Bloem et al. (1980)	PCSA radiometric	25 × 75 μm @ 70°		50 μm spot location		632.8 nm	Automatic wafer handling and positioning
Cohn et al. (1988b), Cohn and Wagner (1989c)	PHSCA radiometric	10 × 30 μm @ 70°	50 s per frame		±0.1° rms Δ ±0.025° rms ψ	632.8 nm	
Erman et al.	Spectroscopic	10 × 20 μm	10 to 60 min (200 point spectrum), 1 to 20 h (4000 point image)	10^{-3}	10^{-5}	0.7 to 4.2 eV	Energy (λ) resolution 1 meV
Gaertner L116B[a]	PCSA rotating analyzer	25 × 86 μm @ 70°	4 s per point	±0.3 nm	±0.1 nm	632.8 nm	
Holzapfel and Riss (1987)	PCSA microspot *transmission*	100 μm		±0.0006° Δ ±0.0003° ψ	±0.005° Δ ±0.004° ψ	632.8 nm	8-zone averaging used
Lauer and Marxer (1986b)	PM(QSA Faraday modulator 500 Hz	20 μm		±0.05° Δ ±0.02° ψ		632.8 nm	3 operational modes, variable AOI
Photo Acoustic Tri-Beam[b]	PCSA, three detectors per beam	10 μm @ 70°	5 ms per point			780 and 830 nm	4 beams, 12 detectors, for high-speed wafer scanning
Rudolph AutoEl[c]	Automatic null	25 × 76 μm @ 70°	Average 12 s per point	0.1° Δ 0.05° ψ		632.8 nm plus options	
SOPRA[d]	PSA spectroscopic, rotating polarizer	100 × 150 μm	0.3 up to 9999 s		cosΔ ± 0.0015° tanψ ± 0.001° @ 600 nm and 45°	230 to 930 nm plus IR options	λ resolution 0.05 nm @ 313 nm
Sugimoto and Matsuda (1983)	PCSA null	10 μm @ 59.65°				446.1 nm	Detector is mechanically scanned

[a] "Auto Gain Ellipsometers," Bulletin EE, Gaertner Scientific Corp., Chicago.
[b] Tri-Beam, Photo Acoustic Technology Inc., Westlake Village, Calif.
[c] Rudolph Research Technical Bulletins 613 and SAL.PUB.616 through 618, Rudolph Research, Flanders, N.J.
[d] MOSS Multilayer Optical Spectrometric Scanner, SOPRA, Bois-Colombes, France.

tures (and thereby spot sizes) in the focusing optics. Errors are shown to be sensitive to wavelength and to decrease for increasing spot size. At $\theta_0 = 70°$, $g = g'$, an angular aperture of 4.6°, and with a GaAs substrate, errors for a 10×20-μm spot are shown to be less than 1%. This value is on the order of the combined inherent errors (for many experiments) caused by misalignments, noise, imperfect sample preparation, and the ambient environment, demonstrating that microspot ellipsometry can be reliably performed and interpreted.

Erman and Theeten (1986) proceed to analyze laterally nonhomogeneous samples, concentrating on the case of a step discontinuity between bare and film-coated surface regions. When the microspot is contained within a homogeneous portion of the surface, the expected ellipsometric values are measured. However, as the film transition is scanned, contributions from both bare surface and film are optically combined, resulting in potentially large deviations in measured values which the authors refer to as an "optical resonance" caused by interference. Both theoretical and experimental transition scans are presented, leading to the development of the trajectory concept of data analysis. Figure 11.5 shows both thickness trajectories [the paths found in the $\cos\Delta$, $\tan\psi$ plane as the oxide layer thickness (d) increases from an initial to a final value] and spatial trajectories [the path assumed as a weighting function (X) for coverage of a spot straddling the transition boundary increases from zero to one]. Measured data points are then plotted over the resulting grid. Because possible trajectories for both film thickness variations and microspot coverage are found to follow different paths, their effect on the experimental results may directly be graphically evaluated. Examples were given of experimental measurements of narrow gold lines (3–6 μm wide) deposited on a GaAs substrate. The measured values are shown to fall along a line in the $\cos\Delta$, $\tan\psi$ plane corresponding to a weighting value of 0.3, clearly separable from the measured results on the larger (50 μm) structures. This approach provides a framework for analyzing results from any microellipsometer when measuring transitions or small features.

Another approach to the analysis of errors caused by using a focused beam in ellipsometry has been developed by Barsukov et al. (1988). They present derivations for correction factors for the error in Δ and ψ caused by a range of incidence angles resulting from finite aperture focusing with a beam of uniform irradiance. In a second paper (Barsukov et al., 1989), the reduction in Δ and ψ sensitivity in null ellipsometry caused by incomplete beam extinction is analyzed. Extinction is degraded because a range of incidence angles is included in the focused beam. The error in the ellipsometric values is evaluated, then the improvement obtained from the inclusion of a limiting aperture at the analyzer is determined. The use of the aperture also decreases the mag-

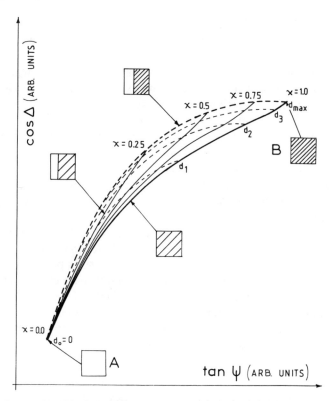

Figure 11.5. "Possible domain in the (cosΔ, Tanψ) plane obtained in SRE measurement (using an incoherent source) under the assumption that the region A is a clean substrate and region B a film deposited on the substrate A. If d_{max} is the maximum thickness of the film then all possible experimental points, whatever the thickness of the layer and the weighing coefficient between A and B, are situated between two limiting curves: the spatial trajectory obtained for $d = d_{max}$ and χ variable, and the thickness trajectory obtained for χ = 1 and variable from 0 to d_{max}. Some other spatial trajectories (broken lines, d constant) and thickness trajectories (solid lines, χ constant) are plotted." (From Erman and Theeten, 1986.)

nitude of the required correction factors as derived in the first paper (Barsukov *et al.,* 1988).

Holzapfel and Riss (1987) present a microspot ellipsometer with two-dimensional scanning designed for transmission mode imaging. A theoretical development is presented for an eight-zone measurement method which derives the local optic axis angle and a pseudoactivity angle in addition to Δ and ψ. The determination of these additional quantities is necessary because transmission ellipsometry at perpendicular incidence lacks a defined plane of incidence as in reflection ellipsometry for use as an absolute reference. Also, the orientation of the optic axis in general varies locally.

The report by Lauer and Marxer (1986b) includes a detailed discussion of their automatic null ellipsometer [also summarized in a research paper (Lauer and Marxer, 1986a)] with a 20-μm spot diameter, and incorporating a Faraday modulator. A Mueller calculus analysis is presented together with a description of the instrument design and experimental results. The instrument contains three operational modes: (1) simultaneous Δ and ψ determination in which 30 to 40 analyzer, polarizer angle pairs are used in the calculation, thus limiting the speed, (2) ψ determination alone from direct analyzer readout (modes 1 and 2 both use no compensator), and (3) Δ determination alone with minimum calculation using a quarter-wave plate compensator. Modes 2 and 3 are generally applied because they maintain a significantly higher measurement speed. However, two scans are required to determine both Δ and ψ.

Stevens (1980) presents a design for a perpendicular-incidence microellipsometer used to determine low levels of preferred crystalline orientation in polycrystalline carbons. A Muller calculus analysis is presented. Also, an oblique-incidence operational mode called a return path ellipsometer is described, in which an isotropic mirror returns light reflected by the sample, along the original path, back through the illumination optics to a beam splitter and a detector. This mode is more closely related to conventional ellipsometry, with the ability to radiometrically determine Δ and ψ.

11.3.2. Full-Field Imaging Ellipsometers

Full-field ellipsometers capture a one- or two-dimensional image directly at the detector plane, thereby eliminating or reducing the need for mechanical scanning. This approach greatly speeds image capture, with several microellipsometers offering the real-time imaging capability necessary for kinetic ellipsometry. Furthermore, these designs have the potential of being modified using high-speed sensors and polarization modulation to achieve far greater temporal resolution. In general, these microellipsometers trade off some of the accuracy and sensitivity of a single-point instrument for the capability of rapidly acquiring an image of high spatial resolution. To paraphrase an old adage, "a picture is worth a hundred thousand individual data points." The importance of obtaining an image of sample variations during data analysis cannot be minimized. Furthermore, when viewing an image, the eye spatially averages over many individual measurement points or *pixels*, thereby significantly reducing the effective noise level relative to the single pixel noise (Cohn and Wagner, 1989a). To gauge the relative magnitude of the measurement time difference, it has been noted (Cohn *et al.*, 1988b) that one commercial microspot ellipsometer* will require over 11 h to acquire a 100×100 point image, while a relatively slow full-field instrument (Cohn *et al.,* 1988b) can acquire 480×512 data points in ~ 55 s and process the Δ and ψ maps within 5 min.

It should be noted, however, that while researchers desire ever-increasing temporal and spatial resolution, serious consideration must be given to data storage and processing. Several of the systems, which partially determine the ellipsometric values, directly record the instrument output on videotape, avoiding this problem. However, full determination of Δ and ψ values requires at minimum three irradiance images. Using a standard video-based 480 $\times$ 512-pixel spatial resolution and 8-bit irradiance resolution, image storage for full determination requires at minimum 0.74 Mbyte per frame. In kinetic studies, when monitoring a sample over an extended period, an enormous volume of data can be acquired rapidly. For such applications, limiting the area of interest under study, or the development of real-time data reduction (allowing efficient analog image recording), may be required.

One problem which affects most two-dimensional imaging microellipsometers is the defocusing encountered when observing at a 45 to 70° angle of incidence. Mishima and Kao (1982a) and Stiblert and Sandström (1983) have addressed this issue, and the degree of reduction in resolution which can be tolerated is situationally dependent. This problem is resolvable through techniques such as the careful design of the optical system, the inclusion of wedge prisms for focus correction, or stepwise mechanically scanning, recording only lines that produce acceptable resolution.

11.3.2.1. Partial Ellipsometric Parameter Imaging

Several ellipsometer designs do not attempt to fully compute the Δ and ψ parameters, but rather to create images in which the contrast in irradiance is a direct function of these parameters. These images may then be captured on a film or video camera, thus only the image recording speed limits the temporal resolution.

Multiple spatial points with different ellipsometric values cannot be nulled simultaneously. Therefore, null ellipsometers cannot be easily used to fully characterize Δ and ψ values when the full field is imaged. However, in a sample with widely ranging ellipsometric values, regions within the image will be nulled at a fixed ellipsometer setting. This technique was used by Löschke (1979) in a microscopic imaging arrangement to record an oxide step on a GaAs wafer. The technique of Mishima and Kao (1982a) called

* "Auto Gain Ellipsometers," Bulletin EE, p. 13, Gaertner Scientific Corp., Chicago, Ill.

"ellipso-interferometry" also applies this principle, reporting a thickness sensitivity of 10 nm and a spatial resolution of 1.5×0.75 μm. An analysis of the blurring caused by the depth-of-field variations when imaging on an angle is included. Thickness resolution for an SiO_2 film has been demonstrated to be better than 15 nm (Mishima and Kao, 1982b). Images are presented with the polarizer, analyzer, and angle of incidence adjusted to different settings. This shifts the locations of the nulls to different features of interest on the specimen, thereby obtaining local thickness estimates. An analysis and experimental study of the effects on Δ and ψ, caused by a uniformly tapered film, has shown that the resulting thickness estimates are within 5% of the true value, for a 1° taper (Mishima and Okuno, 1984). A study by Hurd and Brinker (1988b) has developed a further variant of this technique for real-time studies. A collimated beam with no focusing optics was used. A screen was located near the sample to image the surface with a reported resolution of 50 μm, and the image was recorded using a video camera. Because obtaining an exact null at an arbitrary ellipsometer setting is not possible, a polarizer was continuously rotated, ensuring that null conditions would be periodically obtained. Nulls were identified by measuring minima in the optical irradiance. Then from theoretical computations, the precise film thicknesses at the null sites were established. Intervening film thickness must be inferred from the known points together with the physical constraints of the experiment.

A different technique has been developed by Stenberg et al. (1980), called a "comparison ellipsometer." This approach uses a polarizer, reference sample, test sample, and analyzer configuration. The reference and test samples are arranged so that the s and p components of the optical beam are reversed. When linearly polarized light is incident on the reference sample, covered by a uniform film of known thickness, an elliptical polarization state results. If a region on the test sample has the same thickness and index, the effect of the first reflection will be counteracted, returning the beam to a linearly polarized state. With the analyzer crossed relative to the polarizer, all regions from the test film which match the reference will be extinguished. Optimum results were achieved when the refractive indices of the reference and test samples were matched, but useful results were obtained from differing refractive indices. Using an SiO_2 reference in monochromatic light together with a graded and calibrated SiO_2 film wedge, extinction was achieved at multiple orders of thickness variation. However, the zero-order null was uniquely identified using white light, at which it was the only order to fully extinguish. At a 45° angle of incidence (sacrificing some sensitivity) the geometrical resolution of the comparison ellipsometer was reported to reach 2 μm when objects of limited thickness variations were imaged (Stiblert and Sandström, 1983). The limiting factors were found to be scattering from feature edges and the averaging of properties, for surface features of less than a wavelength in dimension. A 10-μm resolution was reported for a 1-nm thickness sensitivity.

A system using a PCSA system with an objective lens, a CCD camera, and image processor was reported by Beaglehole (1988). Fringe images of $|\rho|$ sinΔ were produced by processing images captured by rotating the quarter-wave plate compensator at two different analyzer settings. The capability of achieving a spatial resolution of 1–2 μm was reported.

11.3.2.2. Full-Field Δ and ψ Imaging Capability

A technique combining the optical system of a radiometric ellipsometer with focusing optics, solid-state image sensors, and digital image processing has been described by Cohn and Wagner (1989c), Cohn et al. (1988a,b) and Hurd and Brinker (1988a). The technique developed by Cohn and Wagner (1989c), called "dynamic imaging microellipsometry" (DIM), has been shown to generally perform best in the PHSCA (polarizer, half-wave plate, specimen, compensator, analyzer) optical configuration as shown in Fig. 11.4. This places the rotating component, the half-wave plate, on the illumination side, minimizing image translation and distortion and, by maintaining a fixed polarizer, irradiance variations. Also, error caused by detector polarization sensitivity is eliminated. Coherent interference in the image is reduced through the use of a spinning ground glass disk to perform speckle averaging. A design variant developed by Allemeier and Wagner (1991) replaces the laser source with a light-emitting diode (LED), eliminating the need for the ground glass disk [which introduces additional spatial image error (Cohn and Wagner, 1989b)]. Random errors, a primary concern with the limited dynamic range of a video camera and 8-bit A/D converter, have been demonstrated to be acceptable over a wide range of Δ, ψ space (Cohn and Wagner, 1989a,b). The DIM system's minimum error range may be tuned by adjusting the quarter-wave plate compensator and analyzer azimuth angles, so that full coverage may be maintained of Δ, ψ space. A four irradiance image algorithm ($P = -45, 0,$ 45, and 90°) has been used in all published experimental studies, but an analytical evaluation of three, four, and Fourier-based algorithms has been published (Cohn, 1990).

Hurd and Brinker (1988a) have used radiometric detection at three polarizer angles to compute Δ and ψ values for one-dimensional line traces. A fast technique for obtaining the required images and eliminating the need for background calibration, called "differential detection," is also described. The first derivative of the optical intensity with respect to polarizer angle may be estimated by differencing two images at closely spaced polarizer angles of up to 5°. Formulating Eq. (11.15) in terms of a rotating polarizer, then taking the first derivative yields

$$\frac{dI(P)}{dP} = \frac{I_0}{2} K_{PAC} |r_p| [-v \sin2P + w \cos2P] \qquad (11.19)$$

in which the value u is eliminated, and therefore only two unknowns (v and w) require solution. When implementing this approach the polarizer only need make small excursions and the computation requirement is minimized, suggesting the possibility of real-time operation.

Itoh et al. (1989) describe a method used to determine linear birefringence and its azimuth angle through the capture of three digital images using a rotating crossed polarizer/analyzer pair at an angle of incidence of 0°. Experiments were performed using a simplified implementation in which the polarizer/analyzer pair were rotated in six 15° steps and the image processor determined the maximum irradiance value at each pixel, from which the azimuth angle was estimated.

11.4. Applications of Microellipsometry

Applications of microellipsometry are appearing with increasing frequency in the literature. In large part, this is attributable to the growing availability of commercial automated ellipsometers with microspot capability. Measurement problems requiring high spatial resolution, range throughout the full spectrum of ellipsometric applications. For example, studies of passive film breakdown, in the corrosion of metals subject to pitting attack, must separate film changes at localized sites from the overall background. Or, the imaging of the film thicknesses on processed semiconductor structures which requires the determination of ellipsometric parameters at a very high spatial resolution. Measurement problems such as these require microellipsometric imaging capability. The following examples illustrate the applications of microellipsometers reported in the literature.

11.4.1. Passive Film Breakdown Studies

A microellipsometer constructed by Sugimoto and Matsuda (1983) was applied to the *in situ* study of passive films on austeno-ferritic stainless steel. The corrosion resistance is known to be dependent on the ratio of the ferrite phase (α) to austenite phase (γ) within the steel. The specimen was maintained under potentiostatic control in an electrochemical cell, enabling the determination of film thickness on individual grains over a range of potentials. Significant differences were found between passive film thicknesses on α grains compared with γ grains. Pitting corrosion is known to begin at the α/γ grain boundaries, and it was suggested that the large thickness mismatch could create a weak point in the passive film protection layer, thus leading to pitting attack.

A subsequent paper by Sugimoto *et al.* (1985) studied the formation and growth of pits in 18Cr-8Ni stainless steel. The microellipsometer was adjusted for a 25 × 50-μm spot size and scanned with horizontal steps of 50 μm and vertical steps of 25 μm, over a 500 × 700-μm region. An 18Cr-8Ni stainless steel specimen was maintained under potentiostatic control. The sample was first passivated, then a NaCl solution was added, to initiate pitting. The pits were allowed to grow for selected time periods, then the polarization current was stopped and the area surrounding the pit measured using the microellipsometer. Surface plots of the film thickness are presented, showing regions of both film thinning around the pit and thickening below the pit. Secondary pits were found to occur at the boundary between the thinned and original thickness regions in the oxide film.

11.4.2. Tribology

A tribology study of pin-on-disk wear of carbon overcoats on rigid magnetic recording media was reported by Karis and Novotny (1989). Scanning microellipsometry with a 20 × 60-μm spot size, moving in 10-, 25-, or 50-μm steps, with a thickness resolution of 0.1 nm, was used to profile wear tracks generated by a 100-mm-radius test probe, prior to penetration of the carbon overcoat. The ellipsometric technique has the advantage that it is insensitive to plastic deformation of the disk, measuring only the actual removal of carbon. Figure 11.6 presents a comparison between ellipsometer and profilometer traces of a wear track. The differences in the traces are caused by the larger spot size of the microellipsometer (1000 μm^2 versus 1 μm^2 for the profilometer tip) which averages out localized surface structure variations, and the sensitivity of profilometer to both overcoat thickness and local surface topography. The microellipsometer has the greater depth sensitivity, resulting in a track depth detectability of 0.2 nm compared with 200 nm for the profilometer. Therefore, the ellipsometer was the preferred measurement technique, despite its lower spatial resolution, because most wear tracks were too shallow to be detected by the profilometer.

Novotny *et al.* (1989) used a commercial microellipsometer* to measure wear tracks on silicon oxide/silicon surfaces coated with cadmium arachidate Langmuir–Blodgett (LB) layers. Low-speed wear tracks (1 to 10 cm/s) were

---

* Rudolph Research Technical Bulletins 613 and SAL.PUB.616 through 618, Rudolph Research, Flanders, N.J.

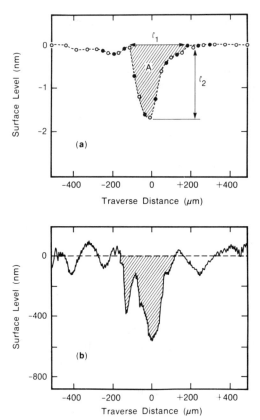

Figure 11.6. Example of wear track measurement by (a) the ellipsometer and (b) the profilometer. Regions outside the wear tracks indicate the typical surface roughness (Karis and Novotny, 1989).

produced with a pin on oscillating slider arrangement and high-speed tracks (10 to 10^4 cm/s) with a pin on rotating disk. Normal loads and frictional forces were measured using semiconductor strain gauges. Ellipsometric scans of the 100-μm-wide wear tracks were performed at a wavelength of 632.8 nm, 70° AOI, and with a 25×75-μm spot size, moving in 10-μm steps. A monolayer LB film on the disk alone was not found to improve durability. A multilayer LB film on the disk did improve durability and reduce the dynamic friction coefficient, after an initial period of higher friction. The surface coating was found, from microellipsometric measurements, to be worn to a single monolayer, and a disorganized coating had transferred to the pin's surface. When both surfaces were covered with an LB monolayer, the dynamic friction coefficient remained low (0.12) and the system ran for several million cycles before forming a visible wear track. Wear rates when both surfaces were coated were up to five orders of magnitude lower than for the uncoated case.

Lauer and Marxer (1985, 1986) and Lauer *et al.* (1984, 1985) measured the oxide thickness in wear tracks on M-50 steel (an Fe, Cr, V, Mo, C alloy) with a base lubricant of Trimethylol Propane Triheptanoate and three different additives: (1) benzotriazole (BTZ), (2) dioctyldiphenylamine (DODPA), and (3) tricresyl phosphate (TCP). A steel ball was loaded on the plate surfaces and rotated at 220 rpm. Δ and ψ traces across the 400-μm-wide wear track were measured independently, for increased ellipsometer speed, as described in Section 11.3.1. Large variations were found in Δ and ψ resulting from wear track surface roughness and oxide patchiness. An average film thickness of 30 to 40 nm and the major oxide component, Fe_3O_4 (70% mol fraction), were both determined from the ellipsometric results. TCP and BTZ were shown to form patchy nonuniform oxide films. The oxide acts to protect the steel surfaces from welding; therefore, oxide-free regions lead to increased scuffing, and thereby increased wear.

Lauer and Marxer (1986a,b) also used their microellipsometer to determine the thickness of carbon films sputtered on metal-coated disks, the thickness of evaporated metal films on metal-coated disks, and stearic acid films on metal. They found that the carbon coating was best measured by assuming a value for the real part of the refractive index (n) and determining the extinction coefficient (k) and the thickness from the ellipsometric data and the model. Δ and ψ traces are presented (Stevens, 1980) from a disk coated with a plastic containing magnetic particles and lubricant, showing that both n and k increase with increasing distance from the periphery. The suggested explanation was that the lubricant concentration increased with radial distance.

Hu and Talke (1988) studied lubricant depletion in slider wear tracks of magnetic recording disks, using a microellipsometer with a 25×75-μm spot size, moving in 75-μm steps across the wear tracks. The disks consisted of an aluminum substrate and magnetic film, which are coated with a sputtered carbon layer and a liquid lubricant. Both the liquid layer and the carbon film were optically transmitting, forming a multilayer film optical model. Ellipsometric measurements of Δ and ψ were made at angles of incidence ranging from 60 to 80° in 5° steps, for the film thickness analysis. Nominal lubricant thickness was approximately 8 nm. After 12,000 cycles of reciprocating motion of a 15.4-mm stroke length with a 0.084-N load, the depletion of lubricant was shown, using a microellipsometer scan, to form a trough 3 nm deep. Lubricant depletion was found to vary relative to slider stroke length, number of cycles, materials, and applied force.

11.4.3. Semiconductors

A kinetic ellipsometry study by Moritani and Hamaguchi (1985) applied high-speed microellipsometry to the direct measurement of cw laser-induced solid-phase epitaxial growth in arsenic-implanted silicon wafers. A microellipsometer with 60×90-μm spatial resolution, at a $60°$ angle of incidence, a 632.8-nm wavelength, 8-μs data point acquisition times, at intervals of 320 μs was used. The phase growth at a single point, created by a 20-W Ar-ion laser with a spot size of 260×320 μm on the wafer surface, was monitored *in situ,* in realtime. The laser-induced reaction was shown to reach completion in ~30 ms.

Erman *et al.* (1987) used their SRE technique to measure the uniformity in GaInAs metal–insulator–semiconductor structures at each processing stage. The surface preparations consisted of:

Step 1. As-grown, thick (4 μm) GaInAs layer
Step 2. Oxide removal with HF solution
Step 3. Chemical etching with H_3PO_4 solutions
Step 4. Second oxide removal with HF

A variant of the previously described trajectory approach (Erman and Theeten, 1986) was employed to separate rough and amorphous interface effects from changes in the oxide thickness. Analysis of the wafer produced cosΔ and tanψ SRE maps at each stage, using a $66°$ angle of incidence, optical energy of 3 eV, and 100-μm spatial resolution, over a 6×6-mm region. The cosΔ and tanψ data points are mapped onto the trajectory plot shown in Fig. 11.7. The mesh in Fig. 11.7 is comprised of trajectories for a constant roughness thickness d_i (a layer which is a mixture of substrate and oxide), and a constant total thickness ($d_i + d_o$ = constant), where d_o is the thickness of the oxide layer. This analysis enables the evaluation of both oxide layer thickness and interface roughness. Processing steps (S1 through S4) were shown to increase the overall roughness and surface nonuniformity, as observed in the figure. An array of electrodes, providing connections to the PECVD Si_3N_4 and GaInAs junction, were deposited on the surface, then capacitive hysteresis values were measured at several frequencies. Equations for the computation of surface roughness and surface roughness plus oxide thickness of the final GaInAs/Si_3N_4 interface are presented. The surface roughness plus oxide thickness variations were shown to correlate directly with the measured capacitive hysteresis values.

Spectroscopic SRE was employed by Erman *et al.* (1988) to study the transition in a multiple-quantum-well structure, from single quantum well to superlattice behavior, as the multilayer thicknesses of GaAs and GaAlAs decrease. A structure consisting of 25 periods of GaAs and GaAlAs layers was grown on a GaAs substrate with decreasing layer thickness relative to distance along the surface. SRE spatial scans were made over a 46-mm line at 100-μm resolution, with spectral scans from 1.4 to 4.0 eV, at a resolution of 5–10 meV. These data are first analyzed using multilayer modeling together with linear regression, determining the quantum well thickness which varied

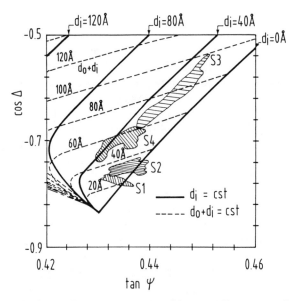

Figure 11.7. Analysis of processing stages 1 through 4 of GaInAs MIS structures using the trajectory approach (Erman *et al.*, 1987). d_o is the oxide thickness and d_i is the roughness thickness.

between 9.5 and 1.0 nm. Optical transitions are observed by tracking excitonic absorption lines in the ellipsometric spectrum relative to the decreasing well thickness. Dielectric functions are also computed to track these changes, and the experimental results are confirmed through the development of a theoretical model. Erman (1989) in a survey paper includes brief summaries of these two studies, together with overviews of spectroscopic, kinetic, and spatially resolved (micro-) ellipsometry.

A commercial microspot ellipsometer* was one technique applied by Armgarth *et al.* (1984) to the study of hydrogen-induced surface charging in a palladium-gate metal-oxide semiconductor (PdMOS). Hydrogen adsorbs at the palladium gate and is believed to give rise to surface charges, which then diffuse, creating a thin inversion layer at the surface. A 100×200-μm spot was located at two points, 50 and 230 μm from the Pd gate. Changes in the ψ reading over an hour period were plotted, showing gradual increases of 0.6 and 0.2°, respectively, thus confirming that surface diffusion occurs.

---

* Rudolph Research Technical Bulletins 613 and SAL.PUB.616 through 618, Rudolph Research, Flanders, N.J.

Laser-annealed oxide films on GaAs were mapped by Dunlavy *et al.* (1981) using an automatic microspot scanning microellipsometer. This instrument employed a photoelastic modulator as the active element, used to modulate the polarization state of the optical beam. Line scans of the oxide thickness on the GaAs wafer are presented, and compared with a standard laboratory profile measurement. A GaAs film was etched to a wedge by Dmitruk *et al.* (1989) and then scanned using an LEM-2 ellipsometer, with angle of incidence of 45°, at 632.8 nm, with a 50-μm focused spot, and an accuracy of ±0.03° for Δ and ±0.01° for ψ. Δ and ψ profiles over a 30-mm scan through the wedge were used to determine changes in the oxide refractive index relative to the local oxide thickness.

Mapping of SiO_2 oxide uniformity on Si wafers has been demonstrated by Cohn and Wagner (1989c). A wafer with a 55- to 110-nm oxide step was imaged, producing Δ and ψ maps and surface plots. The 55-nm layer alone was observed with higher resolution, imaging a depression with a 2° Δ variation and an inverse 1° ψ variation, corresponding to a total 3-nm (5%) variation in oxide thickness. Löschke (1979) presents measurement results of a sharp 2.6 to 2.3 ± 0.1 nm oxide step etched onto a GaAs wafer. The microellipsometer clearly differentiated between the two regions of only 0.3-nm thickness difference in the presented photomicrograph. Stenberg *et al.* (1980) used a continuously varying SiO_2 film on Si to calibrate their comparison ellipsometer. A film of 1 μm was initially grown, then dipped at a controlled rate into a HF etch producing a wedge of 0 to 1-μm film thickness. A scale was added using a conventional ellipsometer, and the fringes produced by the comparison ellipsometer could then be directly calibrated.

Mishima and Kao (1982a) applied their ellipso-interferometer to the imaging of a silicon transistor at two angles of incidence. Good correspondence was found between thicknesses determined by nondestructive ellipso-interferometry and destructive SEM measurements. Stiblert and Sandström (1983) also used a semiconductor circuit as a test pattern, together with 50-μm holes etched into an oxide layer on GaAs, and a test pattern of silicon nitride on silicon dioxide. The test pattern established that the instrument had a 3-μm spatial resolution.

11.4.4. Bulk Materials

Moy (1981), using a commercial ellipsometer,[*] describes a technique for preventing errors caused by surface oxide formation when measuring bulk material properties. A hemispherical lens and index matching fluid are used

---

[*] "Auto Gain Ellipsometers," Bulletin EE, p. 13, Gaertner Scientific Corp., Chicago, Ill.

to cover a section of surface immediately after cleaning. The focal spot is placed at the center of the curvature of the lens, setting all rays normal to the lens surface, and thereby minimizing any beam deviations. The oxidation rate is sufficiently slowed for reliable readings of bulk properties to be performed.

Holzapfel and Riss (1987) and Riss and Holzapbel (1988) used their eight-zone averaging transmission microellipsometer to produce two-dimensional maps of retardation (Δ), polarization-dependent loss ($|1 - \tan\psi|$), fast axis orientation, and pseudoactivity over a 2-mm^2 region of a laser window. Anisotropy maps showing the effect of a 2-μm-wide channel are presented (Riss and Holzapfel, 1988), together with measurements of relative retardation Δ and fast axis angle in a mechanically stressed glass sample.

11.4.5. Sol–Gel and Liquid Films

Hurd and Brinker (1988a,b) and Brinker *et al.* (1990) used their imaging ellipsometer to generate thickness and index profiles of a drying film. TiO$_2$ and SiO$_2$ sols were imaged during a dip coating process, where an adjusted steady rate of substrate drawing allowed continuous observation of the drying front. Theoretical predictions of film thickness and index, based on the mechanisms of evaporation, viscoelastic properties, and hydrodynamic draining, were compared with the measured profiles. Gravitational flow alters film thickness but not refractive index, while evaporation changes both index and thickness. After the drying front passes, the void structure in the film collapses because of surface tension, thereby increasing the refractive index and decreasing the thickness. Thickness profiles were found to be consistent with a combination of the evaporation and gravitational flow draining mechanisms.

Surface wetting properties are important to many industrial applications such as painting, dying, lubricating, and coating. Léger *et al.* (1988a,b) used microellipsometry to study the precursor film spreading of polydimethylsiloxane on horizontal Si wafers. The microellipsometer described by Erman and Theeten (1986) was used, with a 10×25-μm spot size and 10-μm steps. The liquid film ($n = 1.43$) and the 2-nm SiO$_2$ layer ($n = 1.45$) were treated as a single dielectric layer. Drops with different characteristics were measured, and precursor films were found to develop far more rapidly than current theory predicts. In some cases, a wafer surface roughness of 0.4 to 0.5 nm was believed to obscure identification of the film extremity. A discussion and image of a precursor film around a droplet obtained through "ellipsocontrast" microscopy, and surface plots ("ellipsometric cartography") of the Δ values of the precursor film are presented (Léger *et al.,* 1988b).

The migration of liquid polymers has been studied by Novotny (1990). This study was reported to be the first to investigate the microscopic migration

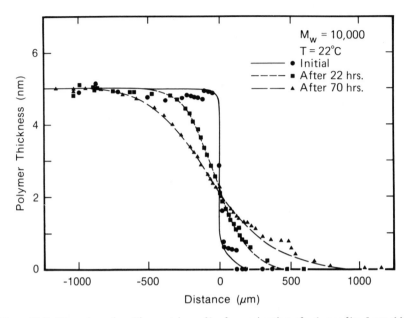

Figure 11.8. "Scanning microellipsometric profiles from migration of polymer film 5 nm thick. The solid line represents the initial profile and the dashed lines are subsequent migration profiles." (From Novotny, 1990.)

of films within a factor of 10 of the thickness of the polymer size in bulk. A commercial microellipsometer* was used, with a 20 × 60-μm spot size, 3-s acquisition time, and a 10-μm overlapping spatial resolution. Typical scans consisted of at least 100 points and required 5 min to complete. Changes in polymer film thickness were minimal during this period. Scanning microellipsometer profiles were made of the edges of the polymer film, before and after migration, as shown in Fig. 11.8, from which the thickness-dependent diffusion coefficient was determined. The surface diffusion coefficient was shown to increase as the film thickness decreases, down to 1 nm, below which diffusion becomes independent of polymer film thickness. A further study by Novotny and Marmur (1991) used the scanning microellipsometer to study wetting autophobicity (low-surface-tension liquids which do not completely spread on high-energy surfaces). The study showed that the droplets were surrounded by an adsorbed film of monolayer or submonolayer molecular

---

* Rudolph Research Technical Bulletins 613 and SAL.PUB.616 through 618, Rudolph Research, Flanders, N.J.

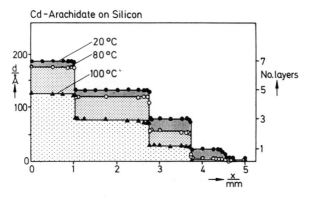

Figure 11.9. Ellipsometric thickness profile of an LB staircase of 1, 3, 5, and 7 monolayers of CdA on a silicon wafer at room temperature (●), after about 1 h at 80°C (○), and after about 1 h further at 100°C (▲) (Rabe *et al.*, 1988).

thickness, extending over several millimeters. The film was also shown to be adsorbed from the vapor phase, rather than through surface diffusion.

A search for the prewetting line in a binary liquid system at the vapor–liquid interface was performed by Schmidt and Moldover (1986) using an imaging ellipsometer design based on the instrument developed by Beaglehole (1988). The imaginary part of ρ was monitored and converted into thickness, but the expected prewetting line was not detected.

Microellipsometry was used in a study of the diffusion of cadmium arachidate Langmuir–Blodgett layers by Rabe *et al.* (1988). A commercial automatic nulling microellipsometer* was used, with a 25 × 75-μm spot and 0.01° measurement uncertainty. A staircase of 1, 3, 5, and 7 LB monolayers was deposited on a silicon wafer, then the profile, shown in Fig. 11.9, was measured with a 10-μm step size, both before and after heating. The minimal migration of the step boundaries revealed that the lateral diffusion coefficient was small, $D \le 10^{-10}$ cm^2/s.

Mate *et al.* (1989) compared the profile of a perfluoropolyether polymer liquid film measured with an atomic force microscope (AFM) to that measured with a microellipsometer* to demonstrate the capability of the AFM on polymeric liquid films.

Beaglehole (1988) imaged a spreading oil drop on a mica substrate with 3-μm resolution over a 0.7 × 0.5-mm surface area. The profile of the precursor was followed, by observing contours of $|\rho|$, to a thickness of 5 nm where the

---

* Rudolph Research Technical Bulletins 613 and SAL.PUB.616 through 618, Rudolph Research, Flanders, N.J.

contours can no longer be observed, because of noise and random substrate variations.

11.4.6. Immunology and Biomedicine

Stenberg *et al.* (1980) applied the previously described comparison ellipsometer to study the reaction between bovine serum albumin (BSA) and rabbit anti-BSA immune serum, using the diffusion-in-gel technique developed by Elwing and Stenberg (1981). A silcon wafer was coated with BSA, then a layer of agar gel. Anti-BSA was placed in a small well in the agar gel and allowed to diffuse for 40 h. The gel was then rinsed off and the thickness profile on the silicon surface was measured, using both a conventional ellipsometer and the comparison ellipsometer. The comparison ellipsometer had 1-nm thickness resolution and better than 100-μm spatial resolution. This diffusion-in-gel technique has several advantages: only one plate is required for several assays because each test requires only several square millimeters, a dose–response relationship can be established between dose and distance, and nonspecific binding events are reduced (Place *et al.*, 1985).

Bille *et al.* (1989) describe a laser scanning tomography system for imaging the retina that includes a Fourier ellipsometry mode with PCSCA optics in which the quarter-wave plates synchronously rotate. A Muller calculus analysis is used to evaluate the thickness of the retina's nerve fiber layer.

11.4.7. Metallic Oxides

A study of the growth of cuprite (Cu_2O) on Cu was performed by Allemeier and Wagner (1991) using the DIM technique (Cohn *et al.*, 1988b) to verify its capability. Two copper grains on the order of 5 mm in diameter were monitored over several hours at 100°C in air. Film growth over time was obtained, which exhibited the expected rate variations relative to grain crystal orientation.

Kimball (1991) used a DIM instrument developed by Cohn *et al.* (1988b) to measure the profiles of argon-ion laser (514.5 nm, 2.5 W, and a spot diameter of 2.7 mm)-generated Cu_2O spots on a Cu substrate, over increasing exposure periods in air. Figure 11.10 shows a surface plot of a typical spot with an exposure time of 5.4 s. The thickness of the oxide film was found to be a critical parameter in determining its growth rate. Rapid increases in growth rate were demonstrated when the air/oxide/substrate system approached a thin film transmission peak for the 514.5-nm wavelength, thereby coupling a greater percentage of the incident radiation into the substrate. Consideration of thin film transmission must be included in laser heat treating and drilling computations.

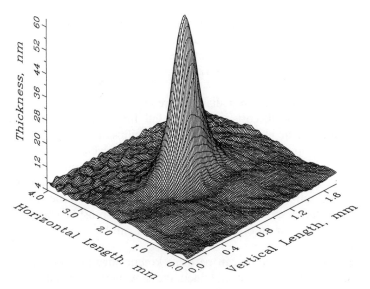

Figure 11.10. Surface plot of a Cu_2O on Cu oxide spot generated by a 5.4-s pulse from a 2.5-W argon-ion laser (Kimball 1991).

11.5 Conclusions

Microellipsometry is a relatively new branch of the well-established field of ellipsometry. Its development has been spurred by the growing availability and sophistication of automated computer-based technology. In the case of microspot instruments, its increasingly common application is facilitated by the availability of fast automatic ellipsometers, particularly those with automated specimen scanning capability. Most commercial manufacturers now offer a microspot option with their automatic ellipsometers, opening the field of microellipsometry to researchers and industries that are not inclined to construct their own instruments. Full-field instrument development has been advanced by the availability of solid-state image sensors and cameras, together with digital image processing.

Microellipsometric studies have spanned the range of fields in which conventional ellipsometry has found application. As Section 11.4 demonstrates, microscopic studies of surface properties and films have measured the uniformity of oxides on semiconductor wafers, the precursor film of liquid droplets, localized passive oxide changes associated with pitting corrosion, diffusion in semiconductors and immunology, film thickness changes caused by wear, and properties in semiconductor physics. In general, microellipsometry is applicable when localized interface or film properties must be measured independently of their immediate surroundings. Line or area imaging is appropriate when the exact location of interest is not known *a priori* (e.g., the site at which a corrosion pit will form), or when variations over a region must be determined.

Looking toward the immediate future, as hardware and processing capabilities increase, while costs continue to fall, further combinations of spectroscopic and kinetic ellipsometry with microellipsometry can be expected. While spectroscopic capability is most directly applicable to microspot ellipsometers, where it enables multilayer film analyses and optimization of imaging sensitivity through the selection of a fixed wavelength, the potential for application to full-field instruments also exists. In full-field spectroscopic ellipsometry, either a more limited set of wavelengths must be chosen to minimize data set size, or an optimum wavelength can be selected to maximize sensitivity.

Microspot instruments can perform kinetic measurements at a single spot, but when images must be obtained, the slow mechanical scanning speeds limit their applicability. Herein lie the major applications for full-field instruments, which rapidly capture area measurements at the justifiable sacrifice of some precision. Indeed, the speeds of the current full-field microellipsometers are more commonly limited by cost, rather than technological constraints. Electro-optic polarization modulation and high-speed image sensors, rather than 30 frame per second video, would greatly accelerate data collection times. Multiple image sensor designs can completely eliminate the need for time-consuming polarization scanning in radiometric systems. Furthermore, real-time data processing may be accomplished using lookup tables stored directly in semiconductor memory, allowing images of Δ, ψ, refractive index, thickness, etc. to be directly recorded in either analog or digital format. Subsecond image capture and processing are obtainable. This will open microellipsometric applications to studies of fast-changing phenomena, such as found in liquid diffusion or corrosion attack on passivated surfaces.

While not all ellipsometric measurements require either microscopic resolution or imaging capability, the widespread availability of such capability will lead to new directions not currently envisioned. With the continuing development of theoretical models for analysis of spectroscopic and kinetic ellipsometric data, together with further improvements in the hardware and processing capability, the range of applications of microellipsometry can be expected to undergo continued expansion.

References

Allemeier, R. T., and Wagner, J. W. (1991). Real-time thickness measurement of metal oxide film growth, in *Nondestructive Evaluation of Materials IV*, (C. O. Ruud, J. Bussiere and R. E. Green, eds.). Plenum Press, New York. 41–47.

Alterovitz, S. A., Woollam, J. A., and Synder, P. G. (1988). Variable angle spectroscopic ellipsometry, *Solid State Technol.* **31**(3), 99–102.

Archer, R. J., and Gobeli, G. W. (1965). Measurement of oxygen adsorption on silicon by ellipsometry, *J. Phys. Chem. Solids* **26**, 343–351. (Reproduced in Azzam, 1990).

Armgarth, M., Nylander, C., Svensson, C., and Lundstöm, I. (1984). Hydrogen-induced oxide surface charging in palladium-gate metal-oxide-semiconductor devices, *J. Appl. Phys.* **56**(10), 2956–2963.

Aspnes, D. E. (1971). Measurement and correction of first-order errors in ellipsometry, *J. Opt. Soc. Am.* **61**(8), 1077–1085.

Aspnes, D. E. (1983). The characterization of materials by spectroscopic ellipsometry, *SPIE* Vol. 452 (F. H. Pollack and R. S. Bauer, eds.).

Aspnes, D. E. (1985). The accurate determination of optical properties by ellipsometry, *Handbook of Optical Constants of Solids,* Academic Press, New York.

Aspnes, D. E., Theeten, J. B., and Hottier, F. (1979). Investigation of effective-medium models of microscopic surface roughness by spectroscopic ellipsometry, *Phys. Rev. B.* **20**(8), 3292–3302. (Reproduced in Azzam, 1990).

Azzam, R. M. A. (1976). Use of different periodic analyzers for the frequency-mixing detection of polarization-modulated light, *SPIE* **88**, 11–18.

Azzam, R. M. A. (1977). Fourier photoellipsometers and photopolarimeters based on modulated optical rotation, *SPIE* **112**, 54–57.

Azzam, R. M. A. (ed.) (1990). *Selected Papers on Ellipsometry,* SPIE Milestone series, **MS 27,** SPIE Optical Engineering Press, Bellingham, Wash.

Azzam, R. M. A., and Bashara, N. M. (1974). Analysis of systematic errors in rotating-analyzer ellipsometers, *J. Opt. Soc. Am.* **64**(11), 1459–1469.

Azzam, R. M. A., and Bashara, N. M. (1987). *Ellipsometry and Polarized Light,* Elsevier, Amsterdam.

Barsukov, D. O., Gusakov, G. M., and Komarnitskii, A. A. (1988). Precision ellipsometry based on a focused light beam, *Opt. Spectrosc. (USSR)* **64**(6), 782–785. (Reproduced in Azzam, 1990).

Barsukov, D. O., Gusakov, G. M., and Komarnitskii, A. A. (1989). Precision ellipsometry based on a focused light beam 2: Analysis of sensitivity, *Opt. Spectrosc. (USSR)* **65**(2), 237–240.

Bashara, N. M., and Azzam, R. M. A. (eds.) (1976). *Ellipsometry: Proceeding of the Third International Conference on Ellipsometry,* North-Holland, Amsterdam.

Bashara, N. M., Buckman, A. B., and Hall, A. C. (eds.) (1969). *Proceeding of the Symposium on Recent Developments in Ellipsometry,* North-Holland, Amsterdam.

Beaglehole, D. (1988). Performance of a microscopic imaging ellipsometer, *Rev. Sci. Instrum.* **59**(12), 2557–2559.

Bille, J. F., Grimm, B., Liang, J., and Mueller, K. (1989). Imaging of the retina by scanning laser tomography, *SPIE* **1161**, 417–425.

Bloem, H. H., Goetz, W. E., Jackson, R. N., and Kern, R. W. (1980). Development of an automatic ellipsometer, *Electro-Opt. Syst. Des.* **12**(3), 38–45.

Brinker, C. J., Hurd, A. J., Frye, G. C., Ward, K. J., and Ashley, C. S. (1990). Sol-gel thin film formation, *J. Non-Cryst. Solids* **121**, 294–302.

Budde, W. (1962). Photoelectric analysis of polarized light, *Appl. Opt.* **1**(3), 201–205.

Cohn, R. F. (1990). Evaluation of alternative algorithms for dynamic imaging microellipsometry, *Appl. Opt.* **29**(2), 304–315.

Cohn, R. F., and Wagner, J. W. (1989a). Absolute and random error analysis of the dynamic imaging microellipsometry technique, *Appl. Opt.* **28**(15), 3187–3198.

Cohn, R. F., and Wagner, J. W. (1989b). Current accuracy limits of dynamic imaging microellipsometry, *SPIE* **1036**, 125–134.

Cohn, R. F., and Wagner, J. W. (1989c). Nondestructive mapping of surface film parameters with dynamic imaging microellipsometry, in *Review of Progress in Quantitative NDE* (D. O. Thompson and D. E. Chimenti, eds.), Vol. 8B, Plenum Press, New York, pp. 1219–1226.

Cohn, R. F., Wagner, J. W., and Kruger, J. (1988a). Dynamic imaging microellipsometry: Proof of concept test results, *J. Electrochem. Soc.* **135**(4), 1033–1034.

Cohn, R. F., Wagner, J. W., and Kruger, J. (1988b). Dynamic imaging microellipsometry: Theory, system design, and feasibility demonstration, *Appl. Opt.* **27**(22), 4664–4671. (Reproduced in Azzam, 1990).

Collins, R. W. (1987). In situ ellipsometry comparison of the nucleation and growth of sputter and glow-discharge a-Si:H, *J. Appl. Phys.* **62**(10), 4146–4153.

Collins, R. W. (1988). In situ study of p-type amorphous silicon growth from B_2H_6:SiH_4 mixtures: Surface reactivity and interface effects, *Appl. Phys. Lett.* **53**(12), 1086–1088.

Collins, R. W. (1990). Automatic rotating element ellipsometers: Calibration operation and real-time applications, *Rev. Sci. Instrum.* **61**(8), 2029–2062.

Collins, R. W., and Cavese, J. M. (1987). Surface structures of glow discharge a-si : H: Implications for multilayer film growth, *J. Non-Cryst. Solids* **97/98**, 1439–1442.

Dmitruk, N. L., Mayeva, O. I., and Antonyuk, V. N. (1989). Ellipsometry of anodic oxide films on a GaAs surface, *Phys. Chem. Mech. Surf.* **4**(12), 3609–3616.

Drude, P. (1989). Ueber oberflächenschichten. II. Theil, *Ann. Phys. Chem.* **36**, 865–897. (Reproduced in Azzam, 1990).

Dunlavy, D. J., Hammond, R. B., and Ahrenkiel, R. K. (1981). Scanning microellipsometer for the spatial characterization of thin films, *SPIE* **288**, 390–394.

Elwing, H., and Stenberg, M. (1981). Biospecific bimolecular binding reactions—A new ellipsometric method for their detection, quantitation and characterization, *J. Immunol. Methods* **44**, 343–349.

Erman, M. (1989). Optical techniques of surface analysis for microelectronics: Ellipsometry and photoluminescence, ECASIA 89, Antibes-Juan les Pins, France.

Erman, M., and Theeten, J. B. (1986). Spatially resolved ellipsometry, *J. Appl. Phys.* **60**(3), 859–873.

Erman, M., Renaud, M., and Gourrier, S. (1987). Spatially resolved ellipsometry for semiconductor process control: Application to GaInAs MIS structures, *Jn. J. Appl. Phys.* **26**(11), 1891–1897.

Erman, M., Alibert, C., Theeten, J. B., Frijlink, P., and Catte, B. (1988). Continuous transition from multiple quantum-well regime to superlattice regime in GaAlAs/GaAs system as observed by spectroscopic ellipsometry with high lateral resolution, *J. Appl. Phys.* **63**(2), 465–474.

Erman, M., Theeten, J. B., and Le Bris, J. Spatially resolved ellipsometry, Laboratoires d'Electronique et de Physique appliquée, BP15, 94451 Limiel Brévannes Cedex, France.

Greef, R. (1984). Ellipsometry, in *Experimental Methods in Electrochemistry,* Vol. 8, Plenum Press, New York, pp. 339–371.

Hauge, P. S. (1980). Recent developments in instrumentation in ellipsometry, *Surf. Sci.* **96**, 108–140. (Reproduced in Azzam, 1990).

Hauge, P. S., and Dill, F. H. (1973). Design and operation of ETA, an automated ellipsometer, *IBM J. Res. Dev.* **Nov.**, 472–489. (Reproduced in Azzam, 1990).

Hauge, P. S., Muller, R. H., and Smith, C. G. (1980). Conventions and formulas for using the Mueller–Stokes calculus in ellipsometry, *Surf. Sci.* **96**, 81–107.

Hecht, E. (1987). *Optics,* Addison–Wesley, Reading, Mass.

Holzapfel, W., and Riss, U. (1987). Computer-based high resolution transmission ellipsometry, *Appl. Opt.* **26**(1), 145–153.

Hu, Y., and Talke, F. E. (1988). Lubricant depletion in a slider wear track of a magnetic recording disk measured by ellipsometry, *Am. Chem. Soc. Div. Polym. Chem. Polym. Prepr.* **29**(2), 277–278.

Hurd, A. J., and Brinker, C. J. (1988a). Ellipsometric imaging of drying sol-gel films, *Better Ceramics through Chemistry III*, MRS.

Hurd, A. J., and Brinker, C. J. (1988b). Optical sol-gel coatings: Ellipsometry of film formation, *J. Phys. (Paris)* **49**, 1017–1025.

Itoh, K., Chichibu, T., and Ichioka, Y. (1989). Microscopic digital imaging ellipsometry, *SPIE* **1166**, 301–307.

Karis, T. E., and Novotny, V. J. (1989). Pin-on-disk tribology of thin-film magnetic recording disks, *J. Appl. Phys.* **66**(6), 2706–2711.

Kimball, B. R. (1991). *A Dynamic Imaging Microellipsometric Analysis of Laser-Induced Thermal Oxidation of Polycrystalline Copper,* Master's Thesis, Worcester Polytechnic Institute, Worcester, Massachusetts.

Kruger, J. (1973). Application of ellipsometry to electrochemistry, in *Advances in Electrochemistry and Electrochemical Engineering*, Vol. 9 (P. Delahay and C. W. Tobias, eds.), Wiley, New York.

Kruger, J. (1988). Passivity of metals—A materials science perspective, *Int. Mater. Rev. (UK)* **33**(3), 113–130.

Lauer, J. L., and Marxer, N. (1985). Ellipsometric surface analysis of wear tracks produced by different lubricants, *ALSE Preprints presented at the ASLE/ASME Tribology Conference*, pp. 1–8.

Lauer, J. L., and Marxer, N. (1986a). Ellipsometric film thickness, *Instrum. Aerosp. Ind.* **32**, 283–298.

Lauer, J. L., and Marxer, N. (1986b). Polarization Modulated Ellipsometry, NASA Report N86-26599.

Lauer, J. L., Marxer, N., and Jones, W. R. (1984). Optical and other properties of M-50 bearing steel surfaces for different lubricants and additives prior to scuffing, *ASLE Trans.* **29**(1), 13–24.

Lauer, J. L., Marxer, N., and Jones, W. R. (1985). Characterization of lubricated bearing surfaces operated under high loads, *Proceedings of the JSLE International Tribology Conference*, pp. 351–356.

Léger, L., Erman, M., Guinet-Picart, A. M., Ausserre, D., and Strazielle, C. (1988a). Precursor film profiles of spreading liquid drops, *Phys. Rev. Lett.* **60**(23).

Léger, L., Erman, M., Guinet-Picart, A. M., Ausserre, D., Strazielle, C., Benattar, J. J., Daillant, J., and Bosio, L. (1988b). Spreading of non volatile liquids on smooth solid surfaces: Role of long range forces, *Rev. Phys. Appl.* **23**, 1047–1054.

Löschke, K. (1979). Microscopy with an ellipsometric arrangement, *Kris. Tech.* **14**(6), 717–720.

McBee, C. L., and Kruger, J. (1974). Events leading to the initiation of the pitting of iron, *Localized Corrosion*, NACE-3, pp. 252–260.

McMarr, P. J., Vedam, K., and Narayan, J. (1986). Spectroscopic ellipsometry: A new tool for nondestructive depth profiling and characterization of interfaces, *J. Appl. Phys.* **59**(3), 694–701.

Mate, C. M., Lorenz, M. R., and Novotny, V. J. (1989). Atomic force microscopy of polymeric liquid films, *J. Chem. Phys.* **90**(12), 7550–7555.

Mishima, T., and Kao, K. C. (1982a). Detection of the thickness uniformity of thin-film layers in semiconductor devices by laser ellipso-interferometry, *Opt. Eng.* **21**(6), 1074–1078. Also found as: Mishima, T., and Kao, K. C. (1982). Checking of the thickness uniformity of thin-film layers in semiconductor devices by laser ellipso-interferometry, *SPIE* **334**. (Reproduced in Azzam, 1990).

Mishima, T., and Kao, K. C. (1982b). Improved lateral resolution in the thickness measurement of thin films by ellipsointerferometry, *Appl. Opt.* **21**(23), 4203–4204.

Mishima, T., and Okuno, M. (1984). Determination of thickness distribution in non-uniform film sample by ellipso-interferometry, *Conference Digest of the 13th Int. Comm. for Optics,* pp. 556–557.

Moritani, A., and Hamaguchi, C. (1985). High-speed ellipsometry of arsenic-implanted Si during cw laser annealing, *Appl. Phys. Lett.* **46**(8), 746–748.

Moy, J. P. (1981). Immersion ellipsometry, *Appl. Opt.* **20**(22), 3821–3822.

Muller, R. H. (1973). Principles of ellipsometry, in *Advances in Electrochemistry and Electrochemical Engineering,* Vol. 9 (P. Delahay and C. W. Tobias, eds.), Wiley, New York.

Muller, R. H., Azzam, and Aspnes, D. E. (eds.) (1980). *Ellipsometry: Proceeding of the Fourth International Conference on Ellipsometry,* North-Holland, Amsterdam.

Nee, S. F. (1988). Ellipsometric analysis for surface roughness and texture, *Appl. Opt.* **27**(14), 2819–2831.

Novotny, V. J. (1990). Migration of liquid polymers on solid surfaces, *J. Chem. Phys.* **92**(5), 3189–3196.

Novotny, V. J., and Marmur, A. (1991). Wetting autophobicity, *J. Colloid Interface Sci.* to be published.

Novotny, V., Swalen, J. D., and Rabe, J. P. (1989). Tribology of Langmuir–Blodgett layers, *Langmuir* **5**, 485–498.

Passaglia, E., Stromberg, R., and Kruger, J. (eds.) (1963). *Ellipsometry in the Measurement of Surfaces and Thin Films,* National Bureau of Standards Misc. Publ. 256, Washington, D.C.

Place, J. F., Sutherland, R. M., and Dähne, C. (1985). Opto-electronic immunosensors: A review of optical immunoassay at continuous surfaces, *Biosensors* **1**, 321–353.

Rabe, J. P., Novotny, V., Swalen, J. D., and Rabolt, J. F. (1988). Structure and dynamics in a Langmuir–Blodgett film at elevated temperatures, *Thin Solid Films* **158**, 359–367.

Riss, U., and Holzapfel, W. (1988). High precision measurement of glass sample properties by transmission ellipsometry, *SPIE* **970** 48–61.

Schmidt, J. W., and Moldover, M. R. (1986). A search for the prewetting line, *J. Chem. Phys.* **84**(8), 4563–4568.

Stenberg, M., Sandström, T., and Stiblert, L. (1980). A new ellipsometric method for measurements on surfaces and surface layers, *Mater. Sci. Eng.* **42**, 65–69. (Reproduced in Azzam, 1990).

Stevens, D. W. (1980). A perpendicular-incidence microellipsometer, *Surf. Sci.* **96**, 174–201.

Stiblert, L., and Sandström, T. (1983). Geometrical resolution in the comparison ellipsometer, *J. Phys. (Paris)* **44** (Colloq. C10, supplement to No. 12), 79–83.

Sugimoto, K., and Matsuda, S. (1983). Analysis of passive films on austeno-ferritic stainless steel by microscopic ellipsometry, *J. Electrochem. Soc.* **130**(12), 2323–2328.

Sugimoto, K., Matsuda, S., Ogiwara, Y., and Kitamura, K. (1985). Microscopic ellipsometric observation of the change in passive film on 18Cr-8Ni stainless steel with the initiation and growth of pit, *J. Electrochem. Soc.* **132**(8), 1791–1795.

Theeten, J. B., and Aspnes, D. E. (1981). Ellipsometry in thin film analysis, *Annu. Rev. Mater. Sci.* **11**, 97–122.

(1987). Automatic Thin Film Wafer Mapping and Characterization, *Solid State Technol.* **May,** 94–96.

International Conference on Ellipsometry and other Optical Methods for Surface Thin Film Analysis, (1983). *J. Phys.* **44** (Colloq. C10, supplement to No. 12).

Part V

Acoustic Wave Excitation

12

Scanning Acoustic Microscopy

P. Mutti and G. A. D. Briggs

12.1. Introduction

Scanning acoustic microscopy is a form of microscopy based on the generation and detection of elastic waves in solids. The basic mechanism is the interaction of an acoustic wave with a specimen and the consequent generation of acoustic waves inside the material. Such interaction is characteristically different from optical or electronic interactions, and is mainly dependent on the mechanical properties of the specimen. As a consequence, acoustic microscopy provides an important and complementary source of information for the examination of materials. Two major advantages are obtained using an acoustic microscope. First, acoustic waves are capable of penetrating materials that are opaque to other kinds of radiation. As a result, the acoustic microscope can image subsurface characteristics of materials without the necessity of etching or coating the surface of the sample. The second advantage relies on the origin of contrast in acoustic microscopy lying in the interaction of elastic waves with local variations in mechanical properties. Using an acoustic microscope it is therefore possible to study, with high resolution and sensitivity, mechanical properties of the specimen such as density, elasticity, and viscosity. In the field of material science, applications of acoustic microscopy are quite wide and range from the analysis of cracks and other defects in engineering materials to the

P. Mutti • Dipartmento di Ingeneria Nucleare, Palitecnico di Milano, Via Ponzia 34/3, Milano 20133, Italy. *G. A. D. Briggs* • Department of Materials, University of Oxford, Oxford OX1 3PH, England.

Microanalysis of Solids, edited by B. G. Yacobi *et al.* Plenum Press, New York, 1994.

study of integrated circuits and electronic components. Moreover, recent advances in quantitative acoustic microscopy enable the determination of mechanical material parameters on a microscopic scale.

In this chapter, the basic principles of the technique will be reviewed. Descriptions of acoustic images and of quantitative results will be given to demonstrate the potential of acoustic microscopy for material microanalysis. Section 12.2 will outline the general instrumental characteristics of a high-resolution acoustic microscope. In Section 12.3 the basic features of acoustic contrast, together with a description of subsurface imaging, will be presented. Section 12.4 will provide a general idea of the propagation of surface acoustic waves in solids. As will be emphasized later, these kinds of waves are of fundamental importance in the generation of the contrast in the acoustic microscope. The last two sections, 12.5 and 12.6, will deal with quantitative applications of the acoustic microscope as well as acoustic images in the area of the science of materials.

12.2. The Scanning Acoustic Microscope

Although the first acoustic microscope operated in transmission (Lemons and Quate, 1974), and some applications (Derby *et al.*, 1983) of this instrument to material evaluation have been made, the most typical configuration for high-resolution work is the reflection scanning acoustic microscope (SAM) (Briggs, 1992). In this instrument (Fig. 12.1) an acoustic pulse is generated by means of a piezoelectric transducer and is focused by an acoustic lens onto

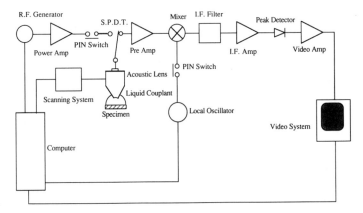

Figure 12.1. Schematic block diagram of a reflection scanning acoustic microscope for high-resolution work.

the surface of the specimen. A coupling medium, usually a liquid, is used to transmit the sound wave generated by the lens to the object to be imaged. The acoustic properties of the couplant play a vital role in the overall resolution of the microscope. For most applications in materials science, distilled water is used. In the reflection configuration, the lens is also used to receive the acoustic pulse reflected from the object which, in turn, produces a proportional voltage signal at the transducer. This electrical signal is detected, processed, and then used to control the brightness of a point on the screen of a TV display system. The lens is electromechanically scanned in a "raster" pattern over the entire field to be viewed: the image can be recorded and displayed using a framestore.

12.2.1. The Acoustic Lens

The basic element of a high-resolution acoustic microscope is the transducer–lens assembly shown in Fig. 12.2. This consists of a spherical cavity ground in the center of one face of a cylinder of sapphire oriented with its long axis parallel to the c-axis of the crystal. The choice of this material for a high-frequency acoustic lens results from the high velocity of the longitudinal wave parallel to the c-axis ($v \approx 11,000$ ms^{-1}) and from the low acoustic absorption at high frequencies ($\alpha \approx 0.5$ dB/cm at 1 GHz). The transducer is made of ZnO sputtered epitaxially onto the other face of the lens rod. Two gold layers deposited before and after the transducer are used as electrical contacts. The lens–transducer assembly is usually mounted in a metallic housing.

Despite the simplicity of the focusing element, the acoustic lens is able to provide a diffraction-limited focus. This is because of the high index of

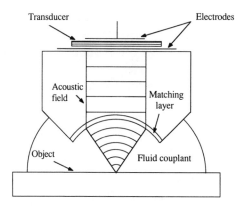

Figure 12.2. The acoustic lens assembly.

refraction of acoustic waves at the solid/liquid interface ($n = 7.4$ for the sapphire/water interface). In a lens system, spherical aberrations are inversely proportional to the square of the refractive index (Born and Wolf, 1980), so using acoustic waves, a single surface lens can lead to very small spherical aberrations. In addition, the high index of refraction allows a high numerical aperture and therefore a high spatial resolution. Unfortunately, the high velocity in the solid leads to a large impedance discontinuity at the solid/liquid interface (the acoustic impedance Z of a material is equal to the product of the sound velocity v and the density ρ). As a result, only a small portion of the incident signal would be transmitted across the boundary. A quarter-wavelength matching layer of SiO_2 (Kushibiki *et al.*, 1980) or chalcogenide glass (Kushibiki *et al.*, 1981a) is coated on the lens surface to reduce the transmission losses because of the impedance mismatch.

12.2.2. The Coupling Medium

The spatial resolution of an acoustic microscope is determined by the diameter of the spot size at focus which is proportional to the wavelength λ. This, in turn, is determined by the excitation frequency f used and by the sound velocity in the couplant v_w through the relation

$$\lambda = v_w/f \tag{12.1}$$

High spatial resolution is therefore obtained using a high excitation frequency and a coupling fluid with a low sound velocity. There are requirements which limit the maximum utilizable frequency in an acoustic microscope. In order to have an adequate signal-to-noise ratio (S/N), the coupling medium must possess a low coefficient of attenuation for ultrasonic waves. In the majority of cases, acoustic absorption in liquids is proportional to the square of the frequency, and an attenuation coefficient per unit distance can therefore be written as $\alpha = \alpha_0 f^2$ (Pinkerton, 1949). Therefore, for fixed acceptable attenuation α_{acc} and focal length q, the constraint on the frequency is

$$f \leq (\alpha_{acc}/2\alpha_0 q)^{1/2} \tag{12.2}$$

In principle, to reduce the loss suffered by the acoustic wave in the fluid it would be possible to use a lens with the smallest possible radius of curvature in order to reduce the path between the lens and the specimen. A limitation, however, is imposed on the smallest focal length that can be achieved in an acoustic lens. In practice, the echo from the surface of the lens can never be entirely eliminated, and because the reflection from the specimen must pass through the coupling fluid it will suffer attenuation that may well make it smaller than the lens surface echo. As a result, these two reflections must be separated in time. To provide adequate time isolation of the specimen echo from the lens surface echo, it is necessary that the temporal separation between these two pulses be greater than the pulse time-width. There is a limit to the shortest radio frequency (r.f.) pulse that can be achieved with adequate isolation. If the time interval between echoes is required to be t_0, then the minimum focal length is

$$q = v_0 t_0 / 2 \tag{12.3}$$

The shortest wavelength λ_{min} that can be used is therefore

$$\lambda_{min} \equiv v_0 / f_{max} = (v_0^3 \alpha_0 t_0 / \alpha_{acc})^{1/2} \tag{12.4}$$

To assess the performance of a couplant for acoustic microscopy, a resolution coefficient R is defined by calculating the minimum attainable wavelength for fixed losses α_{acc} and transit time t_0 in the fluid couplant. The resolution coefficient R may be defined as

$$R \equiv (v_0^3 \alpha_0)^{1/2} \tag{12.5}$$

since this gives the quantity $\lambda_{min}(\alpha_{acc}/t_0)^{1/2}$, and therefore enables the resolution available from different fluids to be compared directly. In Table 12.1 the

Table 12.1. Acoustic Parameters of Various Fluids

Fluid	T (K)	Velocity v_0 (μm ns^{-1})	Impedance Z (Mrayl)	Attenuation α_0 (dB μm^{-1} GHz^{-2})	Resolution coefficient R (μm$^{3/2}$ ns$^{-1/2}$)
Water	298	1.495	1.5	0.191	0.799
	333	1.550	1.5	0.095	0.595
Methanol	303	1.088	0.866	0.262	0.581
Ethanol	303	1.127	0.890	0.421	0.776
Acetone	303	1.158	0.916	0.469	0.853
Carbon tetrachloride	298	0.930	1.482	4.67	1.94
Hydrogen peroxide	298	1.545	2.26	0.087	0.566
Carbon disulfide	298	1.310	1.65	0.087	0.442
Mercury	297	1.449	19.7	0.050	0.391
Gallium	303	2.87	17.5	0.0137	0.570
Air (dry)	273	0.33145	0.4286×10^{-3}	—	—
	293	0.34337	0.4137×10^{-3}	1.6×10^5	80
	373	0.386	—	—	—

acoustic properties and the resolution coefficient of fluids relevant for acoustic microscopy are calculated; distilled water is taken as the reference liquid. We see that the selection of efficient coupling media for a room temperature acoustic microscope is limited to some organic solvents, some liquid metals, and distilled water. Considerations of the physicochemical stability of the medium and the chemical compatibility with the specimen further restrict the choice, and most room temperature acoustic microscopy is performed with distilled water. In a typical acoustic microscope at 1 GHz with water as the coupling fluid, a lens with a radius of curvature of 100 μm is used; the water is usually heated to 60°C to reduce attenuation. Using a smaller lens radius of 40 μm, the microscope can be operated at 2 GHz and the resultant spatial resolution is 0.7 μm.

Higher spatial resolution can be obtained using cryogenic liquids (Foster and Rugar, 1985). The highest resolution (15 nm) for an acoustic microscope was achieved using pressurized liquid helium as the coupling medium and operating the acoustic microscope at 15.3 GHz (Muha et al., 1990). Pressurized superfluid helium was used because below 0.7 K the attenuation of this medium decreases as the fourth power of the temperature.

12.2.3. The Radio Frequency System

An important aspect of the design of the SAM, which has a crucial effect on the quality and reproducibility of the measurements, is the design of the electronic system which is used for the detection of the acoustic pulses reflected from the specimen. The problems encountered concern the isolation of the signal generated by the specimen reflection from other pulses generated by the internal reflections inside the lens or from spurious signals produced by switching transients and electrical cable reflections. Early versions of SAM used switches at the input of the video detector in order to gate the signal of interest. Considerable distortion in the detected signal was introduced by gating transients at the switch output. For high-resolution work, a different version of the r.f. detection system, based on heterodyne detection, has been implemented (Weaver, 1986, Briggs, 1992) (Fig. 12.1). In this system, an r.f. continuous wave signal is gated by a fast PIN diode switch to produce pulses of short time-width (typically 60 ns), which are then amplified and used for the excitation of the piezoelectric transducer. A single-pole-double-throw (S.P.D.T.) microwave switch, placed at the lens input, is used to change between the transmitted and reflected pulses from the lens. The switching time and the isolation of these switches are fundamental parameters for the overall resolution of the microscope. The signal reflected by the specimen is then fed to a broadband low-noise preamplifier. This amplifier must possess a fast recovery time, so that it can recover quickly from leakage signals from the

S.P.D.T. switch and from any switching spikes. The amplified signal is then mixed with a gated local oscillator (LO) signal and filtered by an intermediate-frequency (IF) filter at the output of the mixer. Gating the LO signal, instead of the signal, and using an IF filter greatly improves the suppression of switching spikes. After further amplification, with various controls to set gain and offset, and after additional filtering, the signal is fed to a peak detector to measure the signal for each point in the image.

12.3. Contrast Theory

The importance of the SAM as an instrument for material characterization lies in its combination of resolution and capability of imaging subsurface features in materials which are opaque to other kinds of radiation such as light. But an unambiguous interpretation of surface and sub-surface material signatures requires a knowledge of the mechanisms responsible for the generation of contrast in acoustic microscopy. The contrast theory, often referred to as the "$V(z)$ theory", is developed through the analysis of the output signal V generated by the transducer as a function of the amount of defocus z, i.e., the distance between the object plane and the focal plane. By convention, the focal plane is designated as $z = 0$ and displacement of the object toward the lens is considered as negative. An experimental $V(z)$ curve for a glass sample is shown in Fig. 12.3. The key features of this curve, typical of most solid materials, are the maximum at $z = 0$ and the oscillatory behavior in the negative z region. A complete description of all of the $V(z)$ features should consider a diffraction model based on Fourier optics (Atalar, 1978), but a more direct insight into the physical mechanisms which are responsible for

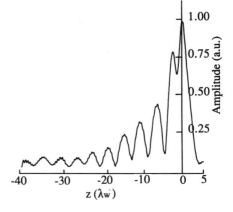

Figure 12.3. Experimental $V(z)$ curve of a glass sample; operating frequency $f = 350$ MHz. Defocus distance is in units of acoustic wavelength in water (λ_w).

these features is given by a simple ray interpretation (Bertoni, 1984). A striking characteristic of this model, valid only far from the focal region, is the description of the oscillations in the $V(z)$ as resulting from the interference between two principal rays when the specimen is negatively defocused. As depicted in Fig. 12.4, ray 1 propagates parallel to the lens axis and is reflected back along the same path. Ray 2 is incident on the specimen at a critical angle associated with the excitation of a Rayleigh wave, or surface acoustic wave (SAW), in the surface of the specimen. The surface wave excites longitudinal waves in the fluid at the same critical angle. Only rays which propagate along a symmetrical path with respect to the lens axis contribute to the signal at the transducer, because the transducer sums all waves incident on it with respect to amplitude and phase so that there is a phase cancellation for all those wave fronts which arrive at other than normal incidence. Interference between rays 1 and 2 generates oscillations in the signal, with maxima whenever the two rays are in phase.

Relative to the phase of the axial ray when the specimen is at focus, the phase variations ϕ_1 of ray 1 and ϕ_2 of ray 2, when the specimen is negatively defocused by an amount z, are

$$\phi_1 = -2\, \overrightarrow{OF} \cdot \overrightarrow{k_w} = -2k_w z \qquad\qquad (12.6)$$

$$\phi_2 = -2 \cdot \overrightarrow{OB} \cdot \overrightarrow{k_w} + \overrightarrow{BC} \cdot \overrightarrow{k_R} = -2k_w z/\cos\theta_R + 2K_R z \tan\theta_R. \quad (12.7)$$

where $k_w = 2\pi f/v_w$, $k_R = 2\pi f/v_R$ are the wave-vector magnitudes of the wave in water and SAW, respectively. The critical angle for the excitation of a SAW is easily given by Snell's law, i.e., $\theta_R = \sin^{-1}(v_w/v_R)$. The quantities v_w, v_R, and f are respectively the sound velocity in water, the phase velocity of the SAW, and the excitation frequency. A phase change of 2π in the phase difference $\Delta\phi = \phi_1 - \phi_2$ is associated with the dip interval in the $V(z)$ curve oscillations; the corresponding spacing Δz is given by

$$\Delta z = \pi/k_w(1 - \cos\theta_R) = v_w/2f(1 - \cos\theta_R) \qquad (12.8)$$

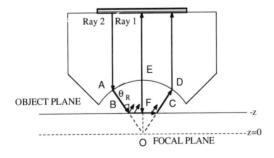

OBJECT PLANE

O   FOCAL PLANE

Figure 12.4. Ray model of an acoustic lens defocused below the surface of a specimen.

Although this simple picture is not able to describe the characteristics of the $V(z)$ curve near the focal region, where diffraction effects are dominant, it emphasizes the importance of Rayleigh waves in the generation of contrast in acoustic microscopy.

12.3.1. Subsurface Imaging

There are two major mechanisms which allow subsurface information using an acoustic microscope to be obtained. In the first mechanism, contrast arises from the reflection of longitudinal and shear waves from local variations of mechanical properties in the specimen. A clearer idea of this kind of contrast can be gained if we consider the interaction of the sound wave launched by the transducer at the couplant–specimen interface (Fig. 12.5). Three different waves appear after the reflection of the incident ray: the reflected wave, which propagates through the liquid with velocity v_w at an angle θ_w; the refracted wave, which propagates through the solid as a longitudinal wave of velocity v_L at an angle θ_L; and the refracted wave, which propagates through the solid as a shear wave with a velocity v_S at an angle θ_S. The angles can be easily calculated by applying Snell's law, i.e., $\sin\theta_w/v_w = \sin\theta_L/v_L = \sin\theta_S/v_S$. When the appropriate boundary condition for the wave displacement is applied for the liquid–solid interface, the reflection coefficient $R(\theta)$ as a function of the angle of incidence is found. This function, usually called the reflectance function of the specimen, is given by (Atalar, 1979)

$$R(\theta) = [Z_L \cos^2(2\theta_S) + Z_S \sin^2(2\theta_S) - Z_w]/$$

$$[Z_L \cos^2(2\theta_S) + Z \sin^2(2\theta_S) + Z_w] \quad (12.9)$$

where the acoustic impedance, for the case of oblique incidence at an angle θ, is defined as $Z = v\rho/\cos\theta$. For an isotropic homogeneous solid where there

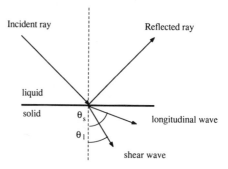

Figure 12.5. Wave model of an incident ray focused at the liquid–solid interface.

are no local variations in elastic parameters, the reflectance function will not vary across the specimen, hence the resulting constant voltage across the transducer will produce a featureless image. If, however, the material has local elastic inhomogeneities (e.g., different grains or phases, defects, voids), then there will be associated variations in the reflectance function which will appear as contrast in the acoustic micrograph. There are, however, different problems which limit the possibility of imaging the interior of a solid object using longitudinal and shear waves. The first problem arises from the fact that the large velocity discontinuity at the liquid–solid interface imposes a high spherical aberration on the focal spot generated by the incoming ray (Fig. 12.6a). In addition, there will be two different foci for the acoustic lens associated with the excitation of longitudinal and shear waves (Fig. 12.6b). It is therefore necessary to use a lens with a sufficiently small semiangle to minimize the generation of shear waves in the solid. Further restrictions in the capability of subsurface imaging using bulk waves stem from the large discontinuity of impedance at the couplant–specimen interface, which results in high loss by reflection and leads to a poor returned signal. For instance, in aluminum, only 25% of the power of the normally incident ray is transmitted through the solid.

A second method of obtaining subsurface information is to consider the contrast generated by Rayleigh waves when the lens is defocused toward the specimen. Indeed, using negative defocus it is possible to image all of those

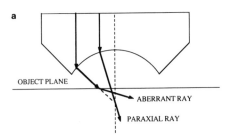

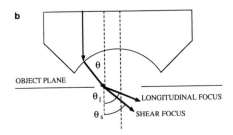

Figure 12.6. Rays focusing below the surface of a solid: (a) aberrations arising from refraction at the solid–liquid interface; (b) shear and longitudinal foci.

factors that affect the propagation of Rayleigh waves. Such factors include changes of velocity and attenuation from one phase to another or from one grain to another within a polycrystalline material (Somekh *et al.*, 1984), or in regions of high strain or deformation (Weaver and Briggs, 1985); and also reflection, scattering, and attenuation of these waves by interfaces and defects such as cracks (Ilett *et al.*, 1984). As will be shown in the next section, the Rayleigh wave is confined to a depth near the surface of the specimen approximately one wavelength thick. The Rayleigh wavelength is related to the frequency f through the equation $\lambda = v/f$. Hence, by varying the frequency it is possible to sample different near-surface layers of material. The acoustic microscope can therefore be utilized as a technique to image mechanical properties of thin film and subsurface layers for materials which support Rayleigh waves.

12.4. Surface Acoustic Waves in Solids

Surface acoustic waves (Farnell, 1970) are the solutions of the wave equation for the displacement in an infinite half-space in which the displacement amplitude decays exponentially with depth below the surface. Unlike a bulk wave, approximately all of the elastic energy is confined within a wavelength or two beneath the surface. These waves were first derived theoretically by Lord Rayleigh for a free, isotropic half-space in 1885, and they are known as Rayleigh waves (Auld *et al.*, 1985) after him. They consist of a superposition of a longitudinal and a shear wave, the shear wave being polarized normal to the plane of the surface. Both of these waves propagate parallel to the surface with the same velocity. The displacement amplitude of each component decays in an exponential fashion below the surface, and the particle motion is elliptical with eccentricity varying with depth. The phase velocity is given by the unique, real, and positive solution of a sextic equation:

$$\beta_S^3 - 8\beta_S^2 + 8(3 - 2\beta_L)\beta_S - 16(1 - \beta_L) = 0 \qquad (12.10)$$

where $\beta_S = (v_R/v_S)^2$, $\beta_L = (v_R/v_L)^2$ and v_R, v_L, v_S are respectively the Rayleigh, longitudinal, and shear wave velocities. An approximate expression for the magnitude of the Rayleigh wave velocity, dependent on the Poisson ratio σ, is given by

$$v_R = v_S(0.87 + 1.13\sigma)/(1 + \sigma) \qquad (12.11)$$

In the acoustic microscope, surface waves are excited through a coupling medium, so that the surface of the specimen cannot be regarded as a free surface. The presence of the liquid has the effect of slightly increasing the velocity of the Rayleigh wave to an extent proportional to the ratio of the liquid density to the density of the solid (Viktorov, 1969). More important is that the Rayleigh wave becomes leaky, as it can lose energy by radiating back into the coupling medium (Bertoni and Tamir, 1973). Because of this behavior, leaky modes are not proper modes of wave propagation and travel along the surface with attenuation. Rayleigh waves are not the only surface modes for a solid, and it is possible to find higher surface and interface modes in layered and plate structures (Brekhovskikh and Godin, 1990) as well as other Rayleigh-type modes in anisotropic solids (Farnell, 1970). In deriving the velocity of the surface wave, we have so far ignored the attenuation suffered by the wave when it travels along the surface of a real solid. The causes of attenuation of elastic waves in solids have various origins (Truell *et al.*, 1969); these could be, for example, the dislocation damping of ultrasound waves or the scattering of stress waves caused by local variations in elastic properties. When attenuation is present, we can write the wave vectors as complex quantities, i.e., $k = k(1 + i\alpha)$, where the imaginary component represents an attenuation factor per unit wavelength. Since the Rayleigh wave is a combination of two partial bulk waves, its attenuation α_R will be related to the loss suffered by each partial bulk component. If the attenuation coefficients of longitudinal and transverse waves are small, i.e., α_L, $\alpha_S \ll 1$, it is possible to show (Viktorov, 1969) that the attenuation factor for a Rayleigh wave α_R becomes a linear combination of the same factors for each partial bulk wave, namely,

$$\alpha_R = A\alpha_L + (1 - A)\alpha_S \qquad (12.12)$$

where the factor A depends only on the Poisson ratio. For most solid materials the attenuation of Rayleigh waves is related primarily to its transverse component.

12.5. Quantitative Acoustic Microscopy

As emphasized previously, the potential of acoustic microscopy to image mechanical properties of materials can be exploited to make quantitative measurements of elastic parameters with high spatial resolution. Different techniques have been developed which are capable of obtaining quantitative information from the experimental data, i.e., the $V(z)$ curve (Weglein, 1979; Briggs *et al.*, 1988; Liang *et al.*, 1985). The most striking feature of the $V(z)$ curve is the oscillatory behavior in the negative defocus region, associated with the excitation of Rayleigh waves. By inverting the formula for the periodicity of these oscillations, as predicted by the simple ray model, it is possible to derive an expression which gives the Rayleigh velocity in terms of Δz:

$$v_R = v_w/[1 - (1 - v_w/2\Delta z f)^2]^{1/2} \qquad (12.13)$$

Therefore, by measuring the periodicity of the oscillations and knowing f and v_w, it is possible to deduce the Rayleigh velocity.

The most precise method which exploits this simple ray description for measuring both the velocity and the attenuation of leaky surface acoustic waves propagating on the liquid–sample interface is represented by the line-focus-beam acoustic microscope technique (LFB) (Kushibiki and Chubachi, 1985). The instrument is based on a typical reflection acoustic microscope in which an acoustic lens with a cylindrical surface (Kushibiki *et al.*, 1981b) is used instead of the traditional point focus lens. This lens generates an acoustic field which is linearly focused along the axis of the cylindrical surface and excites surface waves which propagate perpendicular to this axis. Using this configuration it is possible to show that the component $V(z)_R$ of the acoustic signal associated with LSAW excitation is small compared with the component $V(z)_G$ which is specularly reflected. It is then possible to write the transducer signal $V(z)$ as an arithmetical sum:

$$V(z) = V(z)_G + V(z)_R \cos(\phi_1 - \phi_2) \qquad (12.14)$$

where ϕ_1 and ϕ_2, defined in Section 12.3, are respectively the phase variations of $V(z)_G$ and $V(z)_R$. The $V(z)_G$ component depends entirely on the spatial distribution of the acoustic field generated by the lens, and it could be calculated knowing the pupil function of the lens and using a diffraction model. It is determined experimentally by measuring the $V(z)$ for a material, such as lead or Teflon, in which Rayleigh waves are not excited (Fig. 12.7b). As shown in Fig. 12.7c, the subtracted $V_R(z)$ curve, away from focus, has the shape of an exponentially damped sinusoidal curve:

$$V(z)_R = V(0)_R \exp(-2\alpha z) \sin[(2\pi/\Delta z)z] \qquad (12.15)$$

The oscillatory component contains all of the information regarding the velocity of the Rayleigh wave, i.e., Δz. The exponential decay of the Rayleigh oscillation is expressed in terms of the loss suffered by the principal Rayleigh ray when it propagates in the coupling medium and along the surface of the solid; with reference to Fig. 12.4, it is easy to show that

$$\alpha = (\alpha_0 - \alpha_R \sin\theta_R) \sec\theta_R \qquad (12.16)$$

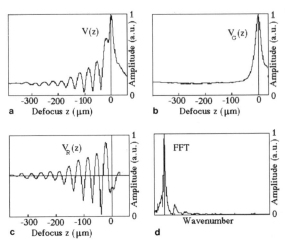

Figure 12.7. Steps in the analysis of an experimental $V(z)$ as obtained by an LFB microscope. (a) $V(z)$ of a glass sample. (b) $V_G(z)$ for Teflon $V(z)$. (c) $V_R(z) = V(z) - V_G(z)$. The curve was filtered to remove short-period ripple caused by lens reverberations and long-period error in $V_G(z)$. (d) Power spectrum of $V_R(z)$ from which the Rayleigh wave velocity and attenuation are found.

where θ_R is the critical angle for the excitation of Rayleigh waves, and the quantities α_0 and α_R are the attenuation per unit of length in the fluid and in the solid. This last factor takes into account both the loss relating to the water loading effect and the acoustic bulk absorption in the solid. To analyze simultaneous surface modes present along the couplant–specimen interface, a spectral analysis is performed to derive velocity and attenuation from the $V(z)_R$ curve; peaks in the power spectrum correspond to LSAW propagating on the specimen surface. Using filtering techniques in the frequency domain, it is possible to isolate the particular SAW of interest and to determine with high accuracy its velocity (0.01%) and attenuation (1–2%). An important characteristic of the cylindrical configuration is the possibility to analyze samples with anisotropic characteristics; by varying the rotation angle between the lens and the specimen it is possible, for instance, to determine experimentally velocity and attenuation of LSAW as a function of propagation direction in single crystals (Kushibiki *et al.*, 1983). Figure 12.8 is an example of this kind of application; here the surface modes present on the {100} surface of a single crystal of GaAs are detected.

The possible applications of this instrument for material characterization are quite promising and cover a broad spectrum from the analysis of residual surface stresses (Obata *et al.*, 1990) to the evaluation of thin film thickness (Kushibiki *et al.*, 1990), or to the analysis of subsurface damage in semiconductors (Logitech Ltd., personal communication). An example of this kind

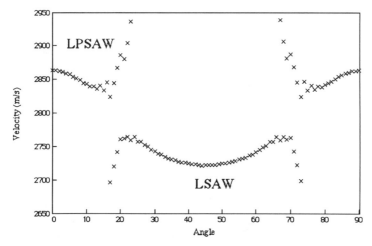

Figure 12.8. Velocities of surface acoustic modes on a GaAs {001} surface measured by the LFB microscope (LPSAW, leaky pseudosurface acoustic wave; LSAW, leaky surface acoustic wave); the angle of propagation θ in the plane is relative to a (110) direction.

of application is shown in Fig. 12.9, where the LFB technique was applied to characterize texture during the deformation of a metal–metal composite. The angular dependence of Rayleigh wave velocity was measured in different samples of rolled metal–metal (Cu–20% Nb) composites (Thompson *et al.,* 1990) as a function of the draw ratio η. This quantity is defined as $\eta = \ln(A_0/A)$, where A_0 and A are respectively the initial and final cross-sectional areas of the specimen during the rolling procedure. Interpretation of experimental data in terms of deformation-induced changes of the texture in the copper matrix gave a good quantitative agreement with theoretical values of standard deformation texture.

12.6. Acoustic Images of Materials

In the introduction, it was mentioned that the high-resolution SAM finds a number of applications in various areas of materials science and a wide literature exists in this regard (Briggs *et al.,* 1982,1989; Kulik *et al.,* 1990): imaging cracks in ceramics (Lawrence *et al.,* 1990), electronic and optoelectronic components and materials (Miller, 1985), different phases and grains in polycrystalline materials (Hoppe and Bereiter-Hahn, 1985). As it would be impossible to describe the full range of applications, we will present a limited number of figures taken in the last few years in our laboratory. The

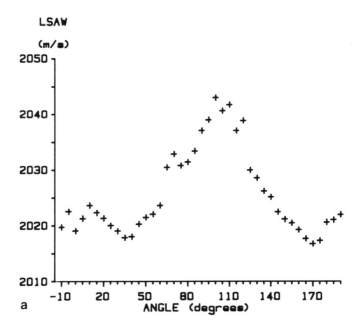

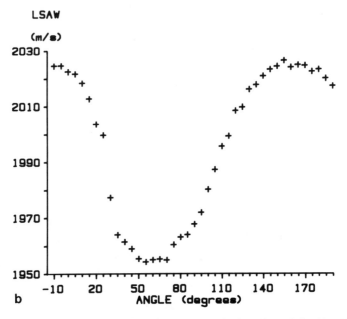

Figure 12.9. Angular dependence of Rayleigh wave velocity in deformed Cu–20% Nb: (a) η = 3.6, (b) η = 5.4.

aim will be to illustrate the major advantages of the SAM for material analysis as well as the specimen requirements for acoustic microscopy.

Acoustic microscopy requires great care and experience in the preparation of the surfaces for observation. The major requirement is that the flatness of the surface must be less than the wavelength to be used for imaging in such a way as to reduce topographical effects. Long- and short-range local topographical variations result in random scattering of the acoustic waves and produce degradations of the contrast or dark fringes in the images (Fig. 12.10).

12.6.1. Defects

Because of the particular origin of contrast, the acoustic microscope is extremely sensitive to any surface or subsurface defects that can affect the propagation of Rayleigh waves. These could be grain boundaries, voids, or inclusions. Among all of these, the experimental evidence suggests that the most striking characteristic of the SAM is its remarkable sensitivity to surface and subsurface cracks (Briggs, 1989; Briggs *et al.*, 1990). In an acoustic picture, the marked contrast around a crack arises from several mechanisms. If the lens is at focus, the contrast is the result of the strong reflection of bulk waves from the high change of impedance over the crack. A large enhancement in contrast and the appearance of characteristic fringes of periodicity of half a Rayleigh wavelength along the crack length ("Yamanaka ripple"; Yamanaka and Enomoto, 1982) are produced if the image is taken at negative defocus. Since Rayleigh waves travel parallel to the surface, they hit the crack broadside and are therefore strongly scattered even when the crack width is less than the wavelength of the incident radiation. The fringes are generated by the interference of the incident surface wave with waves scattered by the crack (Somekh *et al.*, 1985). Using an acoustic microscope, detection of cracks of width less than the resolution limit is thus possible.

Experimental confirmation of the contrast generated by a crack is given in the following figures (Briggs *et al.*, 1990), where crack lengths in different materials were measured using a high-frequency (800 MHz–2

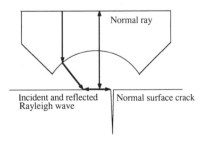

Figure 12.10. Fringe formation caused by normal crack.

GHz) Ernst Leitz SAM (ELSAM) and a Hitachi S-530 scanning electron microscope (SEM).

Figure 12.11 shows an acoustic micrograph (panel a) and SEM image (panel b) of cracks in ceramic materials. The specimens consisted of partially stabilized zirconia (PSZ) and silicon carbide ceramics. Both specimens had crack patterns formed by indenting the surfaces using a Vickers indenter. In the acoustic micrograph of the PSZ ceramic, an irregular fringe pattern is seen, characterized by a periodicity of half a Rayleigh wavelength. This is caused by multiple scattering of Rayleigh waves from grain boundaries, since at the frequency used for imaging (1.5 GHz), the Rayleigh wavelength (2 μm) was comparable with the grain size. The same crack was imaged at a similar magnification using the SEM; the length of the crack was estimated from both of these images to be 85 μm from the apex of the indent. Figure 12.12 shows an acoustic and a SEM micrograph of the radial crack at the right corner of the indent in a sample of reaction-bonded silicon nitride. The high contrast in the acoustic micrograph is generated by the high porosity of this material. The measured length of the crack was in both cases nearly 100 μm.

In addition to the ability to measure crack lengths with high accuracy, acoustic microscopy can furnish additional information about the particular nature of the crack. Figure 12.13 is an acoustic and a SEM image of a fatigue crack in a specimen of stainless steel. With negative defocus ($z = -4.4$ μm) the crack can be seen with its characteristic fringe pattern, and the grain structure of the stainless steel is also revealed. The SEM picture revealed only details concerning the dimension of the crack. With the SAM it was thus possible, without etching, to see the relationship between the crack path and the grain structure; in this case the crack seemed to be largely intragranular.

A final illustration is provided in Fig. 12.14, which shows acoustic micrographs of cracks in two specimens of WC–6% Co hardmetal. Both specimens had precracks formed by wedge indenting, and both were then subject to high-temperature loading for different periods of time. The specimen in Fig. 12.14a was held under load for a very short time (much less than a minute); the specimen in Fig. 12.14b was held under load for 12 min. In Fig. 12.14a the crack appeared straight along its whole length, and the characteristic fringes with half Rayleigh wavelength spacing are nicely visible. The micrograph shows the radial precrack generated by indentation which grew to a limited extent during the very short loading time. In Fig. 12.14b the crack had grown by void nucleation and coalescence, and secondary microcracking is evident. Rayleigh wave interference is apparent, with characteristic half-wavelength periodicity, but the pattern is irregular.

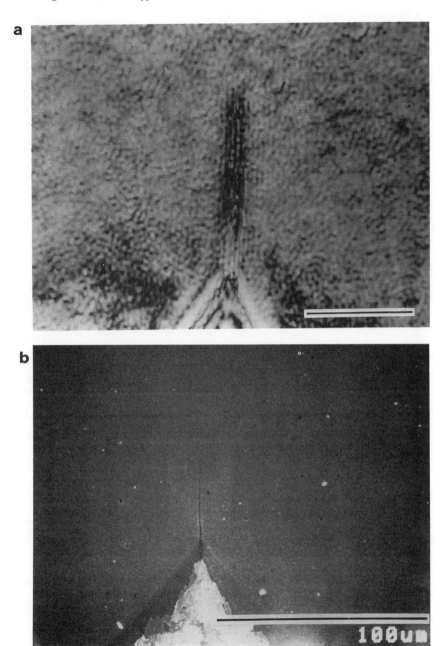

Figure 12.11. Micrographs of cracks formed from indents in partially stabilized zirconia (PSZ) ceramic. (a) Acoustic micrograph, 1.5 GHz, $z = -2.4$ μm; bar = 200 μm. (b) SEM micrograph of the same area.

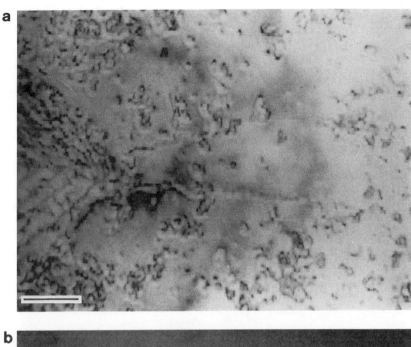

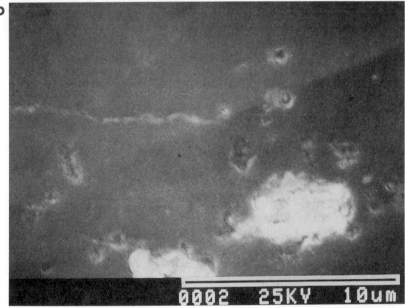

Figure 12.12. Micrographs of radial crack generated by microhardness indent in reaction-bonded silicon nitride. (a) Acoustic micrograph, 1.4 GHz, $z = 0$ μm; bar = 50 μm. (b) SEM micrograph of the same crack shown in (a).

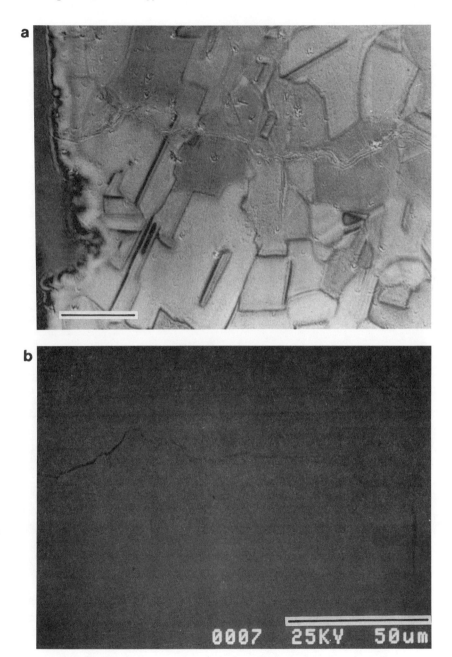

Figure 12.13. Micrograph of fatigue crack in stainless steel. (a) Acoustic micrograph, 1.5 GHz, $z = 4.4$ μm, bar = 25 μm. (b) SEM micrograph.

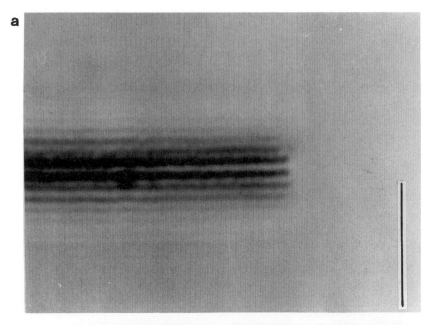

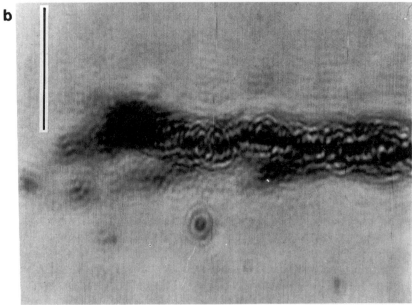

Figure 12.14. Acoustic micrographs of cracks in two specimens of WC–6% Co hardmetal held under load at 850°C. (a) Held under load for less than 1 min. 250 MHz, $z = -66$ μm. (b) Held under load for 12 min. 370 MHz, $z = -34$ μm. Bars = 100 μm.

12.6.2. Microstructure

We have shown how acoustic microscopy can be used to detect very small cracks in a broad class of materials. Other information about microstructural features of materials can be obtained using the same technique.

A high-frequency SAM (ELSAM) was used in our laboratory to characterize microstructural features of a series of PSZ ceramics (Fagan *et al.,* 1991). These ceramics present higher values of fracture toughness when they are deformed because of the tetragonal-to-monoclinic phase transformation which occurs in these constrained metastable composites. For the specimen analyzed, a sub-eutectoid aged ZrO_2–2 wt % MgO (MPSZ) ceramic, the stress-induced phase transformation was initiated by indenting the surfaces using a Vickers indenter. The polished surface was observed using a high-frequency Ernst Leitz SAM. Since the acoustic microscope is sensitive to local variations of elastic properties, it was possible to obtain images from regions of polymorphic zirconia where contrast was produced by areas of differing elastic properties. In the SAM micrograph of Fig. 12.15, the background contrast variations were the result of intergranular changes in orientation of the crystal elastic tensors, and to a lesser extent, of the local elastic modulus variations in regions of the tetragonal-to-monoclinic phase change. In these pictures, higher contrast is produced by Rayleigh wave interaction with cracks, pores, and grain boundaries. Phase-transformed materials were revealed as associated with shear banding in high-stress regions surrounding indentations. Indeed, in addition to fringes associated with odd reflections at the topographical contour of the indent ("water ripple") and the Yamanaka ripple, two new types of fringes were identified. The additional fringes, indicated as types C and D in the micrographs, were generated either by an interference phenomenon between incident Rayleigh waves and waves scattered from shear bands or from cracks generated in regions of high deformation. It was therefore concluded that these new fringes had a microstructural origin, and occupied regions where the tetragonal-to-monoclinic phase change had occurred.

As a last application of acoustic microscopy for microstructural analysis, we present some acoustic micrographs of a sample of borosilicate glass matrix reinforced with silicon carbide fibers (Lawrence, 1990). Figure 12.16 shows acoustic and optical micrographs of a cross section of a glass matrix composite, and that the microstructure is composed of the silicon carbide fibers and of a matrix containing two different phases. This last evidence was clearly observable only in the acoustic pictures, where about 70% of the matrix area appears homogeneous, while the remaining 30% appears granular, with a grain size less than 10 μm. The interpretation of this image, confirmed using an electron microprobe and an x-ray diffractometer, was that the bright and featureless region was amorphous borosilicate glass, while the crystalline phase

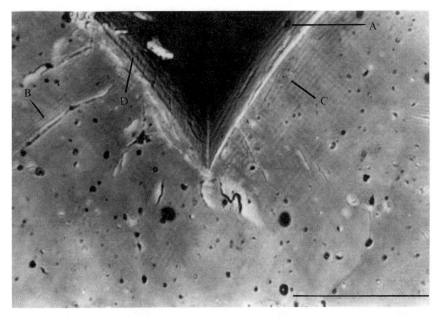

Figure 12.15. Acoustic micrograph of an indent on a manganese partially stabilized ceramic (MPSZ). Interference fringes: A, "water ripple"; B, "Yamanaka ripple"; C, cross-hatched fringes caused by scattering from shear bands; D, fringes on the indentation faces different from "water ripple." Imaging frequency 1.3 GHz, $z = -1$ μm, scale bar = 100 μm.

was identified as being crystobalite, the cubic allotrope of silica. An acoustic micrograph of a section through a single filament (Fig. 12.17) showed that the crystobalite grew at the fiber–matrix interface to form a layer 3–5 μm thick around the fiber, but there was also an extensive nucleation inside the borosilicate glass. The precipitation of crystobalite, known as devitrification, was caused by the high temperature and pressure conditions employed for the fabrication of these materials.

12.7. Conclusions

In this chapter we have examined some general aspects of acoustic microscopy and described some applications of this technique for materials microanalysis. It has been recognized that the usefulness of the acoustic microscope lies in the unique origin of contrast, whereby a probing elastic wave interacts with the specimen. The dominant mechanism of image formation

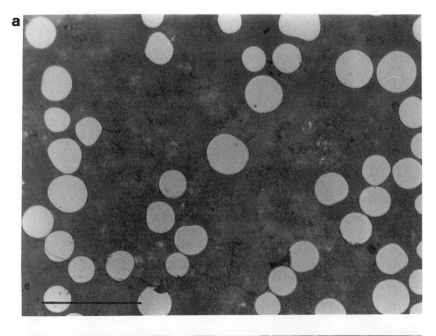

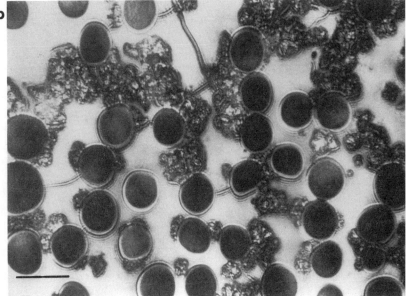

Figure 12.16. Micrographs of a 7740/SiC glass matrix composite showing general microstructure. (a) Optical micrograph; (b) acoustic micrograph, 1.9 GHz, $z = -2.8$ μm. Bars = 50 μm (a), 25 μm (b).

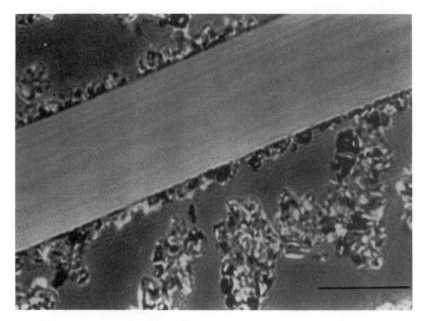

Figure 12.17. Acoustic micrograph showing crystobalite formation at the fiber–matrix interface; 1.9 GHz, $z = -1.0\ \mu$m, bar = 20 μm.

relates to the excitation and perturbation of leaky surface acoustic waves. These waves penetrate the specimen to a depth of approximately one or two wavelengths and strongly interact with surface and subsurface mechanical features in the material. There exist different ways whereby the propagation of surface waves can be affected. These could be changes in elastic properties or, if the material is anisotropic, changes in the orientation of the stiffness tensor, elastic inhomogeneities, or plastic zones. As a consequence, compared with other techniques, the SAM offers a unique combination of advantages that render it an important tool for the study of materials. First, the microscope may be used to image specimens which are optically opaque without the necessity of sectioning or etching the sample. This means that no external modifications are introduced in the preparation of the sample, and therefore reliable subsurface information of the specimen can be obtained. Moreover, taking into account its high resolution and sensitivity, the SAM extends the application of nondestructive testing (NDT) to small samples where higher resolution is required. This is especially important in the utilization of new high-strength engineering materials, such as ceramics and composites, where small defects generally cause failure and reliability problems. The second advantage resides in the possibility of using surface acoustic waves to extract quantitative information on mechanical properties of the sample. By determining propagation characteristics, i.e., velocity and attenuation, of these waves, characterization of elastic parameters of materials can be performed. Different applications confirming the feasibility of quantitative acoustic microscopy for material characterization have already been performed. From these first encouraging results, we expect that this technique will play, in the near future, an important role as an analytical method for material science.

ACKNOWLEDGMENTS. We express our thanks to the following for providing equipment and support: Logitech ltd. and SERC for providing their financial support during P. Mutti's research in Oxford University. Particular thanks to Mr. Max Robertson and Dr. Scott McMeekin of Logitech for their helpful discussions and encouragement. Dr. J. Kushibiki of Tohoku University, Sendai, Japan, for the line focus beam acoustic microscope. Wild Leitz Gmbh., Wetzlar, Germany, for the Elsam scanning acoustic microscope. All of the members of the acoustic microscopy group in Oxford for their constant assistance, in particular Z. Sklar, Dr. A. Fagan, and Dr. C. Lawrence for supplying the pictures and for the revision of the text. Professor Sir Peter Hirsch for the provision of the laboratory facilities. Finally, special thanks to Manuela for helping with the proofreading and typing of the text.

References

Atalar, A. (1978). An angular spectrum approach to contrast in reflection acoustic microscopy, *J. Appl. Phys.* **49**, 5130–5139.

Atalar, A. (1979). A physical model for acoustic signatures, *J. Appl. Phys.* **50**, 8237–8239.

Auld, B. A. (1985). In *Rayleigh-Wave Theory and Applications* (E. A. Ash and E. G. S. Paige, eds.), Springer-Verlag, Berlin, pp. 12–28.

Bertoni, H. L. (1984). Ray-optical evaluation of V(z) in the reflection acoustic microscope, *IEEE Trans.* **SU-31**, 105–116.

Bertoni, H. L., and Tamir, T. (1973). Unified theory of Rayleigh angle phenomena for acoustic beams at liquid–solid interfaces, *Appl. Phys.* **2**, 157–172.

Born, M., and Wolf, E. (1980). *Principles of Optics: Electromagnetic Theory of Propagation, Interference and Diffraction of Light,* Pergamon Press, Elmsford, N.Y.

Brekhovskikh, L. M., and Godin, O. A. (1990). *Acoustics of Layered Media I: Plane and Quasi-Plane Waves,* Springer-Verlag, Berlin.

Briggs, G. A. D. (1992). *Acoustic Microscopy,* Clarendon Press, Oxford.

Briggs, G. A. D. (1989). How sensitive is acoustic microscopy? Proc. EUROMAT '89.

Briggs, G. A. D., Somekh, M. G., and Ilett, C. (1982). Acoustic microscopy in materials science, *SPIE* **368**, 74–80.

Briggs, G. A. D., Rowe, J. M., Sinton, A. M., and Spencer, D. S. (1988). Quantitative methods in acoustic microscopy, *IEEE 1988 Ultrasonics Symposium,* pp. 743–749.

Briggs, G. A. D., Daft, C. M. W., Fagan, A. F., Field, T. A., Lawrence, C. W., Montoto, M., Peck, S. D., Rodriguez-Rey, A., and Scruby, C. B. (1989). Acoustic microscopy of old and new materials, in *Acoustical Imaging* (H. Shimizu, N. Chubachi, and J. Kushibiki, eds.), Vol. 17, Plenum Press, New York pp. 1–16.

Briggs, G. A. D., Jenkins, P. J., and Hoppe, M. (1990). How fine a surface crack can you see in a scanning acoustic microscope? *J. Microsc. (Oxford)* **159**, 15–32.

Derby, B., Briggs, G. A. D., and Wallach, E. R. (1983). Nondestructive testing and acoustic microscopy of diffusion bonds, *J. Mater. Sci.* **18**, 2345–2353.

Fagan, A. F., Briggs, G. A. D., Czernuszka, J. T., and Scruby, C. B. (1991). Microstructural observations of two deformed P.S.Z. ceramics using acoustic microscopy, *J. Mater. Sci.* in press.

Farnell, G. W. (1970). Properties of elastic surface waves, in *Physical Acoustics VI* (W. P. Mason and R. N. Thurston, eds.), Academic Press, New York, pp. 109–166.

Foster, J. S., and Rugar, D. (1985). Low-temperature acoustic microscopy, *IEEE Trans.* SU-32, 139–151.

Hoppe, M., and Bereiter-Hahn, J. (1985). Applications of scanning acoustic microscopy—Survey and new aspects, *IEEE Trans.* SU-32, 289–301.

Ilett, C., Somekh, M. G., and Briggs, G. A. D. (1984). Acoustic microscopy of elastic discontinuities, *Proc. R. Soc. London Ser. A* **393**, 171–183.

Kulik, A., Gremaud, G., and Satish, S. (1990). Acoustic microscopy as a polyvalent tool in materials science, *Trans. R. Microsc. Soc.* **1**, 85–90.

Kushibiki, J., and Chubachi, N. (1985). Material characterization by line-focus-beam acoustic microscope, *IEEE Trans.* SU-32, 189–212.

Kushibiki, J., Sannomiya, T., and Chubachi, N. (1980). Performance of sputtered SiO_2 film as an acoustic antireflection coating at sapphire/water interface, *Electron. Lett.* **16**, 737–738.

Kushibiki, J., Maehara, H., and Chubachi, N. (1981a). Acoustic properties of evaporated chalcogenide glass films, *Electron. Lett.* **17**, 322–323.

Kushibiki, J., Ohkubo, A., and Chubachi, N. (1981b). Linearly focused acoustic beams for acoustic microscopy, *Electron. Lett.* **17**, 520–522.

Kushibiki, J., Horii, K., and Chubachi, N. (1983). Velocity measurement of multiple leaky waves on germanium by line-focus-beam acoustic microscope using FFT, *Electron. Lett.* **19**, 404–405.

Kushibiki, J., Ishikawa, T., and Chubachi, N. (1990). Cut-off characteristics of leaky Sezawa modes and pseudo-Sezawa modes for thin-film characterization, *Appl. Phys. Lett.* **57**, 1967–1969.

Lawrence, C. W. (1990). Acoustic microscopy of ceramic fibre composites, D. Phil. thesis, Oxford University.

Lawrence, C. W., Scruby, C. B., Briggs, G. A. D., and Dunhill, A. (1990). Crack detection in silicon nitride by acoustic microscopy, *NDT Int.* **23**, 3–10.

Lemons, R. A., and Quate, C. F. (1974). Acoustic microscope—scanning version, *Appl. Phys. Lett.* **24**, 163–165.

Liang, K. K., Kino, G. S., and Khuri-Yakub, B. (1985). Material characterisation by the inversion of V(z), *IEEE Trans.* SU-32, 213–224.

Miller, A. J. (1985). Scanning acoustic microscopy in electronics research, *IEEE Trans.* SU-32, 320–324.

Muha, M. S., Moulthrop, A. A., and Kozlowski, G. C. (1990). Acoustic microscopy at 15.3 GHz in pressurized superfluid helium, *Appl. Phys. Lett.* **56**, 1019–1021.

Obata, M., Shimada, H., and Mihara, T. (1990). Stress dependence of leaky surface wave on PMMA by line-focus-beam acoustic microscope, *Exp. Mech.* **March,** 34–39.

Pinkerton, J. M. M. (1949). The absorption of ultrasonic waves in liquids and its relation to molecular constitution, *Proc. Phys. Soc.* **62,** 129–140.

Somekh, M. G., Briggs, G. A. D., and Ilett, C. (1984). The effect of anisotropy on contrast in the scanning acoustic microscope, *Philos. Mag.* **49,** 179–204.

Somekh, M. G., Bertoni, H. L., Briggs, G. A. D., and Burton, N. J. (1985). A two-dimensional imaging theory of surface discontinuities with the scanning acoustic microscope, *Proc. R. Soc. London Ser. A* **401,** 29–51.

Thompson, R. B., Li, Y., Spitzig, W. A., Briggs, G. A. D., Fagan, A. F., and Kushibiki, J. (1990). Characterization of the texture of heavily deformed metal–metal composites with acoustic microscopy, in *Review of Progress in Quantitative Nondestructive Evaluation 9* (D. O. Thompson and D. E. Chimenti, eds.), Plenum Press, New York, pp. 1433–1440.

Truell, R., Elbaum, C., and Chick, B. B. (1969). *Ultrasonic Methods in Solid State Physics,* Academic Press, New York.

Viktorov, I. A. (1969). *Rayleigh and Lamb Waves, Physical Theory and Applications,* Plenum Press, New York.

Weaver, J. M. R. (1986). The ultrasonic imaging of plastic deformation, *D. Phil. Thesis,* Oxford University.

Weaver, J. M. R., and Briggs, G. A. D. (1985). Acoustic microscopy techniques for observing dislocation damping, *J. Phys. (Paris)* **12 C10,** 743–750.

Weglein, R. D. (1979). A model for predicting acoustic materials signatures, *Appl. Phys. Lett.* **34,** 179–181.

Yamanaka, K., and Enomoto, Y. (1982). Observation of surface cracks with scanning acoustic microscope, *J. Appl. Phys.* **53,** 846–850.

Part VI

Tunneling of Electrons and Scanning Probe Microscopies

13

Field Emission, Field Ion Microscopy, and the Atom Probe

J. J. Hren and J. Liu

13.1. Introduction

With the rapid development of scanning tunneling microscopies in the last decade, several older techniques from which they sprang have been nearly ignored. Yet these techniques offer similar resolution and data that are valuable in themselves. Field electron emission, which dates to the 1930s, is recognized today as a phenomenon inherent to STM or it is associated with very bright electron sources for electron microscopes (SEMs and TEMs). Recently, field electron emission has also been linked to a generation of "new" devices for vacuum microelectronics (although this is hardly a new application). If field electron emission is a relatively obscure, but useful phenomenon, field electron microscopy (FEM) is viewed as a technique for aficionados which has little to offer that can't be better measured in other ways. The related technique of field-ion microscopy (FIM) and the atom probe (AP) are recognized as having highly specialized applications, at best.

Given this background, this chapter has several purposes: (1) general enlightenment about the related phenomena of field electron emission, field ionization, and field desorption (or field evaporation) which are at the heart

J. J. Hren and J. Liu • Department of Materials Science and Engineering, North Carolina State University, Raleigh, North Carolina 27695.

Microanalysis of Solids, edited by B. G. Yacobi *et al.* Plenum Press, New York, 1994.

of FEM, FIM, and AP techniques; (2) an evaluation of the experimental utility and limitations of these techniques; (3) a description of several related by-products, such as intense electron sources, vacuum microelectronic devices, and intense ion sources; and (4) an introduction to the relationship between FEM, FIM, and AP and the several common scanning probe techniques covered in the next chapter.

The treatment throughout is intended to be introductory and tutorial, not exhaustive or definitive. A small number of comprehensive references have been cited for the serious reader to explore the major topics at greater depth.

13.2. Field Electron Emission

Experimental evidence for field emission of electrons was published by Eyring *et al.* in 1928, but classical theoretical predictions used to explain it were not successful. That same year, the quantum-mechanical treatment of the same problem by Fowler and Nordheim (1928) appeared. By 1936 it was clear that their treatment successfully explained the main features of the field emission process and a working field emission microscope (FEM) was introduced by E. W. Mueller (1936a,b). These accomplishments are worth serious reflection even now, half a century later! At the time, Mueller was in Berlin and interested in developing an *electron microscope*. Indeed he did just that. The FEM is a point projection electron microscope with ~2nm lateral resolution and even better vertical resolution! This latter feature was not truly exploited until the development of STM. In another sense the FEM was truly the first modern instrument used for surface science. It was immediately obvious that stable field emission images required vacuums far better than those in common use at that time. Field emission was so extremely sensitive to adsorbates that the effects of even a few molecules were detectable. Indeed, the conjugate effects are used today in STM to probe single adsorbates or small clusters. But the interpretation of images from single adsorbate molecules by FEM remains controversial (Brodie, 1978; Giaever, 1972).

13.2.1. The Fowler–Nordheim Equation

Although there have been many refinements to the Fowler–Nordheim equation, including more realistic boundary conditions, it remains the basis for understanding field emission phenomena. In thermionic emission, an electron must acquire sufficient thermal energy to overcome the surface potential barrier ϕ we call the work function. A classical treatment of the problem gives the current density J:

$$J = \alpha T^2 \exp[-\phi/kT] \tag{13.1}$$

where α and k are constants and T is temperature. When suitable values are inserted, J is 10 A/cm^2 at most.

In the presence of a sufficiently large electric field the surface barrier of a conductor will be distorted such that the probability for electron tunneling directly through it will become significant. Field emission of electrons at low temperatures (where direct thermal activation over the barrier can be neglected) was one of the first direct manifestations of quantum-mechanical tunneling with no classical analogue. The field strength F required for tunneling can be simply estimated from the Heisenberg uncertainty principle: appreciable tunneling will occur only when the uncertainty in the position of the electron is comparable with the width of the barrier (Fig. 13.1). If the uncertainty in kinetic energy of the electron is taken to be the barrier height, ϕ, and the width of the barrier is ϕ/Fe (where e is the electron charge), then:

$$\Delta x \Delta p = (\phi/Fe)(2m\phi)^{1/2} \approx \hbar/2 \tag{13.2}$$

or

$$F \approx (8m\phi^3/\hbar^2e^2)^{1/2} \cong 10 \text{ volts/nm} \tag{13.3}$$

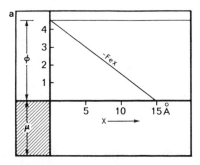

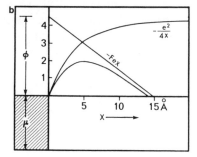

Figure 13.1. Potential energy diagram of electrons at a metal surface in the presence of an applied field: (a) without electron image potential; (b) with electron image potential.

where $\hbar$ is Planck's constant and m is the rest mass of the electron. The field strength predicted is about an order of magnitude too large compared with experiment. The major difficulty lies with the oversimplified triangular barrier which does not include the contribution of the coulombic attraction of the electron to its "image charge" in the metal, $-e^2/4x$. When an image potential is included, the shape of the barrier is considerably lowered. The classical expression must be reasonably correct since it predicts the current–voltage characteristic of field-emitted electrons to reasonable accuracy.

Several other simplifying assumptions are usually made:

1. The metal is one-dimensional and obeys Fermi–Dirac statistics at $T = 0$.
2. The emitter surface is approximated as a smooth, infinite plane.
3. The work function is uniform and isotropic over the emitter surface.

Under these assumptions, if the barrier penetration probability (calculated using the WKB approximation) is multiplied by the arrival rate of electrons, the result is the Fowler–Nordheim equation:

$$J = (AF^2 \exp[-6.83 \times 10^7 e^{3/2} \phi^{3/2} f(y)/F])/e\phi t^2(y) \qquad (13.4)$$

where A is a constant and $t(y)$ and $f(y)$ are tabulated functions of $y = (eF)^{1/2}/_\phi$, and the electric field strength F is given by V/Kr, where r is the emitter radius and $K = 1$ for an isolated sphere (in reality, $4 \leq K \leq 6$). It is found that a field strength of about 0.3 V/Å is required for field emission, so that a potential of one or more kilovolts must be applied for $r \approx 200$ nm. For simple comparison with thermionic emission Eq. (13.1), the Fowler–Nordheim equation, may be written as

$$J = BV^2 \exp[-\phi^{3/2}/CV] \qquad (13.5)$$

where B and C are constants. Note the similarity between Eqs. (13.1) and (13.5), except for the $\phi^{3/2}$ dependence. If suitable numerical values are inserted, J may have values as high as 10^4 A/cm^2 or about 10^3 times greater than for thermionic emission. In a crude sense, we may think of the energy provided by the electric field as replacing thermal energy. If the logarithm of I/V^2 is plotted against $1/V$, a straight line is obtained (Fig. 13.2) and from the Fowler–Nordheim equation, the slope of this line is found to be:

$$S = -6.8 \times 10^7 e^{3/2} \phi^{3/2} f(y) Kr \qquad (13.6)$$

Assuming that the tip radius, r, is known (e.g., from a micrograph of the tip profile), and $K \approx 5$, an average work function of the emitter, ϕ, can be obtained from Eq. (13.6).

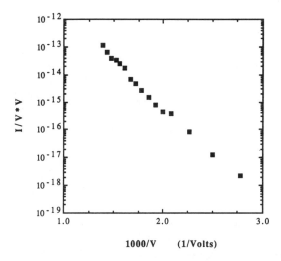

Figure 13.2. Field emission data from a silicon tip following a Fowler–Nordheim curve.

13.2.2. Point Projection Microscopes and the FEM

A fortuitous combination of properties makes field electron emitters ideal specimens for point projection microscopes. First, in order to achieve field emission at applied voltages of a few kilovolts, a radius of curvature $\leq 1\ \mu$m is essential. For example, using $r \approx 100$ nm and $K \approx 5$ yields an approximate voltage of 1.5 kV to achieve the required field strength. Practically speaking, such small radii can be achieved with either very thin wires, sharp edges, or needle-shaped points. As will become apparent, the needle shape is preferred to avoid widely differing magnification in different directions. Imagine an idealized point projection microscope comprised of a small spherical emitter of radius r surrounded by a large glass sphere of radius R. Coat the inside of the large sphere with a transparent conductive coating (e.g., doped SnO_2) and a suitable phosphor (e.g., Zn_2SiO_4:Mn or willemite) and evacuate it to a vacuum of 10^{-11} torr. Now apply a sufficient potential difference V between the emitter and screen to achieve field electron emission. (Neglect for the moment the obvious practical difficulties.) The resulting pattern on the spherical screen should be a pure point projection of electron emission from the surface of the small sphere. Since there are no lenses, there are no lens distortions. The resolution of the image is limited only by contrast from the intrinsic detail of the surface reflected in the electron emission pattern. The magnification M, given to first approximation by $M = R/r$, must also be adequate. If the values of R and r are respectively 10^{-1} m and 10^{-7} m, a magnification of one million timcs is attainable. In other words, 1 nm^2 on the specimen will project onto 1 mm^2 on the screen, sufficient to resolve detail of atomic dimensions.

Mueller first achieved a workable approximation to the idealized point projection microscope in 1936 (Mueller, 1936a,b). The microscope (first known as the field electron emission microscope, or FEEM, since shortened to field emission microscope (or FEM) is shown schematically in Fig. 13.3. The field-emission cathode consists of a fine tungsten wire, formed into a triangular loop with the protruding end etched to a fine point. One side of the loop bridges two supporting electrodes which could be outgassed by passing an electric current through them. A transparent anode or vapor-deposited tin oxide is coated with zinc oxide or willemite (Zn_2SiO_4:Mn) in order to make the emission pattern visible. Side arms contained sublimation sources aimed at the tip and a barium getter. After bakeout, the glass microscope body was isolated from a trapped, mercury diffusion pump. Flashing a getter then permitted an ultimate vacuum below 10^{-11} torr. No hot-filament pressure gauge was needed because the pressure in the microscope body could be deduced from the change in appearance of the emission pattern, caused by adsorption of the residual gas contaminants on the tip. This method was most reliable at the very low pressures required to produce a stable image.

How did Mueller's FEM perform relative to his expectations or that of our idealized instrument? A typical result is shown in Fig. 13.4. To begin with, the magnification was somewhat altered because he used a needle and not a sphere; that is, something geometrically approaching a hemispherical cap on a cone. This resulted in some image compression β, such that:

$$M = R/\beta r \tag{13.7}$$

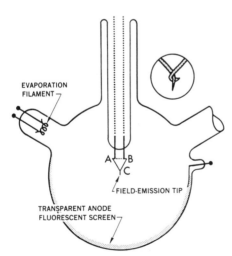

EVAPORATION
FILAMENT

A⟨⟩B
C
FIELD-EMISSION TIP

TRANSPARENT ANODE
FLUORESCENT SCREEN

Figure 13.3. Schematic diagram of a projection-type field emission microscope.

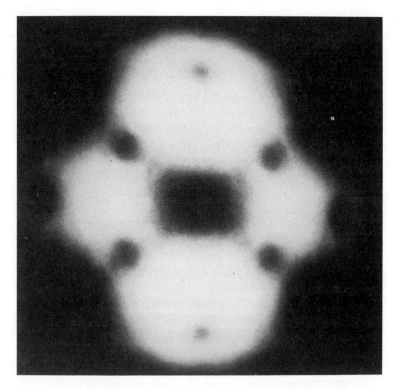

Figure 13.4. Field emission pattern of the tungsten (110) plane.

where $\beta \approx 1.5$ which depends primarily on the cone angle. The decreased magnification could be easily compensated by increasing the distance from tip to screen. However, it was found that detail finer than 2 nm still could not be resolved, except under very special circumstances. In fact, the resolving power was found to be limited by the spread in electron velocities directed parallel to the surface. This can be understood as follows. Consider electron emission from a point on the surface. If it takes time for the electron to travel to the fluorescent screen and the transverse velocity component of the electron is v_t, then the electrons will be spread over a distance $v_t t$ when they reach the screen. That is, every point on the surface will project in an extended cone to the screen; thus, the inherent microscope resolution cannot be better than the resulting "circle of confusion" at the screen. That is, two emitting points on the surface will not be resolved if they are separated by a distance less than δ, given by:

$$\delta = 2v_t t \beta r / R \qquad (13.8)$$

where $R/\beta r$ is the magnification. Since an electron acquires its full kinetic energy very close to the tip surface, its travel time, t, is approximately equal to the drift time from tip to screen at voltage V, and the resolution becomes

$$\delta = v_t (2m\beta^2 r^2/eV)^{1/2} \qquad (13.9)$$

During field emission, there are two contributions to v_t: the average transverse velocity with which electrons leave the surface and the uncertainty caused by their finite de Broglie wavelength (i.e., diffraction). Mueller (1953) combined both contributions in quadrature to obtain a typical resolution of 2 nm, in good agreement with experiment. Subsequently, it has been argued that under special conditions, near-atomic resolution ($\delta \approx 4$ Å) may be possible.

As we noted earlier, a crucial component in image formation is the contrast. Even if the inherent resolution is adequate to achieve 2 nm, we must ask: is there sufficient variation in electron current from point to point that features can differentiate? To answer this question requires a further understanding of the specimens themselves and the likely variations in electron emission to form contrast. The fundamental model in considering the imaging mechanisms giving rise to contrast is the same as that used to derive the Fowler–Nordheim equation. The two major contributions are: the field strength and the strength of the barrier to electron tunneling. However, these parameters must now be modeled on a *local* scale. Consider first the overall specimen. It is normally a crystalline conductor or semiconductor.

The size of the tip is normally smaller than the size of individual crystallites in most polycrystalline materials. Thus, an etched tip will normally expose a single, perfect crystal at its apex. (Occasionally, a well-defined grain boundary will be exposed.) The etching process results in a shape corresponding roughly to a hemisphere on a conical shank and exposes many different crystal planes of differing orientations joined smoothly into a surface. Individual crystal planes have different atomic packing densities and inherently different charge distributions which results in a corresponding distribution of tunneling barriers. In turn, this will be reflected in the field-emission image because the probability of tunneling depends so strongly on the local surface barrier. Since surface adsorbates change the charge distribution on a clean surface, their presence will also alter the field-emission pattern and give localized contrast variation. This sensitivity of FEM to local changes in tunneling allowed Mueller to study adsorption and desorption phenomena on clean metal surfaces in ultrahigh vacuum as early as 1937, the first true surface science as thought of today (Mueller, 1937).

13.2.3. Adsorption Studies and Probe Hole Measurements

If the average work function of a clean emitter is known, the change in work function caused by adsorption can also be determined. This is achieved by measuring the ratio of the Fowler–Nordheim slope before and after adsorption and assuming that the product Kr remains constant. However, the accuracy of this result depends on the assumption that the relative anisotropy of the field-emitted current does not change appreciably over the surface before and after adsorption. Since this is often not true, it is safer to measure work function changes in small regions of the emitter which are each single-crystal facets. An apertured photomultiplier can be used for this purpose, but more commonly a "probe-hole" technique is used. The preferred method is to manipulate the tip such that the field-emission image can be shifted with respect to a small aperture (or probe hole) in the fluorescent screen, the dimensions of which are much smaller than a facet. In order to measure the field-emission current, a Faraday collector or electron multiplier is placed behind the probe hole. If the aperture is much smaller than the region being examined, a reasonably accurate measurement of local work function can be made; however, one cannot completely ensure that all electrons from adjacent regions of the emitter are completely excluded. Even with these limitations, the probe-hole technique has made major contributions to our understanding of the concept of a work function.

Although qualitative information concerning the adsorbate–substrate interaction is easy to obtain from the appearance of the field-emission pattern and its current–voltage characteristic, quantitative information from such measurements, including absolute values of the work function and coverage, is not reliable. To obtain quantitative data it is necessary to determine the kinetic energy distribution of field-emitted electrons. That is, the energy distribution of the field-emitted electrons should be sensitive to both the properties of the clean metal surfaces and modifications of the tunneling process resulting from a local change in surface charge distribution, for example, by adsorption.

13.2.4. Field-Emission Energy Distributions

Although the energy distribution of field-emitted electrons was measured before 1940, it was not until 1959 that Young made measurements of sufficient energy resolution that useful fundamental information resulted. Electrons having both perpendicular *and* transverse energy components to the emitter surface contribute to the measured current, but the transverse component is quickly dominated by the radial kinetic energy acquired in the electric field above the emitter. In order to detect only electrons emitted in a direction normal to the tip surface, either a very small probe hole has to be placed very close to the emitter surface, or the tip surface must be imaged onto the entrance surface of an energy analyzer by a virtually distortion-free lens.

Most analyzers are of the retarding potential type consisting of three electrodes: the field-emitter tip held at potential V_T, an apertured counterelectrode (the anode) held at potential V_A, and a collector located immediately behind the aperture at potential V_C. The various designs differ primarily in their geometry, the method used to manipulate a selected region of the field-emission image over the anode aperture, and the method used to focus the electrons onto the collector. The field strength required for electron emission is established from the potential difference ($V_A - V_T$). Only electrons with sufficient kinetic energy to overcome the retarding potential difference between collector and tip ($V_C - V_T$) will be collected (Fig. 13.5). An optimized lens system maximized luminosity (area times solid angle) and provided energy resolution of 20 mV or less at a signal-to-noise ratio of about 1000:1 (Kuyatt and Plummer, 1972).

Energy distributions having only one peak in the total energy distribution are typical of most clean metal surfaces, but Swanson and Crouser (1966, 1967) were the first to observe two peaks for electrons emitted from the (100) plane of tungsten (Fig. 13.6). One peak at 0.37 eV was shown to be sensitive to the cleanliness of the (110) surface Partial monolayers of a variety of adsorbates (e.g., CO, H_2, or even Ar) were shown to attenuate or remove this structure, which is apparently related to a surface state in the $\langle 100 \rangle$ spin-orbit-split gap of tungsten.

However, experimental aspects of FEED and data interpretation in terms of band structure effects, surface states, and adsorbates are still a topic of research. Several excellent reviews (e.g., Gadzuk and Plummer, 1973; Swanson and Bell, 1973) exist, but unambiguous interpretations of experimental energy

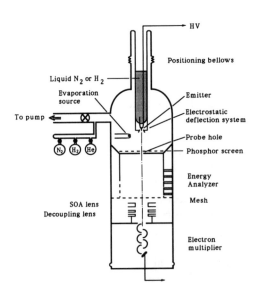

Figure 13.5. Retarding potential energy analyzer for field emission total energy distribution measurement.

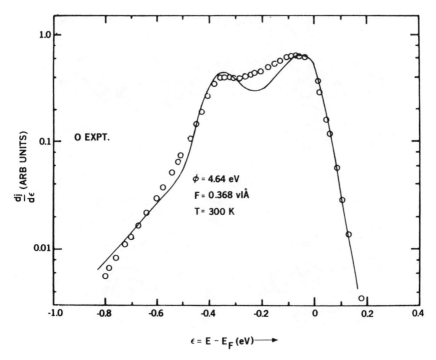

Figure 13.6. Theoretical (solid line) and experimental field electron emission energy distribution from tungsten (100).

distributions are still rare. This method remains a seldom used but potentially very powerful means to obtain fundamental surface and band structure data.

13.2.5. Molecular and Atomic Imaging

The inherent simplicity of a FEM coupled with its excellent resolution led to many attempts to resolve individual adsorbed molecules and atoms. Successful imaging presupposes that an isolated molecule on the surface will change the local work function so as to reflect the "true contour" of the adsorbate. Of course, it is important to ensure that contaminants are minimized, but, in effect, this precludes depositing of molecules *ex situ*. The results of the first *in situ* experiments by Mueller (1950a,b) have remained controversial to this day. He used the organic dye copper-phthalocyanine $(C_{32}H_{16}N_8Cu)$ which sublimed onto the field emitter in high vacuum. Striking high-contrast images were obtained (Fig. 13.7) in which fourfold symmetry, apparently characteristic of the planar phthalocyanine molecule, was seen.

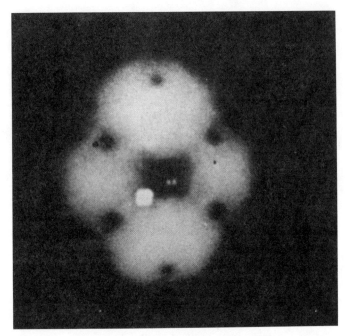

Figure 13.7. Field-emission image of tungsten (110) plane with adsorbates.

Surprisingly, these images were an order of magnitude larger than predicted yielding an apparent resolution an order of magnitude better than expected. These results were explained in terms of a locally enhanced magnification.

Additional support came from images of a twofold symmetric molecule, flavanthrene ($C_{28}H_{12}N_2O_2$), which could be vapor-deposited onto the tip *in situ*. More than 99% of the images were doublets, apparently reflecting the symmetry of this molecule as well. Unfortunately, when a large number of organic molecules of other expected shapes were examined, only quadruplet and doublet images were ever seen. Evidently the FEM images were not displaying the actual molecular contour of an adsorbed molecule, but some other characteristic. Many explanations and a series of "clarifying" experiments have been conducted to explain these spectacular data; however, none have provided convincing evidence of molecular imaging as intended. One explanation by Giaever (1972) is consistent with the hypothesis that adsorbed molecules stack to form an ordered crystallite. He postulates that doublet images are caused by electrons emitted from single, planar molecules standing on edge, while quadruplet, and other less frequently observed patterns, are caused by emission from the ends of small molecular crystallites. He also suggests that under certain conditions, a field-emission pattern can reflect the presence of an individual molecule, but not its actual morphology. The common symmetry in the emission pattern of different molecules he concludes is the result of the common symmetry in the electric field distribution surrounding each adsorbed species.

The possibility that single atoms could be imaged in the FEEM is related to the molecular imaging problem, but was raised even earlier. When the FEM was introduced, Mueller suggested that some image features apparent after barium deposition on tungsten were caused by enhanced emission at individual barium atoms. Ashworth (1948) also found that if hydrogen was introduced into a well-baked FEEM, single image spots could be observed on flat, smooth crystal faces of low work function. Each spot would eventually split into two halves and these remained close together, usually rotating about each other before disappearing or moved randomly away from each other over the surface. If argon was adsorbed in a similar manner, no splitting was observed. Ashworth thus believed that the dissociation of molecular hydrogen was being observed. A smooth, flat surface of low work function was required to see an image, because on such a surface an adsorbate would produce the maximum perturbation in the surface potential. This would lead to enhanced emission and an increase in local magnification and resolution (Brodie, 1978; Rose, 1956).

Despite these tantalizing results, the goal of imaging individual atoms in the FEEM, and proving unambiguously that individual atoms were imaged, was never completely achieved although Brodie (1978) has accumulated some convincing evidence to the contrary. But because of these obvious difficulties, this challenge led Mueller to develop the field-ion microscope in which atomic resolution was first unambiguously achieved.

13.3. Field Ion Microscopy

Since atomic resolution seemed to be just beyond the capability of field electron microscopy, it was natural to turn to other prospects for point projection. Ion sources did not appear attractive at first because the ion currents produced by field-induced desorption were extremely low (as were detection efficiencies). The required field strengths were also higher. However, by applying a dc bias to the tip, any resulting ions would follow trajectories very similar to electrons (of course, neutrals would not). Mueller was able to barely record an image by a rapid succession of adsorption/desorption sequences using barium ions coupled with photographic integration. He attempted to increase the supply of ions by supplying hydrogen from the gas phase to yield a steady supply of adsorbates for subsequent desorption and ionization and did indeed obtain a dramatic image improvement, but not by an adsorption/

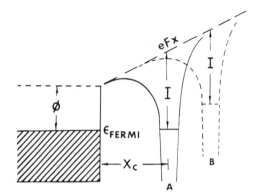

Figure 13.8. Potential energy diagram of a gas atom near a metal surface in the presence of an electric field.

desorption mechanism. We now know that ionization of gas-phase hydrogen in the high fields in free space just above the specimen was actually the source of the steady-state hydrogen ion currents that produced these early images. Fortuitously, the right conditions for field-ion microscopy were discovered.

13.3.1. Field Ionization

As with field electron emission, the physics of quantum mechanics was needed to understand gaseous ionization in free space by an intense electric field. In fact, field ionization of an isolated atom in free space by electron tunneling was first proposed by Oppenheimer (1928). He demonstrated that the lifetime of an electron in an isolated hydrogen atom depended on the magnitude of an external electric field. At fields approaching 1.5 V/Å, ionization would occur in less than 1 s by tunneling through the reduced barrier. Ingram and Gomer (1954) used Oppenheimer's results to develop a theoretical description of field ionization. They clearly showed that hydrogen was not ionized on the emitter surface, but in the free space just above.

In free space, the valence electron of a neutral atom can be considered as bound in a potential well. An energy equal to the ionization potential, I, must be supplied to ionize the atom. But, in an electric field, the potential well will be distorted and the side toward the anode reduced until its width becomes comparable to the de Broglie wavelength of the valence electron in the atom. The probability for tunneling through the potential barrier will then greatly increase. Further, as the distance between the gas atom and the anode is decreased, the barrier width decreases accordingly. However, decreasing this distance to the point of adsorption is counterproductive, since tunneling can occur only into an unoccupied energy state of the bulk substrate material. That is, the ground state of the valence electron in the gas atom must lie above the Fermi level of the metal. This condition is satisfied at some critical distance, x_c, above the surface, as depicted in Fig. 13.8. This simple model leads to the condition:

$$eFx_c = I - \phi \qquad (13.10)$$

The image potential of the tunneling electron and the change in the polarization energy of the atom upon ionization have been omitted because their contributions are usually negligible. Typical conditions for hydrogen, I = 13.5 eV in a field of 2 V/Å above a tungsten surface, where ϕ = 4.5 eV, yield a critical distance for ionization, x_c = 4.5 Å.

A reasonable estimate of the ionization probability of a gas atom at the critical distance can be obtained by multiplying the probability of an electron tunneling from the atom by the frequency with which it strikes the tunneling barrier. The former can be estimated from the WKB approximation using a simplified triangular barrier. The latter can be estimated from the Bohr model using an "effective" nuclear charge appropriate to the atom. The measured field-ion current should then be the product of the ionization probability and the arrival rate of atoms in the ionization zone. Measured field-ion currents were at least an order of magnitude larger than expected, assuming a supply of imaging gas molecules obeying the kinetic theory and an ionization probability of 100%. This discrepancy was explained when it was realized that the supply of gas-phase molecules in the vicinity of the tip was enhanced by a field-induced dipole attraction to the tip.

Measurements of field-ion energy distributions have proved equally as useful for understanding field ionization and FIM as field emission energy distributions have to FEM. A number of interesting physical phenomena were observed including a multiple-peaked resonance associated with the tunneling of an electron between an atom and a nearby substrate (Fig. 13.9). This has led to new models of surface electronic structure, which are sensitive to experiment. At very high fields the imaging gas atom may even be bound to the surface by a short-range field-induced dipole–dipole bond which has been called field adsorption. However, the major contribution of such field-ion energy distribution studies has been to our understanding of the field-ionization mechanism itself and hence FIM imaging.

13.3.2. Gas Supply

Field-ion energy distribution measurements determined that the region of greatest ionization probability is located just above the field-emitter surface. At a certain field strength, the ionization region was found to separate into "disks" of high ionization probability associated with the protruding surface

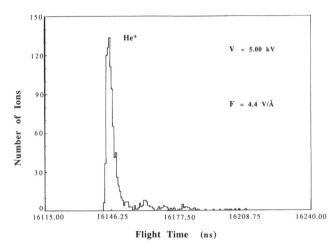

Figure 13.9. Energy distribution of field-desorbed He ions.

atom, where the local electric field strength is greatest. These ionization disks are about 0.2 Å thick, and approximately equal in diameter to an imaging gas atom. Under these conditions, called the best imaging voltage (BIV), the field-ion image will display maximum surface detail over the widest field of view. Although the BIV is subjectively chosen, different observers consistently choose the same value to within a few percent.

Since the supply of imaging is enhanced by gas molecules which are polarized in the field near the emitter, they will strike the emitter with a velocity much larger than kT. If they do not condense on the tip, they will lose some fraction of their energy and rebound, only to be attracted again by polarization forces. In fact, at BIV, a reasonable fraction of the emitter surface will be covered with field-adsorbed imaging gas molecules thereby enhancing considerably thermal accommodation of additional molecules. In effect, the polarized molecules will be trapped in a region of space close to the emitter, while slowly hopping randomly over the surface. Each molecule will pass through the ionization disk above a protruding surface atom a number of times, for each of which it will have a finite ionization probability; when tunneling occurs, the resulting ion will accelerate rapidly away from the tip surface in an almost radial direction (Fig. 13.10). At an imaging gas pressure of several millitorr, about 10^5 ions/s will be created above each ionization disk, which is equivalent to an ion current of about 10^{-14} A. By this measure, the intensity of the resulting field-ion image is roughly equivalent to the Milky Way, viewed with dark-adapted eyes on a moonless night. These conditions immediately suggest the need for image intensification for FIM. Imaging

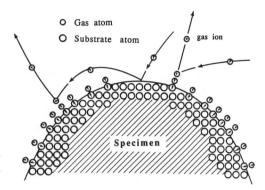

Figure 13.10. Schematic of field ionization processes for gas atoms near a
tip surface.

conditions for FEM, on the other hand, require no special intensification
procedures.

As the imaging molecule thermally accommodates to the tip temperature,
it will lose its field-induced dipole energy $(\frac{1}{2})\alpha F^2$. Therefore, after ionization,
its kinetic energy parallel to the tip surface will be simply

$$(1/2)mv_T^2 \approx kT \qquad (13.11)$$

Although resolution of the FIM is ultimately limited by size of the ionization
zone disk, δ_0, the transverse velocity component of the imaging gas is also
significant in analogy to FEM. The ultimate image resolution is thus

$$\delta = \delta_0 + (4kT\beta^2R/KeF)^{1/2} \qquad (13.12)$$

That is, all things being equal, cryogenically cooling the tip should improve
image resolution roughly as $T^{1/2}$.

In 1956 Mueller described the great improvement in resolution which
could actually be obtained by cooling the tip (Mueller, 1956a,b). Atomic
resolution has been achieved simply by using helium as an imaging gas and
by cooling the tip below 80 K. The field-ion microscope which achieved atomic
resolution is shown schematically in Fig. 13.11. Although many variations of
this basic design have appeared in the literature, this simple prototype remains
the basic research tool of FIM today.

Optimum resolution at any tip radius would seem to require that the tip
temperature be kept as low as possible, the diameter of the imaging gas disk
be chosen to be as small as possible, and that the electric field at BIV be made
as large as possible (Fig. 13.12). Helium would seem to be the best choice as
an imaging gas, since it has a condensation temperature below 5 K, an effective

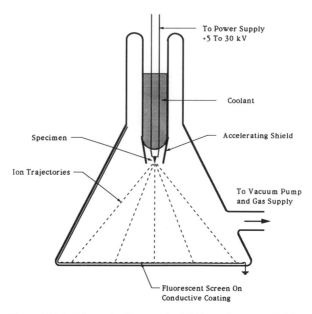

Figure 13.11. Schematic diagram of a field-ion microscope (FIM).

diameter of 1.9 Å, and a best-image field of 4.5 V/Å. However, the stress σ exerted on the emitter by the electric field is very large and given by

$$\sigma = F^2/8\pi \approx 10^{11} \text{ dynes/cm}^2 \tag{13.13}$$

Typical operating conditions at BIV, using helium imaging gas, can elastically deform most materials; in fact, only those with a relatively large elastic modulus, which are free from major lattice defects, will survive. At some sacrifice in resolution, other imaging gases or mixtures can be selected to reduce the field stress. For example, hydrogen produces a field-ion image at about one-half of the helium BIV, reducing the field stress correspondingly. Therefore, by careful choice of imaging conditions a wide range of materials have been successfully imaged.

13.3.3. Field Evaporation

Field evaporation is more than an interesting phenomenon, it is a vital link in the chain of phenomena that make FEM, FIM, and AP techniques possible. Field evaporation has been studied and made use of since the early days of FIM, but it is still not well understood. Although not quantitatively

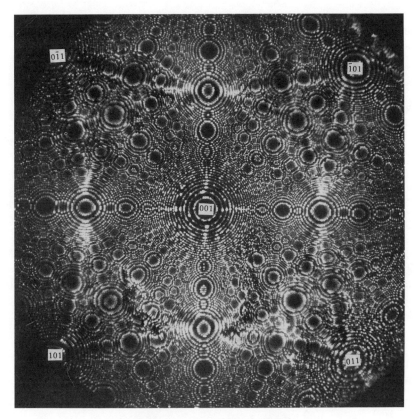

Figure 13.12. FIM image of (001) Pt.

precise, Mueller's original "image force" theory has the virtue of simplicity and consistency with the model of field ionization, while correctly predicting many experimental observations. In the image force model a neutral species is imagined to be desorbed from the surface, and ionized at the critical distance x_c given by Eq. (13.10). The physical considerations inherent to this model are as follows. Removing an atom from its surface site requires a sublimation energy, Λ, assumed to be site independent. At the critical distance, ionization occurs and this requires an additional energy, I_n, equal to the nth ionization potential of the substrate atom. Since the ion now finds itself in an electric field, it can be polarized, which requires an additional energy, $(\alpha_a - \alpha_i)F^2/2$, where α_a and α_i are the polarizabilities of the neutral atom and the ion, respectively. Finally, electrons from the atom must travel back to the metal by tunneling, which yields an energy gain $n\phi$, where ϕ is a crystallographically dependent work function and n is the charge state of the ion. The total energy corresponding to the sum of these contributions is considered the binding energy, Q, so that

$$Q = \Lambda + I_n - n\phi + (1/2)F^2(\alpha_a - \alpha_i) \tag{13.14}$$

The last term in this equation is usually ignored largely because the polarizability of the surface atom, α_a, is not known.

In the presence of an applied field, $-F(ne)x$, the energy barrier is lowered by the amplitude of the "Schottky hump," the magnitude of which can be found by considering the potential V experienced by an ion of charge ne near the surface:

$$V = -F(ne)x - (ne)^2/4x \tag{13.15}$$

where the last term is the same classical image potential used in the Fowler–Nordheim theory of field electron emission. Setting the derivative of Eq. (13.15) equal to zero gives $(n^3e^3F_d)^{1/2}$, which reduces the barrier to desorption accordingly. If we ignore the polarization terms, the activation energy for field desorption (evaporation) is then given by

$$Q = \Lambda + I_n - n\phi - (n^3e^3F_d)^{1/2} \tag{13.16}$$

where F_d denotes the desorption field. The characteristic time τ for thermal activation of the Q is of the form

$$\tau = \tau_0 \exp[-Q/kT] \tag{13.17}$$

where τ_0 is the reciprocal of the attempt frequency of the atom bound at temperature T. Combining Eqs. (13.16) and (13.17) then gives the desorption field strength:

$$F_d = (\Lambda + I_n - n\phi - kT \ln\tau/\tau_0)^2/n^3e^3 \tag{13.18}$$

where the last term is usually negligible.

It is necessary to assume that multiplied charged ions are field-evaporated in order to achieve reasonable agreement with experiment. Indeed, Brandon (1965) found that the calculated evaporation field for most metals was a minimum for $n = 2$. Barofsky and Mueller (1968, 1969) supported the conclusion that $n = 2$ for iron and nickel, but for other materials, they found the charge state depended critically on emitter temperature. Subsequently, it was proposed that normal desorption may be of a singly ionized species but a higher charge state is acquired by subsequent tunneling of one or more electrons to the metal, a process which is field and temperature dependent (and probably also depends on local atomic structure) (Ernst, 1979; Haydock and Kingman, 1980). The theory of field evaporation, particularly the charge state of the evaporated ions, is particularly important in using AP techniques, which are considered separately. However, the field evaporation process itself is important in FIM itself since it establishes the final end form of most specimens.

13.3.4. Specimen Preparation and Image Development

Chemical or electrochemical etching is normally used to prepare a FIM (or FEM) specimen, but before final imaging the specimen is always field-evaporated. It is this final step in specimen preparation that is the most common application of field evaporation. Upon initial insertion of the specimen into a FIM (or FEM), immediately following the usual chemical etch or electrochemical polish, the imaging surface is still rough and contaminated at the atomic scale (Fig. 13.13). Since the local field strength varies as the reciprocal of the local radius of curvature, sharp protrusions are field-desorbed (or evaporated) first. The emitter is thus gradually shaped into a stable end form by

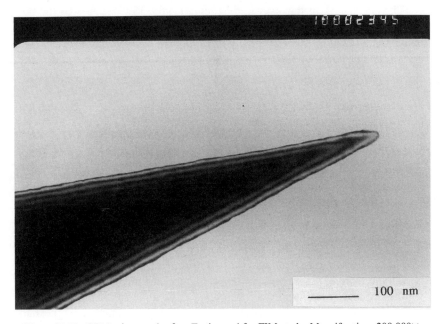

Figure 13.13. TEM micrograph of an Fe tip used for FIM study. Magnification: 200,000×.

the condition that a constant field evaporation rate exists everywhere on the imaging surface. This stable end form varies with a number of the initial conditions, such as the gross specimen shape including the shank, the imaging gas, the specimen temperature, and even which reaction background gases are in the vacuum chamber. In addition, the inherent properties of the specimen (such as orientation, composition, defect state) are also critical. With all of these variables, it is no wonder that precise agreement between a quantitative theory of field evaporation and experimental data does not yet exist. Fortunately, specimen shapes vary through a fairly narrow range usually approximated well by a faceted hemisphere perched onto a cone of small included angle. Note that field evaporation is a unique means for cleaning surfaces, since it can be achieved at cryogenic temperatures in ultrahigh vacuum and leaves atomically smooth facets. It is also worth pointing out that a precise shape usually can be determined by a combination of FIM imaging and subsequent measurements in a transmission electron microscope. Specimen shapes may also be altered by heating, as is sometimes done before FEM studies, but thermal treatments also alter surface and internal atomic structures or microstructures, so they are not always useful.

13.3.5. The Atom Probe

The early work of Barofsky and Mueller (1968, 1969) suggested the possibility of using field evaporation as a unique mass spectrometer capable of identifying a small number of ions. By making the entrance aperture of such a spectrometer equal to the area of a single magnified field-ion image spot, the ultimate microanalytic capability should be possible. A magnetic sector spectrometer would not do because of its limited mass range at any setting. However, time-of-flight (TOF) techniques could be used to overcome this difficulty, since they have a wide mass range and could discriminate between intentionally field-evaporated ions (using a sharp initial pulse) and random events, such as those gaseous ions generated to form a FIM image. An early version of such an instrument is shown in Fig. 13.14. The instrument works as follows. The FIM image of a single atom is positioned over the entrance aperture of a TOF mass spectrometer by manipulating the tip. Field evaporation is initiated by applying a short high-voltage pulse superposed on a constant imaging voltage (BIV). The voltage pulse amplitude should be chosen to field-evaporate only a few atoms from the entire surface layer. In general, these ions will not pass through the aperture, but strike the channel plate assembly in front of the screen. However, by repeated pulsing the atom placed over the probe hole eventually will field-evaporate and pass through the probe hole to the detector. This procedure makes the AP very inefficient (perhaps $1:10^3$ or 10^4). Partly for this reason, the probe hole on the screen is made

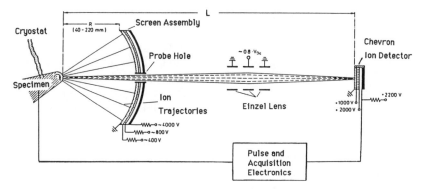

Figure 13.14. Schematic diagram of a linear time-of-flight atom probe.

somewhat larger than the image of a small atom so that one of several neighboring ions may pass through the probe hole.

During the voltage pulse, the field-evaporated ions acquire a kinetic energy:

$$mv^2/2 = neV \qquad (13.19)$$

where V is the total potential and $V = V_{dc} + V_{pulse}$. The steady dc bias makes the image visible, so V_{dc} is usually 80–85% of that required for slow field evaporation. In practice, V pulse must have a rapid rise time and be sufficiently narrow to define the instant of field evaporation.

Because of the intense gradient in an electric field, the field-evaporated ions acquire their full kinetic energy within a few tip radii of the surface. A grounded aperture is normally placed close to the tip to help spatially define the position of the field evaporation event. Ideally, the pulse duration should be just long enough to ensure that the ions pass through the aperture before the pulse terminates. Once past the aperture, the ions enter a field-free drift region of length D en route to the detector. Between the specimen and detector, the ions encounter the fluorescent screen of the field-ion microscope with a small probe hole. Only those surface species whose images are positioned over the probe hole will continue their drift to the detector. If t is the drift time from tip to detector, the mass-to-charge ratio of the ion will be given by

$$m/n = 0.193(V_{dc} + V_{pulse})t^2/D^2 \qquad (13.20)$$

where m/n is expressed in atomic mass units (amu), the sum $(V_{dc} + V_{pulse})$ is in kilovolts, t is in microseconds, and D is in meters for the constant indicated.

A typical mass resolution of $\Delta m/m \approx 1\%$, or better, can be achieved over the mass range 1–100 amu, with the simple TOF AP described. Mass resolution is limited by the spread in kinetic energy of the field-evaporating ions, which can be as large as several hundred electron volts. This energy spread is almost entirely the result of the voltage pulse itself (rise time, pulse shape, and duration) which affects the exact time of evaporation. Very significant improvements in mass resolution are possible without loss in detection efficiency by incorporating energy-compensating electrostatic lenses in the drift region. Mass resolutions better than 1:2000 are common in commercial instruments and 1:20,000 or better has been reported when very elaborate calibrations are made (Fig. 13.15).

Field desorption is a thermally activated process, so field evaporation can be initiated by a short-duration thermal pulse from a laser. Since the tip could be kept at a constant dc potential, the energy spread of the resulting ions should be on the order of kT. Mass resolutions better than 1:5000 can be obtained from this improvement alone. An additional benefit of laser-induced field evaporation is that insulators or high-purity semiconductors can be studied. Otherwise, the resistance of the tip, and its capacitance precludes

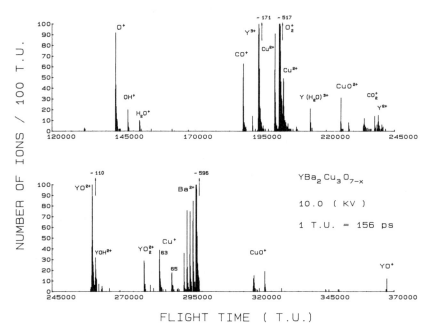

Figure 13.15. A mass spectrum of $YBa_2Cu_3O_{7-x}$ obtained from a high-resolution pulsed-laser time-of-flight atom-probe FIM.

applying the narrow voltage pulses (tens of nanoseconds wide) needed to ensure adequate mass resolution. However, one drawback of laser pulsing is that the surface temperature is elevated during the desorption event and atomic mobility prior to field evaporation may alter the surface structure and/or local composition.

Irrespective of the method of field evaporation, three distinct kinds of measurements can be made with an AP. (1) Determination of the relative abundances of surface species; that is, of those atoms which are weakly bound, such as residual gases. Desorption of this layer is assumed to be possible without disturbing the substrate, although some displaced substrate atoms may be desorbed. (2) The abundance of several species within each layer can be determined by field-evaporating the substrate layer by layer, and a depth profile can be obtained with a depth resolution of one atomic layer. The volume analyzed corresponds to the effective diameter of the probe hole (usually about 2 mm at the screen or 2 nm at the tip for $10^6 \times$ magnification) and the number of layers removed. (3) The average composition for any near-surface volume desired can be determined by summing the data layer by layer in a depth profile. Such data are useful for comparison with other microanalytical techniques (e.g., electron microprobes).

13.3.6. Field Desorption Studies and the Imaging Atom Probe

The development of areal detectors sensitive to single ions, such as the double channel plate (Chevron), made the acquisition of desorption images possible. Repeated adsorption and desorption of gases can yield respectable images of the surface comparable to an FIM image, but at lower applied fields (Fig. 13.16). A camera can be used to integrate field desorption events until sufficient data are recorded to form a satisfactory image of the underlying substrate. Even shadow images of biological molecules adsorbed to a metal substrate have been acquired by uniformly depositing gaseous adsorbates with weak binding to both molecule and substrate and then desorbing them. However, these images remain difficult to reproduce and overall comprise extremely time-consuming experiments. A simpler and more widespread application comes from desorbing (field-evaporating) the substrate atoms themselves with a sharp voltage or laser pulse and placing a Chevron a relatively short distance away (e.g., 10–15 cm). In effect, this device is a very-low-mass-resolution TOF AP ($\Delta m/m \approx 30$–50) with no probe hole and/or a wide collection angle (Fig. 13.17). Compared with a probe hole device the effective collection area is greatly increased (several thousandfold), but at the cost of mass resolution.

An unanticipated benefit of examining such a large surface area was that a species with small abundance still produced an easily identified mass peak. Another advantage was that even species which did not survive field-ion im-

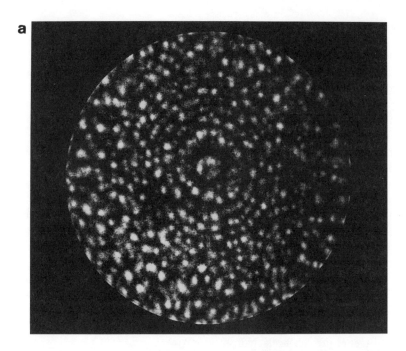

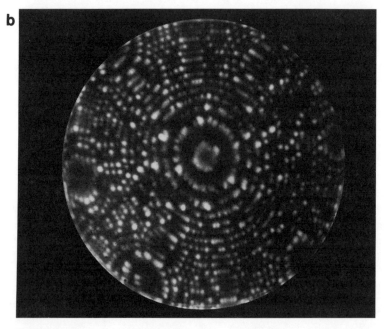

Figure 13.16. Field desorption and field ion images detected from the same tungsten (110) plane: (a) field desorption image; (b) field ion image.

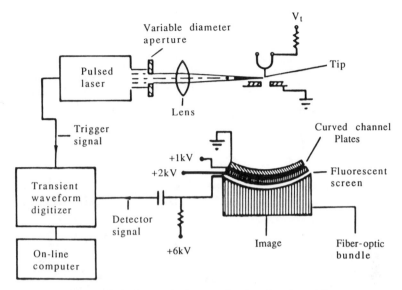

Figure 13.17. Schematic diagram of an imaging atom probe.

aging fields could be identified by recording their mass spectrum and characteristic desorption fields. In fact, such desorption images can be used to deduce the distribution of adsorbates on the surface when they are compared with subsequent FIM images of the substrate. However, the abundance of an adsorbed species is usually small and there is more than one species present, which desorbs simultaneously. As a result, a desorption image often may display the locations of several indistinguishable species. One solution is to activate the Chevron detector only over a time increment corresponding to the expected arrival time of one species ("time gating"). A complication is that the distance from the tip to the detector is so small that a flat Chevron results in a noticeable difference in flight paths for the same species arriving near an edge compared with the center. In turn, such widely differing paths result in significant overlap in arrival times. Panitz (1974) overcame this difficulty by using a spherically curved Chevron, which when combined with ultrafast electronics, provided acceptable mass resolution ($\Delta m/m \approx 50$ and even better). Such an instrument is now called an imaging atom probe (IAP) and can be used to determine the crystallographic distribution of either adsorbed surface species or one species of the underlying lattice over the entire imaged area (Fig. 13.18). The IAP is now used routinely in a complementary manner with the FIM and/or a high-mass-resolution AP with a probe hole all in the same chamber.

The concept of the IAP has been carried one step further by Cerezo *et al.* (1988), by substituting a position-sensitive particle detector (such as a

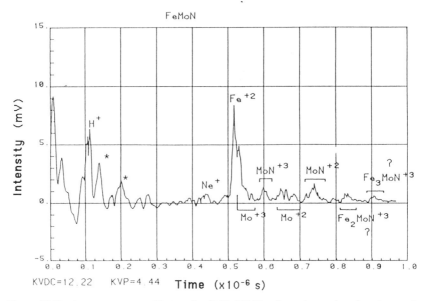

Figure 13.18. A mass spectrum of internally nitrided FeMo alloy using an imaging atom probe.

resistive anode) for the Chevron. The result is a capability to determine both position and flight time for each field-evaporated ion over the entire image. Because the detection and encoding times are relatively long, only one event can be recorded at a time from the entire field. Nonetheless, a three-dimensional map of all species can be reassembled and color-coded to produce a spatial distribution of several species. Mass resolution is reasonable because the flight paths can be corrected after position is recorded and laser pulsing reduces the energy spread. Ultimately, a better spatial detector (i.e., faster and finely segmented into independent pixels) will permit rapid collection of three-dimensional compositional profiles.

Such an instrument could be envisioned as the ultimate FIM/AP and FEM, if the attainable vacuums are UHV. Although we cannot yet achieve such performance, we are close.

References

Ashworth, F. (1948). Ph.D. thesis, University of Bristol, Bristol, England.
Barofsky, P. F., and Mueller, E. W. (1968). *Surf. Sci.* **10**, 177.
Barofsky, D. F., and Mueller, E. W. (1969). *Int. J. Mass Spectrom. Ion Phys.* **2**, 125.
Brandon, D. G. (1965). *Surf. Sci.* **3**, 1.

Brodie, I. (1978). *Surf. Sci.* **70**, 186.
Cerezo, A., Godfrey, T. J., and Smith, G. D. W. (1988). *Rev. Sci. Instrum.* **59**, 862.
Ernst, N. (1979). *Surf. Sci.* **87**, 469.
Eyring, C. F., Mackeown, S., and Millikan, R. A. (1928). *Phys. Rev.* **31**, 900.
Fowler, R. H., and Nordheim, L. W. (1928). *Proc. R. Soc. London Ser. A* **119**, 173.
Gadzuk, J. W., and Plummer, E. W. (1973). In *Solid State Surface Science* (M. Green, ed.), Vol. 3, Dekker, New York.
Giaever, I. (1972). *Surf. Sci.* **29**, 1.
Haydock, R., and Kingman, D. R. (1980). *Phys. Rev. Lett.* **44**, 1520.
Ingram, M. G., and Gomer, R. J. (1954). *J. Chem. Phys.* **22**, 1279.
Kuyatt, C. E., and Plummer, E. W. (1972). *Rev. Sci. Instrum.* **43**, 108.
Mueller, E. W. (1936a). *Z. Phys.* **37**, 838; **102**, 734.
Mueller, E. W. (1936b). *Z. Tech. Phys.* **17**, 412.
Mueller, E. W. (1937). *Z. Phys.* **106**, 132.
Mueller, E. W. (1950a). *Naturwissenschaften* **14**, 333.
Mueller, E. W. (1950b). *Z. Naturforsch.* **5a**, 473.
Mueller, E. W. (1953). *Ergeb. Exakten Naturwiss.* **27**, 290.
Mueller, E. W. (1956). *Z. Naturforsch.* **11a**, 87.
Mueller, E. W. (1956). *J. Appl. Phys.* **27**, 474.
Oppenheimer, J. R. (1928). *Phys. Rev.* **31**, 67.
Panitz, J. A. (1974). *J. Vac. Sci. Technol.* **11**, 206.
Rose, D. J. (1956). *J. Appl. Phys.* **27**, 215.
Swanson, L. W., and Bell, A. E. (1973). *Adv. Electron. Electron Phys.* **32**, 193.
Swanson, L. W., and Crouser, L. C. (1966). *Phys. Rev. Lett.* **16**, 389; (1966). *Phys. Rev. Lett.* **19**, 1179 (1967); *Phys. Rev.* **163**, 622 (1967).

General Reference Books

Gomer, R. (1961). *Field Emission and Field Ionization,* Harvard University Press, Cambridge, Mass.
Hren, J. J., and Ranganathan, S. (1968). *Field-Ion Microscopy,* Plenum Press, New York.
Miller, M. K., and Smith, G. P. W. (1989). *Atom Probe Microanalysis: Principles and Applications to Materials Problems,* Materials Research Society.
Mueller, E. W., and Tsong, T. T. (1969). *Field-Ion Microscopy: Principles and Applications,* American Elsevier, New York.
Tsong, T. T. (1990). *Atom-Probe Field-Ion Microscopy,* Cambridge University Press, London.
Wagner, R. (1982). *Field-Ion Microscopy in Materials Science,* Springer-Verlag, Berlin.

14

Scanning Probe Microscopy

D. A. Grigg and P. E. Russell

14.1. Introduction to Scanning Probe Microscopy

Several techniques have been developed over the past decade using near-field effects to obtain high-resolution microscopic images and surface microanalysis. These techniques are referred to as scanning probe microscopies (SPM). In each of these microscopies, a sharpened probe is used to measure a specific interaction between the probe and the surface of interest. The instrument which began this new and expanding field of microscopy is the scanning tunneling microscope (STM) invented in 1981 by Binnig, Rohrer, and co-workers (1982a–c). The STM measures the electronic interaction between a conductive or semiconductive probe and sample which allows the direct imaging of geometric and electronic surface structure with atomic resolution. Several reviews of the STM literature have been published (Golovchenko, 1986; Hansma and Tersoff, 1987; Binnig and Rohrer, 1987; Kuk and Silverman, 1989; Demuth *et al.,* 1988; Griffith and Kochanski, 1990).

Since the invention of the STM, other near-field interactions between a sharpened probe and sample have been utilized to image a variety of materials with nanometer to subnanometer resolution. The scanning force microscope

D. A. Grigg and P. E. Russell • Department of Materials Science and Engineering, North Carolina State University, Raleigh, North Carolina 27695.

Microanalysis of Solids, edited by B. G. Yacobi *et al.* Plenum Press, New York, 1994.

(SFM) was developed immediately after the STM and is considered the cousin of the STM (Baro *et al.,* 1984). In the SFM, various forces such as van der Waals, adhesive, and capillary forces are measured between the probe and sample. By scanning across a sample at constant force, the topography of both conductive and nonconductive samples can be measured.

In this chapter we describe the history behind the development of SPM and the physics and mechanics of their operation. Examples will be shown from the literature which mark various developmental steps. Also, several other SPM techniques will be reviewed, particularly the SFM.

14.2. The Ultramicrometer

In 1966, Russell Young developed a position sensor, the field emission *ultramicrometer,* which utilized the gap sensitivity of a probe field emitting to a flat metal surface. Young reported that this sensor was capable of reproducing measurements of 10^{-5} to 10^{-8} m gap to within one part in 10^4. Figure 14.1 schematically shows the setup of the ultramicrometer. The tungsten probe can be seen facing a tantalum ribbon which was held in an arch shape by two stainless steel strips. The stainless steel strips were attached to 1-mm-diameter tungsten rods forming a bimetallic strip which bent toward the emitter when heated, thus forcing the tantalum ribbon away from the emitter. Experiments were run in ultrahigh vacuum (UHV). During baking, the tantalum ribbon would flex away from the emitter. The tungsten emitter was also attached to two tungsten rods so that during thermal excitation, relative motion between the emitter and Ta strip would be caused only by the bending of the Ta strip. Measurements were started by adding liquid nitrogen to the cold finger. This started the process of drawing the Ta ribbon closer to the emitter. While the ribbon approached the emitter, measurements of emitter current versus bias voltage were obtained.

Figure 14.2 shows theoretical values of detectable electrode displacement versus emitter–anode spacing. Note that the sensitivity increases as emitter–anode spacing decreases, and the results are not strongly affected by the radius of the emitter, r_0. Conditions are such that the Fowler–Nordheim theory of field emission applies (Fowler and Nordheim, 1928).

To understand the operation of the ultramicrometer, one must first understand the concepts of field emission. When a sharp point is placed opposite an anode in UHV and a negative potential of a few kilovolts is applied, field emission of electrons from the tip to the anode occurs. The emitted current is expressed by the simplified form of the Fowler–Nordheim expression (Fowler and Nordheim, 1928):

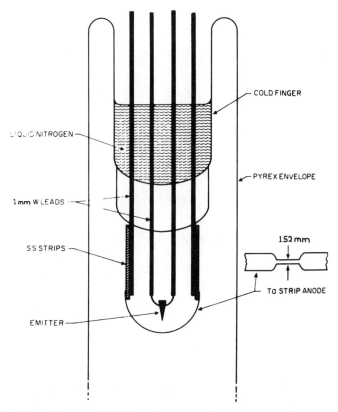

Figure 14.1. Enlarged drawing of the field emission ultramicrometer sensor. The evacuated glass envelope was sealed off from the vacuum system after flashing a Ta getter. (Young, 1966)

$$J = \left(\frac{1.54 \times 10^6 F}{\phi}\right) \exp\left(\frac{-6.83 \times 10^7 \phi v(y)}{F}\right) \qquad (\text{A/cm}^2) \qquad (14.1)$$

$$i = JA \qquad [\text{A}]$$

$$V = \alpha F \qquad [\text{V}]$$

where J is the emission current density, F is the field strength in V/cm, ϕ is the work function of the emitting surface in eV, $v(y)$ is a slowly varying function of ϕ and F, i is the total emitted current, A is the effective area of the emitter, V is the applied potential, and α is the proportionality constant between F and V. Differentiating the above expression results in an expression for the fractional change in applied voltage for a fractional change in emitted current. Using typical parameters for a tungsten emitter ($\phi = 4.5$ eV, $F = 3 \times 10^7$ V/cm):

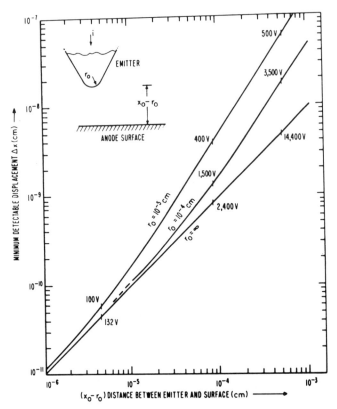

Figure 14.2. Theoretical sensitivity of the ultramicrometer. The emitter is assumed to be tungsten. $dV/V = 10^{-5}$; $F_o = 3 \times 10^7$ V/cm; and for $r_o = 10^{-5}$ cm, $J > 100$ A/cm^2 and $i > 10^{-8}$ A. (Young, 1966)

$$\frac{dJ}{J} = \left(2 + \left(\frac{48 \times 10^7}{F}\right)\right)\left(\frac{dF}{F}\right) \tag{14.2}$$

or

$$\frac{di}{i} = 18\left(\frac{dV}{V}\right) \tag{14.3}$$

Laplace's equation can then be used to solve for the variation of the field between the emitter and anode. The use of Laplace's equation roughly results in a field at the emitter which varies inversely as the 3/2 power of the distance from the emitter surface,

$$F(x) = F_0 \left[\frac{r_0}{x}\right]^{3/2} \tag{14.4}$$

Integrating the field from the emitter surface to the anode results in

$$V(x_0) = -2(r_0)^{3/2}F_0\left(\left(\frac{1}{\sqrt{r_0}}\right) - \left(\frac{1}{(x_0)^{3/2}}\right)\right) \tag{14.5}$$

where r_0 is the radius of curvature of the emitter, x_0 is the distance from the emitter center of curvature to the anode, and F_0 is the constant field strength at the emitter surface associated with a constant current.

In Fig. 14.3, the results of experiments with the ultramicrometer are shown. In the experiment, the apparatus was first equilibrated to room temperature. Then, liquid nitrogen was quickly added to the cold finger to start the process of decreasing the gap distance. A constant-current power supply was used between the emitter and anode, set to approximately 4.5×10^{-7} A.

Therefore, the voltage necessary to maintain a constant current between the emitter and anode was recorded versus time, i.e., gap distance. On the left side of the graph in Fig. 14.3 is a rough scale of emitter–anode spacing. The curve fits the Fowler–Nordheim theory to within 0.3%.

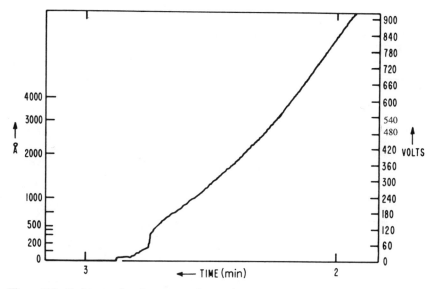

Figure 14.3. Emitter anode voltage versus time as the sensor was cooled to liquid nitrogen temperature. (Young, 1966)

In reporting this technique of position sensing, Young (1966) was already alluding to its use as a topographic mapping instrument. This can be seen in Fig. 14.4 where Young described several possible uses for his ultramicrometer.

14.3. The Topografiner

In 1971, Young constructed the *topografiner* (from the Greek word *topographein*, "to describe a place") and reported the first observations of metal–vacuum–metal (MVM) tunneling (Young *et al.,* 1971) and imaging (Young, 1971). Figure 14.5 shows a schematic representation of the topografiner. X and Y piezo drivers scan the emitter parallel to the specimen surface. The voltage between the emitter and specimen was determined by the constant current passing between them, and their spacing. The Z piezo was controlled by a servo system that holds the voltage, and hence the emitter–specimen spacing, constant. Thus, the Z piezo voltage gave the surface profile directly. The electron multiplier was used to collect secondary electrons and photons emitted from the specimen when the field-emitted electrons strike the surface.

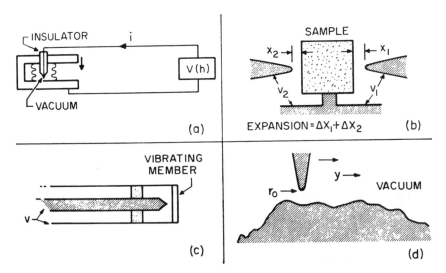

Figure 14.4. Applications of field emission ultramicrometer: (a) strain gauge; (b) differential thermal expansion cell for small samples; (c) mechanical vibration sensor; (d) surface profile delineator. (Young, 1966)

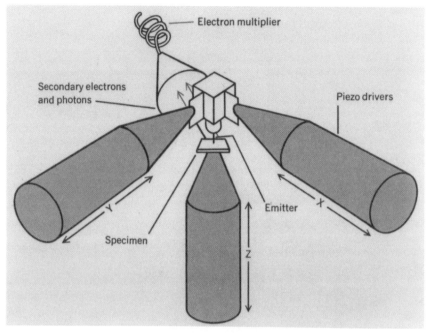

Figure 14.5. The topografiner, shown very schematically. (Young, 1971)

This mode of operation was similar to that of the scanning electron microscope (SEM).

Figure 14.6 shows a more detailed schematic of the topografiner reported in 1972 by Young *et al.* The differential screw was used as a coarse adjustment to bring the specimen close enough to the emitter so that it was within the range of the vertical (Z) piezo. The X-scan piezo deflected the emitter support post so as to scan the emitter in one direction. The orthogonal (Y) piezo is not shown. The specimen was clamped between copper blocks to permit heating of the sample while in UHV.

The topografiner was first used by Young *et al.* (1971) to describe the transition from field emission to MVM tunneling. MVM tunneling is described when electrons from the emitter tunnel to states just above the Fermi energy in the anode material. An expression derived by Simmons in 1963 described this process. Figure 14.7 shows the potential energy diagram defining the parameters used in Simmons's theory. The expression of Simmons describes the current density versus voltage over the full range from close-spaced MVM to field-emission tunneling:

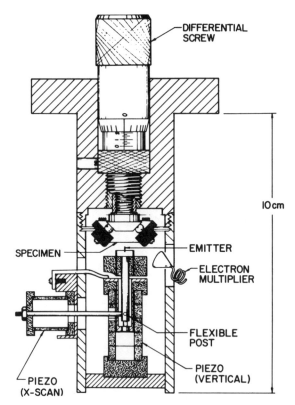

Figure 14.6. Detailed schematic of the topografiner. (Young *et al.*, 1972)

$$J = \left(\frac{e}{2\pi h(\beta \Delta S)^2}\right) [\phi \exp(-A\sqrt{\phi}) - (\phi + eV) \exp(-A\sqrt{\phi + eV})] \quad (14.6)$$

where

$$A = \frac{4\pi \beta \Delta S \sqrt{2m_e}}{h}, \qquad \beta = 1, \qquad \phi = \frac{1}{\Delta S} \int_{S_1}^{S_2} \phi(x)\, dx \quad (14.7)$$

$\phi(x)$ is the potential energy of an electron between the two metal surfaces. S_1 and S_2 are the distances from the first surface to the place where the potential energy equals the Fermi energy near surface 1 and 2, respectively. $\Delta S = S_1 - S_2$, m_e is the electron mass, and V is the potential between the two electrodes. For low-voltage (MVM) range ($eV \ll \phi$),

$$J = \frac{\sqrt{2m_e}}{\Delta S} (e/h)^2 \sqrt{\phi}\, V \exp(-A\sqrt{\phi}) \quad (14.8)$$

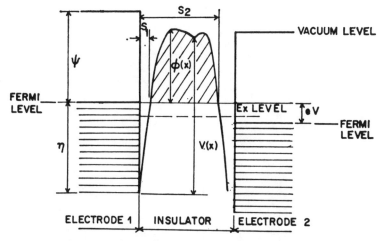

Figure 14.7. Potential energy diagram for an electron between two conducting electrodes (Simmons, 1963). In MVM tunneling the insulator is vacuum.

which gives the expected linear dependence of current density on applied voltage and the exponential dependence on gap distance, i.e., A.

In the Fowler–Nordheim (FN) region, the potential barrier is lowered by increasing the applied voltage, resulting in an exponential increase in current density. The current is not affected by the density of states on the vacuum side of the barrier since all of the states are available. However, in MVM tunneling, there is only a narrow range of available states which causes the current density to increase linearly with applied voltage. Changes in the potential barrier have little effect. Therefore, the two regions are dominated by two different processes. In Fig. 14.8, the calculated voltage necessary to maintain a constant current density of 50 A/cm^2 versus distance is plotted. The FN region is shown as well as the transition region and MVM region predicted by Simmons. The dashed line shows experimental observations by Young *et al.* (1971).

By fixing the emitter at a prescribed height above the surface of the anode and acquiring current (I) versus voltage (V) curves, the three regions could be defined more quantitatively. This was repeated for various gap separations. The results are shown in Fig. 14.9. Notice the linear I versus V shape for the MVM curve predicted by the Simmons equation and the exponential dependence for the intermediate and FN curves in the field emission range predicted by FN and Simmons.

Later in 1971, Young reported some of the first images obtained using the topografiner. A topographic map of a 180 line/mm diffraction grating

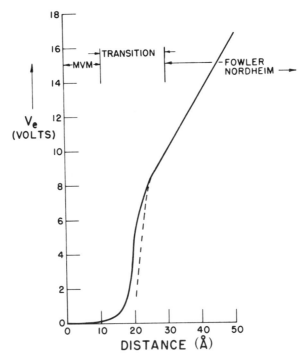

Figure 14.8. Calculated tunnel voltage versus distance for a constant tunneling current density of 50 A/cm². The dashed line depicts attempts to measure the curve experimentally. (Young *et al.*, 1971)

replica obtained with the topografiner is shown in Fig. 14.10a. Figure 14.10b is a map of a disturbed region of the same grating recorded with a memory oscilloscope in place of the X–Y recorder used in (a). The emitter was kept at 200 Å above the surface and the mapping time was approximately 9 min.

Figure 14.11 shows an image acquired using the output of the Channeltron electron multiplier. This work was done in anticipation of using the topografiner as an SEM-type device. When the field-emitted electrons strike the specimen, a large number of secondary electrons are released. The number emitted in a particular direction depends on the slope of the surface, its crystallographic orientation, its work function, and the energy of the incident electrons. Since there is an extremely strong electrostatic field at the surface of the anode (sample), electrons tend to return to the anode, resulting in a strong dependence of the measured yield on the emitter spacing (emitter voltage). The secondary electrons which escaped were collected by a positively biased Channeltron multiplier shown in Fig. 14.6. The number of electrons

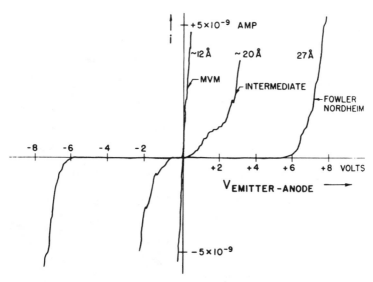

Figure 14.9. Tunneling current versus voltage characteristic for three different emitter-to-surface spacings. (Young *et al.*, 1971)

counted as a function of emitter voltage is shown in Fig. 14.12. When the emitter was scanned over the surface, the output of the Channeltron multiplier was used to modulate the Z signal of an oscilloscope. The size of the electron beam at the surface was believed to be near a few hundred angstroms in diameter. This work has been continued recently in scanning tunneling microscopy to collect secondary signals as well as photon emission signals.

14.4. The Scanning Tunneling Microscope

In 1982, Binnig *et al.* (1982a) announced their first successful tunneling experiments using an apparatus which would soon thereafter be referred to as a scanning tunneling microscope. Previous attempts at MVM tunneling were not adequately founded because of a lack of experimental evidence. Binnig pointed out that it is necessary both to observe an exponential dependence in tunneling resistance and to observe an appropriate work function value for the metals under investigation. Young *et al.* (1971) did show the exponential behavior in the MVM region, but failed to provide adequate measurements of the work function. Both of these measurements are necessary for excluding the existence of ohmic or point contacts or tunneling through contamination layers.

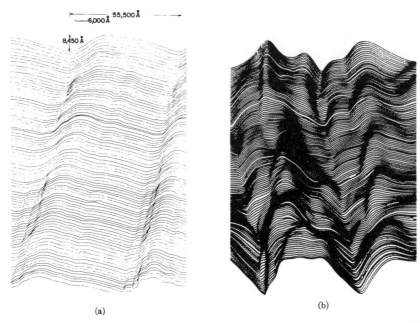

Figure 14.10. (a) Topographic map of 180 line/mm diffraction grating replica. (b) Topographic map recorded on a memory oscilloscope of a disturbed region of the same grating. (Young *et al.*, 1972)

Fowler and Nordheim (1928) predicted that an exponential dependence of the tunnel resistance, $R(s)$, to gap distance, s, would exist as:

$$R(s) \propto \exp(A\sqrt{\phi}\, s) \tag{14.9}$$

which is similar to Simmons's theory (1963). A is given by $(4\pi/h)(2m)^{1/2}$ with m the electron mass in the barrier. Using the free electron mass,

$$A = \frac{1.025}{\sqrt{eV}} \qquad (\text{Å}^{-1}) \tag{14.10}$$

therefore,

$$\sqrt{\phi} \sim \frac{d(\ln R)}{ds} \tag{14.11}$$

This gives a means by which to measure the value of ϕ from curves of resistance versus gap distance. Some typical curves, obtained by Binnig *et al.* (1982a),

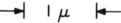

Figure 14.11. Gray-scale image of a Channeltron detector output versus position of the topografiner. (Hansma and Tersoff, 1987)

for the tunneling resistance $R(s)$ between a W probe and a Pt surface are shown in Fig. 14.13. Curves A, B, and C in Fig. 14.13 are resistance measurements made with a moderate vacuum. Curve A was obtained after the first cleaning step, which consisted of putting a 20-Å, 10-kHz dither on the probe, showed a value of $\phi = 0.6$ eV. Curves B and C were after further dithering, but showed no appreciable changes, $\phi = 0.7$ eV. Thus, contamination was believed to be appreciably responsible for the value of ϕ. In a better vacuum (10^{-6} torr), better values of ϕ were obtained; $\phi = 3.2$ eV, closer to that expected for clean Pt and W surfaces [$\phi_{clean} = 1/2(\phi_{Pt} + \phi_W) \sim 5$ eV]. This can be seen in curves D and E.

The apparatus first used by Binnig *et al.* (1982a) is schematically shown in Fig. 14.14. The crucial point for MVM experiments is to suppress vibrations, which are mainly responsible for the failure of previous MVM tunneling experiments. Binnig *et al.* emphasized that this requires excellent mechanical decoupling of the tunneling unit from its surroundings. This was found later to be only half of the problem, since a very stiff mechanical loop between the probe and sample can be made to reject a large portion of the vibration spectrum. Their decoupling was achieved by suspending the tunneling unit by magnetic levitation.

In Fig. 14.14, the vacuum tunneling junction consisted of a platinum plate and a tungsten probe. The tungsten probe was fixed to a support A, which could travel in a semicontinuous fashion in any direction on the bench B. The driving mechanism consisted of a piezoplate PP, resting with three metal feet F_1 and F_2 (F_3 not shown) on a metal plate MP, insulated from each

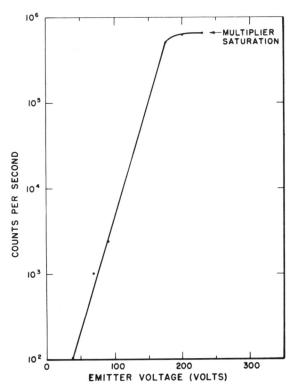

Figure 14.12. Number of secondary electrons detected per second as a function of emitter voltage for a total emission current of about 3×10^{-8} A. (Young *et al.,* 1972)

other by a high-dielectric-constant material D. The feet glided freely on the dielectric, or were clamped in place by applying a voltage between them and the metal plate. Elongation and contraction of the piezo with an appropriate clamping sequence of the feet allowed movements of the support A on the bench in any direction in increments as low as 100 Å. This piezodrive mechanism was given the name *louse.* Piezodrive P allowed fine control of the position of the Pt plate relative to the W probe with a gain of 2 Å/V. The vacuum chamber which housed the system could be pumped down to 10^{-6} torr and rested on inflated rubber tubes for an initial stage of vibration isolation. For further vibration isolation, the tunneling unit was mounted on strong permanent magnets M levitated on a superconducting bowl of lead, Pb, superinsulated and cooled directly by liquid He.

With this apparatus, the first topographic maps of a surface on an atomic scale were obtained. The first surfaces imaged were (110) $CaIrSn_4$ and Au.

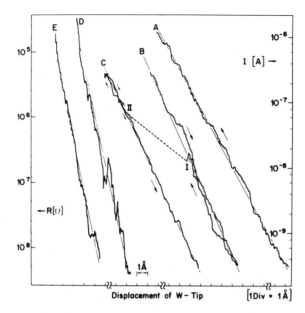

Figure 14.13. Tunneling resistance and current versus displacement of a W probe and a Pt plate for different surface conditions. (Binnig *et al.*, 1982a)

Flux-grown CaIrSn$_4$ crystals were etched in HCl which was believed to stop at the Ir layers. This left an inert surface which was desirable for moderate vacuum levels of 10^{-6} torr. Figure 14.15a shows an STM picture of a (110) surface of CaIrSn$_4$ obtained with their STM at room temperature. On the left side of the STM image, the edge of a spiral growth can be seen. This spiral-type growth has been seen in SEM and optical microscopes. To the right of the spiral growth are single, double, and triple atomic steps. Figure 14.15b shows single line scans over these steps, indicating the single, double, and triple step positions.

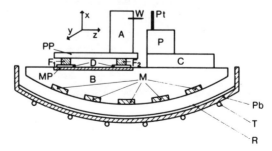

Figure 14.14. Schematic of the tunneling unit and magnetic levitation system used by Binnig *et al.* (1982a).

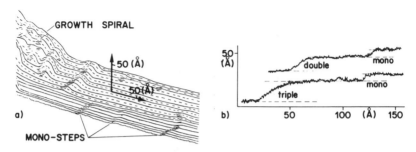

Figure 14.15. Topography of a CaIrSn$_4$(110) surface. (a) Overall view of flat part with single atomic steps (right) and start of a growth spiral. (b) Two individual scans, exhibiting triple, double, and single atomic steps. (Binnig *et al.*, 1982c)

The Au sample was Ar sputtered and subsequently annealed at 600°C in (2–7) × 10^{-10} torr [a standard procedure for inducing reconstructions of Au (110) surfaces]. The surfaces exhibited gentle corrugation in the [001] direction as shown in Fig. 14.16a. Figure 14.16b shows double and single atomic steps observed after rapidly cooling the sample to room temperature after a 300°C anneal.

Later, in 1982, Binnig and Rohrer rebuilt their STM apparatus to go into a UHV chamber. The main change in the apparatus was to suspend the tunneling unit with springs instead of magnetic levitation for vibration isolation. Figure 14.17 shows this new configuration.

With the STM in a UHV environment, surface studies on a subangstrom scale could be performed. Clean surfaces allowed the first STM images of the atomic structure of Si (Binnig *et al.*, 1983). These images were the first true

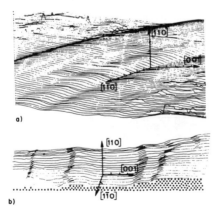

Figure 14.16. Two examples of STM micrographs of a Au(110) surface, taken at (a) room temperature, and (b) 300°C after annealing for 20 h at the same temperature. The scale is 10 Å/div in the figure. (Binnig *et al.*, 1982c)

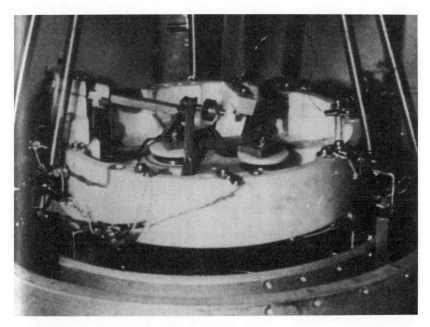

Figure 14.17. Tunneling unit showing piezodrives with a tunneling probe (left) and sample mounted on the "louse." (Binnig and Rohrer, 1982)

examples of the STM's strength in surface science studies and began the SPM revolution.

Figure 14.18 was one of the first images obtained of the Si(111) surface with the STM. The image exhibits monoatomic [−211] steps on a Si(111) surface, up to 50 Å apart. Note that the steps are connected through angles of 60°, which is expected for steps of the same type. STM images were also found having both [−211] and [2-1-1]. This already began to show the importance of the STM to surface scientists. The existence of both types of steps indicated a need to further understand the influence of sample preparation and contamination on Si(111).

Figure 14.19 was the first reported STM image of the famous 7 × 7 reconstruction of the Si(111) surface. The ability to image this surface and to confirm its structure to such detail resulted in Binnig and Rohrer receiving the Nobel prize in 1986 for the STM. The STM image shows two complete rhombohedral 7 × 7 unit cells, clearly bounded by the lines of minima with deep corners. This can be more clearly seen in Fig. 14.20 where a top-down view of the same image is displayed showing the sixfold rotational symmetry. The brighter areas are high positions on the surface and dark areas correspond to low positions. Inside each cell, 12 maxima appear. These were interpreted

Figure 14.18. STM picture of a Si(111) surface exhibiting lines of single atomic step. (Binnig and Rohrer, 1982)

as 12 adatoms, sitting on top of the Si(111) surface in distinct sites. The height of the maxima is uniformly 0.7 ± 0.1 Å with respect to an average level of zero for the unit cell and the depth of the corner minima is 2.1 ± 0.2 Å. Figure 14.21 shows the corresponding adatom model for the 7×7 reconstruction proposed by Harrison in 1979. This imaging of the Si 7×7 reconstruction clearly showed the STM's ability to spatially resolve individual atoms on a surface less than 7 Å apart.

14.5. Instrumentation

The operation of the STM is based on the strong distance dependence of the tunnel current J_T. For two flat, parallel electrodes, J_T is given by

$$J_T \propto (V_T/s) \exp(-A\phi^{1/2}s) \qquad (14.12)$$

where $A \sim 1.025$ (eV)$^{-1/2}$ Å^{-1} for a vacuum gap, ϕ is the average of the two electrode work functions, s is the distance between the electrodes, and V_T is the applied voltage. With work functions of a few electron volts, J_T changes by an order of magnitude for every angstrom change in s. A schematic of the STM is shown in Fig. 14.22. A conductive tunneling probe is securely held onto a piezoelectric mechanism which allows very fine displacement control

Figure 14.19. STM image of the 7×7 reconstruction of Si(111), showing two complete rhombohedral unit cells. (Binnig and Rohrer, 1982)

(~ 10 Å/V). In Fig. 14.22 the tunneling probe is being held to a set of three orthogonal piezoelectric sticks (P_x, P_y, and P_z). The probe can be moved in any direction by applying voltages across the piezoelectric sticks. P_x and P_y are used to raster scan the probe across the sample surface. A voltage V_p is applied to P_z by the control unit CU such that the tunnel current J_T remains constant. Tunneling currents are typically 0.1 nA to 10 nA. Thus, $V_z(V_x, V_y)$ gives the topography of the surface, $z(x, y)$, as illustrated by the surface step at A, i.e., the tunnel probe is moved at constant distance s over the surface. By keeping the tunnel current constant, change in the work function at C induces the apparent surface structure B. Such induced structures, however, can be accounted for by measuring ϕ separately. The signal, $J_s = \Delta(\ln J_t)/\Delta s$ $\sim \phi^{1/2}$, obtained by modulating the tunneling distance s by Δs while scanning, gives ϕ directly, at least in a simple situation as shown in Fig. 14.22. In general, separation of topography and work function is more involved but still possible.

14.5.1. Vibration

Since the probe–sample interaction is highly dependent on the separation distance in the SPM, vibration isolation is a critical factor in the design. For

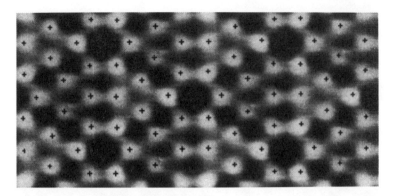

Figure 14.20. Top view of the STM image in Fig. 14.19. Brightness indicates altitude. (Binnig *et al.,* 1983)

many surfaces the atomic corrugations will typically be 0.1 Å, so one must set a vibration isolation goal of better than 0.01 Å.

Figure 14.23 shows a simplified form of a vibration isolation system for an SPM. For a spring and viscous damping system, the resulting vibration amplitude transfer function for a floor-induced vibration input can be described by

$$T = \frac{1 + (2\zeta\nu/\nu_{n})^2}{\sqrt{(1 - \nu^2/\nu_{n}^2)^2 + (2\zeta\nu/\nu_{n})^2}} \qquad (14.13)$$

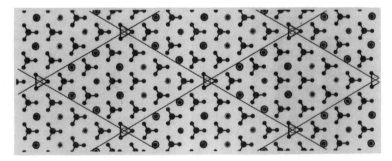

Figure 14.21. Modified adatom model. The underlying top layer atom positions are shown by dots, and the rest atoms with unsatisfied dangling bonds carry circles, whose thickness indicates the depth. The adatoms are represented by large dots with corresponding bonding arms. The empty potential adatom position is indicated by an empty circle in the triangle of adjacent rest atoms. The grid indicates the 7×7 unit cell. (Binnig *et al.,* 1983)

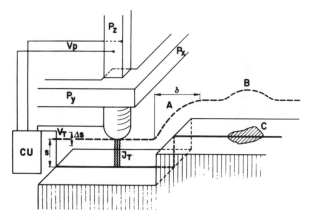

Figure 14.22. Schematic illustration of the principle of operation of the scanning tunneling microscope. (Binnig and Rohrer, 1982)

where ν is the external excitation frequency, ν_n is the resonance frequency, ζ ($=\gamma/\gamma_c$) is the damping ratio, γ is the damping coefficient of the system, and γ_c ($=4m\pi\nu_n$) is the critical damping coefficient. Figure 14.24 shows a family of amplitude transfer curves for the system in Fig. 14.23.

It has been common for many SPM designs to employ several stages of vibration isolation to increase the effectiveness of the isolation system at low frequencies. However, a rigidly constructed SPM does not require many stages of vibration isolation. Vibrations are dissipated by hysteresis loss because of

STM ASSEMBLY

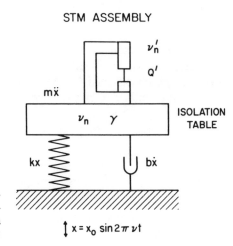

Figure 14.23. Block diagram of an SPM system on vibration isolation system. The isolation system is simplified to a spring and a viscous damper. (Kuk and Silverman, 1989)

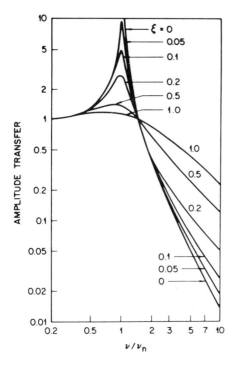

Figure 14.24. Amplitude transfer of one-stage vibration isolation as a function of frequency and damping ratio. (Little, 1973)

the inherent structural damping of a rigid body. The damping transfer function for hysteresis loss, equivalent to amplitude transfer, is given by

$$T^S = \frac{(\nu/\nu'_n)^2}{\sqrt{(1 - \nu^2/\nu'^2_n)^2 + (\nu/\nu'_n Q')^2}} \tag{14.14}$$

where ν'_n is the resonance frequency and Q' is the quality factor of the probe–sample junction. Piezoelectric drivers with resonant frequencies up to 100 kHz can be constructed. Epoxy junctions, joints tightened by screws, and three-point contacts often reduce system resonances to 1–5 kHz. For a system with one-stage vibration isolation and structural damping with $\nu'_n \gg \nu_n$, the resultant transfer function can be expressed by

$$T_{\text{total}} = \sqrt{\frac{1 + (\zeta\nu/\nu_n)^2}{(1 - \nu^2/\nu_n^2)^2 + (2\zeta\nu/\nu_n)^2}} \times \frac{(\nu/\nu'_n)^2}{\sqrt{(1 - \nu^2/\nu'^2_n)^2 + (\nu/\nu'_n Q')^2}} \tag{14.15}$$

The amplitude transfers for four SPMs are shown in Fig. 14.25. The solid line is for a system with $\nu_n = 2$ Hz, $\nu'_n = 2$ kHz, $\zeta = 0.4$, and $Q' = 10$. With

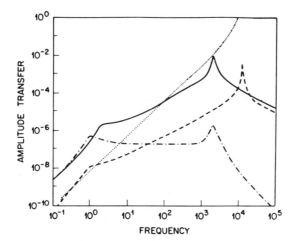

Figure 14.25. Total amplitude transfer for four systems with different vibration isolation and structural dampings. (Kuk and Silverman, 1989)

a floor vibration amplitude of a few thousand angstroms, the gap stability for this system will be worse than 1 Å. For a very rigid system (shown as dotted line where ν'_n = 12 kHz and Q' = 50), the amplitude transfer is still too large for a stable SPM junction. The best case in Fig. 14.25 is the fourth case (shown as a chained line) where a two-stage isolation system was used: one by an external table (ν_n = 1.1 Hz, ζ = 0.5) and another by an internal spring system (ν_n = 1 Hz, ζ = 0.4). The structural damping for the STM in the fourth case is ν'_n = 2 kHz and Q' = 10. In most cases the estimated vibration amplitude is <0.001 Å.

Many applications require large systems capable of handling large samples. As the system increases in its physical size, the resonant characteristics of the system decrease making the system inherently difficult to control. This has become particularly important in the design of scanning probe metrology instruments capable of measuring surfaces on integrated circuit wafers.

14.5.2. Mechanical Designs

It is necessary to fulfill several criteria in the design of an ideal SPM. The basic elements of these criteria are a high mechanical resonance frequency, low Q factor, $x - y$ scan generation of the probe with a range of >1 μm with an accuracy of 0.1 Å, and a z range of >1 μm with an accuracy of ~0.01 Å; also, a mechanism to allow approach of the probe to the surface within the range of the fine z mechanism. The mechanism for reducing the gap to within

z range, known as the coarse approach system, should be smooth and without appreciable backlash, and should move with ~0.1 μm accuracy.

The fine x, y, z motion has been mainly accomplished with the use of piezoelectric elements. Most of the piezoelectric elements used have been of the type PZT [Pb(Zr, Ti)O$_3$]. This material can have gains in the range of 10–10,000 Å/V depending on the material content and the physical shape of the elements. The two types of PZT structures most widely used in SPMs are the orthogonal stick type and the tube type. The orthogonal stick-type scanning element is illustrated in Fig. 14.26. The stick-type piezoelectrics are polarized in the thickness (t) direction. Therefore, by application of an electric field across its thickness, the length changes because of contractions and expansions of the axis orthogonal to the applied field. Displacements on the order of 0.5–1.0 μm are realized with the application of approximately 200–300 V to piezoelectric sticks having a length-to-thickness ratio of approximately 10:1.

In more recent years, tube-type scanners have become widely utilized for SPMs. The tube scanner was first used for STMs by Binnig and Smith in 1986 and was initially considered to be a unique design. However, further investigation revealed that a similar concept had been developed some 30 years earlier by the audio industry as two-channel transducers for use in phonograph cartridges (Germano, 1959). PZT tubes have been recently used to obtain scan ranges as large as 100 μm in x and y and 5 μm in z. The use of tubes instead of sticks reduces the cross talk and raises the resonance frequency of the scanning element (Kuk and Silverman, 1989; Chiang and Wilson, 1986). Figure 14.27 illustrates the basic design of the tube scanner. The PZT tubes are poled radially such that the application of an electric field between the inner diameter electrode and an outer diameter electrode causes the wall thickness of the tube to expand or contract. Thus, because of zero volumetric changes during deformation, the tube also contracts or expands in the axial direction. By separating the outer electrode into four equally spaced quadrants,

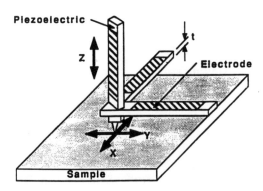

Figure 14.26. Stick-type piezoelectrics used in SPM.

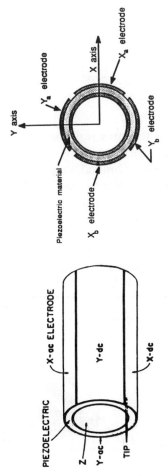

Figure 14.27. (Left) Schematic diagram of a single tube x-y-z scanner. (Right) Axial view of tube scanner design showing electrode placement.

individual sections can be extended or contracted with respect to other sections. Then, by fixing one end of the tube as a cantilever beam, increasing the length of one quadrant of the tube causes the tube to flex, thus deflecting the free end of the tube. Therefore, two mutually orthogonal directions of flexural motion can be created. Also, by applying an equally net positive or negative field to all of the quadrants the tube can be made to move axially for z motion. The resonance frequency in hertz of the single tube design can be estimated by

$$f_r = 1.08 \times 10^5 \left(\frac{\sqrt{r_o^2 + r_i^2}}{l^2} \right) \tag{14.16}$$

where r_i, r_o are the radius of the inner wall and outer wall, respectively, and l is the length of the tube in centimeters.

Since the piezoelectrics have limited range of adjustment in z, $\sim 1~\mu m$, some other mechanism must be used to move the sample and probe within range of the z axis. There have been numerous examples of mechanical systems used to facilitate coarse probe and sample approach. One example of the earlier methods used for coarse positioning was described by Binnig and Rohrer (1982) in their pocket microscope design mentioned earlier. One of the more common sample positioning methods was to use mechanical reduction with lead screws and micrometers. A simple example of this method can be seen in the STM built by McCord and Pease (1987) shown

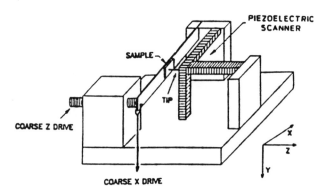

Figure 14.28. Simple mechanical reduction of coarse sample approach designed by McCord and Pease in 1987.

in Fig. 14.28. The coarse mechanical motion in the z direction is provided by a micrometer pushing against a cantilever beam. With the sample mounted approximately halfway between the cantilever end of the beam and the micrometer, the micrometer motion is reduced by a factor of 3.2. Therefore, a micrometer with 1 μm resolution can be used to produce motion of 0.313 μm. This is well within the range of most PZT scanning systems.

Other methods for mechanical demagnification can increase the effective sensitivity of the sample approach by using a method such as that of Smith and Binnig (1986), Fig. 14.29. In this method, the motion of the fine approach adjustment lead screw has been reduced by converting its motion into force through a spring between the lead screw and the center of a beam. The reduction in motion between the lead screw and the sample occurs because of differences in the stiffness of the spring relative to the cantilever beam. For example, by using a cantilever beam with a stiffness 100 times greater than that of the spring, the motion of the lead screw can be reduced by approximately 100 times.

Another popular method used for approaching the probe to the sample over large distances is to use a piezoelectric inchworm. Figure 14.30 illustrates

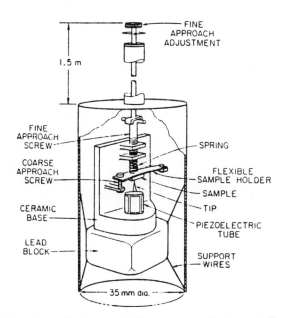

Figure 14.29. Spring demagnification method for coarse sample approach. (Smith and Binnig, 1986)

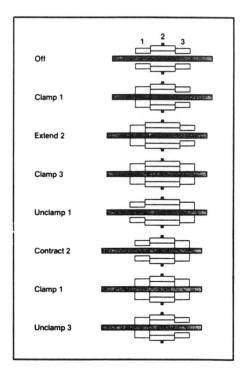

Figure 14.30. Operation of the Burleigh inchworm.

the operation of the inchworm. When a voltage is applied to PZT element 1, it clamps the shaft. Then a variable rate staircase voltage is applied to the center PZT element 2 causing it to change length in discrete steps of approximately 3 nm. The staircase may be stopped or reversed at any step. At the end of the staircase a voltage is applied to PZT element 3 causing it to grip the shaft. Then voltage is removed from PZT element 1 releasing the shaft. The staircase starts downward until it reaches its lower limit at which point PZT element 1 is activated again, PZT element 3 releases, and the staircase starts again. This sequence can be repeated any number of times for a total travel limited only by the length of the motor shaft. Each full extension of the center element moves the shaft about 1 μm. The frequency of the clamp change and extension cycles may exceed 1 kHz, so a maximum speed of 1 mm/s is achieved with a 1.0-kg load capacity.

The inchworm has been used in a number of configurations to approach the sample to the probe, or probe to sample. This type of motor is also very convenient for vacuum conditions since it is made of UHV-compatible materials, can be baked out to 150°C, and only requires standard electrical feedthroughs to operate.

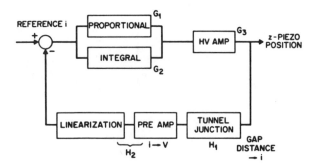

Figure 14.31. Block diagram of STM negative feedback control system. (Kuk and Silverman, 1989)

14.5.3. Electronics

The STM is essentially a servomechanism whose electronic design is strongly influenced by the mechanical design. Figure 14.31 shows the block diagram for a typical STM negative feedback control system. A constant gap separation is maintained by comparing the tunneling current to the set current value. Any difference between the current signal and the set current value will produce an error signal which can then be used in the negative feedback system to compensate the servo position. In linear time-invariant continuous-time systems, the Laplace transform can be utilized in system analysis and design. The Laplace transform is defined as (Phillips and Nagle, 1984)

$$L[f(t)] = \int_0^\alpha f(t)e^{-st}\, dt \qquad (14.17)$$

The resulting transfer function for the feedback system in Fig. 14.31 can be written in the Laplace transform as

$$T(s) = \frac{[G_1(s) + G_2(s)]G_3(s)}{1 + [G_1(s) + G_2(s)]G_3(s)H_1(s)H_2(s)} \qquad (14.18)$$

where G_1, G_2, and G_3 are the transfer functions of the proportional feedback, the integral feedback, and the high-voltage amplifier, respectively. H_1 and H_2 are the transfer functions of the tunneling gap and the preamplifier, respectively. The response function for each component in Laplace transform can be shown in the following form:

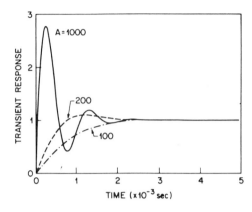

Figure 14.32. Transient response of control system in Fig. 14.31. (Kuk and Silverman, 1989)

$$G_1(s) = \frac{\kappa_p}{(s\tau_p + 1)} \tag{14.19}$$

$$G_2(s) = \frac{\kappa_i}{(s\tau_i + 1)} \tag{14.20}$$

$$G_3(s) = \frac{\kappa_{hv}}{(s\tau_{hv} + 1)} \tag{14.21}$$

$$H_1(s) = \frac{\omega_n^2 \kappa_t}{(s^2 + \omega_n s/Q + \omega_n^2)} \tag{14.22}$$

$$H_2(s) = \frac{\kappa_1}{(s\tau_1 + 1)} \tag{14.23}$$

where κ and τ are the gain and time constant of each system and subscripts p, i, and hv stand for the proportional, integral, and high-voltage amplifier systems, respectively. In the feedback equation for the tunneling junction, H_1, the terms ω_n and Q are the natural frequency and the quality factor of the tunneling assembly, respectively. The quality factor, Q, is related to the damping of the system. For high values of Q, the damping factor is low; therefore, very high amplitudes occur about the resonance of the system. For low values of Q, the damping is high; therefore, the amplitude is reduced substantially near the resonant frequencies. Both ω_n and Q can be measured experimentally.

Typically, this system has three characteristic responses to error compensation: underdamped, critically damped, and overdamped. The feedback electronics are designed such that the gains k_p and k_i can be adjusted for various response speeds and settling times for a variety of systems. Figure 14.32 shows three conditions for gains set at 100 (underdamped), 200 (critically damped), and 1000 (overdamped) for the system in Fig. 14.31 using $\omega_n = 1.2 \times 10^4$ Hz, $Q = 10$, $\tau_1 = 3 \times 10^{-4}$ s^{-1}, and $\tau_i = 0.11$ s^{-1}.

It can be seen in Fig. 14.32 that in the case of underdamping, the rise time is very fast but there is considerable overshoot and the time required for settling is about 2.5 ms. In the case of underdamping, the rise time is very long, but there is no overshoot leading to possible probe crashing. The best case is that of critical damping. The rise time for critical damping is sufficiently high and the settling time is on the order of the rise time, or 1 ms, a factor of 2.5 times better than either of the other cases.

Figure 14.33 shows the basic components used to maintain the probe at a constant height above the sample. The preamplifier converts the current flowing between the probe and sample into a voltage which can be more readily monitored. The current is typically on the order of hundreds of picoamperes to several nanoamperes. The voltage output of the preamplifier is typically between 1 and 10 V. This output can be sent through a logarithmic amplifier whose output is a logarithmic function of the input signal. Such amplification is used because of the exponential relationship between the separation distance and the tunneling current. The logarithmic amplifier may be used, although it is not necessary. For very small corrections in distance from the set current position, the current signal can be closely approximated as a linear signal. This is important because the logarithmic amplifier is typically the slower component within the circuit. Elimination of the logarithmic amplifier can aid in speeding the response time of the acquisition system.

The compensator is built from the above formulation such that a constant current is maintained between the probe and sample, i.e., constant probe-to-sample separation. Deviations from a set current value, predetermined by the

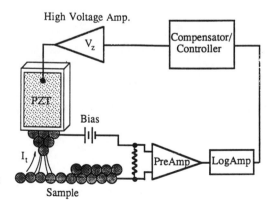

Figure 14.33. Principle of operation of an STM.

operator, result in an error to the compensator. The correction signal from the compensator is sent through a high-voltage amplifier to the z piezoelectric element. The piezoelectric element contracts or expands to adjust the probe position above the sample.

Figure 14.34 shows schematically a typical example of a computer interface to the STM electrical diagram. Using the computer to generate a digital scan for the x and y axis and digital acquisition of voltages applied to the z axis PZT from the controller, three-axis position information is stored and used to display a three-dimensional representation of the surface. Various computer program routines allow topographic views or top-down intensity scales of the surface.

Various methods are used to manipulate the data after acquisition. Some of the most common methods are: slope removal, smoothing, filtering, slope enhancement, and shading. Slope removal is necessary in cases where the slope of the sample is larger than the variations of the topography. In this case, the variations in the topography may be too slight in comparison with the overall slope; therefore, it is necessary to numerically fit a plane to the data. Then, subtracting the plane fit from the data at every point flattens the

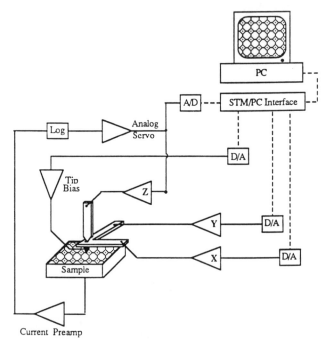

Figure 14.34. Block diagram of computer interface and feedback control circuit.

data such that only the small variations in topography remain. This is the most common type of data manipulation found in presenting SPM data.

Smoothing and filtering are used in cases where the data may be noisy because of inadequate control or instabilities sometimes found in the tunneling current. Other factors contributing to noise in the data may come from such effects as vibration or shock to the system during acquisition. In the case of a continuous vibration influencing the system, filtering can be used to remove a particular frequency superimposed onto the data. Methods for smoothing include such simple schemes as weighted averaging of points around each point in the data. Filtering schemes include one- and two-dimensional Fourier transforms of the surface. The Fourier transforms indicate the direction and frequency of a particular component in the data, which can then be removed or emphasized for data interpretation.

Slope enhancement can be used to emphasize very small changes in slope on the edges of structures. This has been a technique used particularly in the cases where slight changes in slope on step edges of Si and Ge could be emphasized. The application of this technique utilizes a method of measuring the change in the derivative of each line across the sample and plotting a top-down view of the surface where the value of the derivative is used for gray scaling.

Shading is a technique used in a variety of fields from undersea sonar mapping to microwave probing of planetary body surfaces by space probes. This technique uses the three-dimensional data to render a solid model of the surface rather than the typical cross grid representation of the surface used by many STM systems. Computations are then performed to enhance surfaces facing a fictitious light source and deemphasize surfaces facing away from the fictitious light source. This results in an image which looks to the human eye more like a real surface such as a mountain range casting shadows from the rising sun. This method, however, can be quite computer intensive and is typically used for final presentation of the data.

All of the above control and acquisition design considerations can be applied to any of the other probe microscopies. The only difference between the various techniques is the method used to measure the probe–sample interaction of interest. The sensor design and corresponding signal, then, replaces the preamp and log-amp in the above discussion.

14.5.4. Probe Fabrication

The interactions between probe and sample can vary depending on the probe. Therefore, controlling the characteristics of the probe are important in understanding measurements. The material and shape of the probe are the most important variables to control. In STM, the material of the probe is usually a metal such as W, PtIr, or Ir. These can have a wide variety of work functions, conductivity, and elastic moduli. The shape of the probe can have consequences on imaging (Pethica and Sutton, 1988; Kuk and Silverman, 1986; Reiss et al., 1990). A probe which has been fabricated for flat atomic plane imaging may not be an adequate probe for measuring rough surfaces. Therefore, many different procedures have been developed to manufacture probes which will ensure specific properties important to the experiment.

Electrochemical etching procedures are the most common techniques used for probe fabrication. Electrochemical etching of probes was initially developed for use in field ion microscopy (FIM) (Muller and Tsong, 1969). These techniques have been well established and became the starting point of probe development for SPM. However, a variety of probe microscopies have been developed that require many more exotic procedures for probe fabrication. In this section several of these techniques will be discussed which have an impact on different probe microscopies.

14.5.4.1. Electrochemical Etching of Probes

Figure 14.35 illustrates a typical electrochemical etching setup. Only one tip is formed at the surface interface of the wire and the solution. By changing the solution concentrations or the amplitude and frequency of the applied voltage, a variety of probe shapes can be obtained. Other methods have been devised to create two tips at once or create much sharper probes. The trick is to catch the lower half without damaging the tip. A similar procedure can be obtained by suspending the etching solution over a nonconductive solution, such as CCl_4, therefore etching two tips below and above the etchant–CCl_4 interface (Lemke et al., 1990). This basic idea was originally developed as the "drop off method" by Bryant et al. (1987).

The most common probe material used in electrochemical etching is either W or PtIr. For W and PtIr probes a NaOH or $CaCl_2$ etching solution is used, respectively. For W a 0.2-mm-diameter wire of W is lowered 2–5 mm into the NaOH etching solution. Using a gold counterelectrode, a dc or ac (60 Hz) voltage of approximately 10–20 V is applied between the electrode and the W wire. After only 2–3 min, the submerged section of the W wire is etched away completely and a tip remains on the suspended wire with a radius of curvature of approximately 1 μm.

For PtIr or Ir, the electrochemical etching process involves an ac technique using a $CaCl_2$ solution. A saturated $CaCl_2$ solution (74.5 g/100 cm^3 H_2O) is combined with H_2O and concentrated HCl to form the etching solution ($_o$)% saturated $CaCl_2$, 36% H_2O, 4% HCl by volume). A 0.25-mm-diameter wire of Ir is lowered 0.25–0.5 mm into the solution after initial wicking of the solution to the bottom of the cut wire. This is achieved by using a three-axis

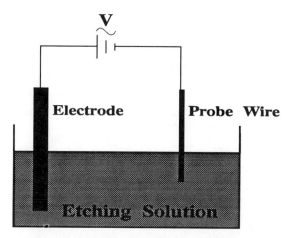

Figure 14.35. Typical electrochemical etching setup.

micrometer stage with a pair of tweezers affixed to the stage to hold the Ir wire vertically in the solution. The counterelectrode is a 1-cm-diameter carbon rod connected to one side of a voltage generator set to produce a 26.5-V, 60-Hz output waveform. With the counterelectrode lowered approximately 1 cm into the solution, the Ir wires are lowered to the appropriate depth 6 cm away from the counterelectrode. The 26.5-V ac signal is then applied and allowed to continue etching until the Ir wire etches out of the solution completely. The 26.5-V signal is then reduced to 5 V, 60 Hz and the tip of the new probe is quickly brought into and back out of contact with the surface of the solution using the z axis of the three-axis micrometer stage. This dunking action is repeated three or four times with a contact duration of no longer than 0.5 s per dunk. A similar process has been reported before by Musselman and Russell (1990) and referred to as "micro-polishing" of the probe.

The current passing between the counterelectrode and the tip can also be monitored (Lemke *et al.,* 1990) to automate the cutoff procedure. This results in a much more controllable process. Through computer control of the process, a threshold value can be set which turns off the process just before the tip has completely etched leaving a very sharp probe, ~ 0.1 μm radius.

14.5.4.2. Mechanical Methods

During the first 4 or 5 years of STM, one of the most common methods for making a probe was simply to cut a W or PtIr wire at an angle with a pair of wire cutters or to grind them on a typical mechanical grinding wheel. These probes are only applicable to atomic imaging since only a small group of protruding atoms at the probe is necessary to obtain atomic resolution. Surface roughness measurements can be greatly influenced by the shape of the probe. A reproducible method of probe fabrication is necessary to increase the measurement reliability and reproducibility. Therefore, these probes were short-lived in the history of probe microscopy because of the increasing demand for sharp well-defined macroscopic probes.

14.5.4.3. Electron and Ion Beam Probe Processing

Two processes which have been developed recently for the fabrication of extremely sharp, contamination-free probes with high aspect ratios are those of electron and ion beam fabrication. Electron beam processing involves the exposure of a material to a focused electron beam, as in a typical scanning electron microscope (SEM). The current density at the focal point of the electron beam can be controlled so as to cause contamination growth. By fixing the electron beam at a specific spot on the end of a blunt tip, a microscopic tip can be grown. Microscopic tips can be grown in this manner which are only slightly larger than the beam diameter, 100 nm (Akama *et al.*, 1990; Ximen and Russell, 1992). Also, by introducing a slight background pressure of known gases in the vicinity of the tip, known materials can be grown. Figure 14.36 shows a SEM micrograph of a tip grown by electron beam induction on a SiN_2 pyramid. Similarly, this process can be achieved with a focused ion beam. However, in this case energy for growth is induced by a focused beam of ions and beam diameters may be only 0.1–0.5 μm.

Focused ion beams (FIB) can be used to grow materials (Vasile and Harriott, 1989), but they are more commonly used to sputter material. Proper control of the beam deflection parameters allows a variety of shapes and sizes to be machined into most materials. The dimensions of machinable features are limited only by the beam diameter (0.1–0.5 μm), the raster size, and time. In a study by Vasile *et al.* (1991), FIB machining of W and Ir probes were performed by rastering a focused ion beam in an annular pattern, such as the shape of a doughnut. An electrochemically etched probe was placed in the FIB such that the axis of the probe was parallel to the FIB. The central axis of the electrochemically etched probe was placed at the center of the annular pattern whose dimensions were: 0.6 μm inner diameter, 7.0 μm outer diameter. The inner diameter was set such that the Gaussian tails of the ion beam would almost overlap at the central axis of the probe. This process creates tips over 5–10 μm long

Figure 14.36. Electron beam-induced growth of a microtip on a SiN$_2$ pyramid. (Ximen and Russell, 1992)

with an end radius of 5–6 nm and a total included cone angle of $\sim 10°$. Figure 14.37 shows an SEM micrograph of W probe before and after FIB machining.

Other ion machining techniques have been used to produce probes for STM. These processes relied on exposing an etched probe to a fixed beam of grazing incidence ions to produce contamination-free (Biegelsen *et al.*, 1987, 1989), sharp W probes. Radii of curvature of approximately 5 nm were produced; however, the cone angles of the probes were no better than 85°. Figure 14.38 shows the results of the work by Biegelsen *et al.* (1989) on single-crystal W. These probes are ideal for atomic-level imaging, but impractical for large-scale imaging, where the goal may be to penetrate down into features requiring much smaller cone angles.

14.5.4.4. Lithographically Shaped Probes

Lithographical methods have become a convenient method for integrating probes into micro-cantilevers for use in scanning force microscopy. Si and Si$_3$N$_4$ pyramids and tetrahedral probes have been successfully fabricated using batch processing lithographic techniques. The first such probes were the py-

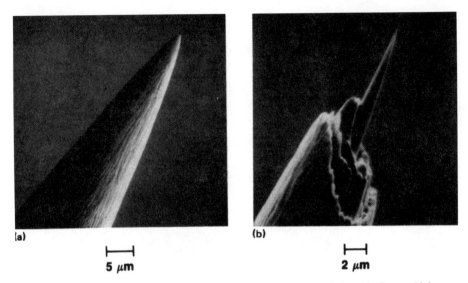

Figure 14.37. Focused-ion-beam-machined W probe. (a) Before machining; (b) after machining. (Vasile *et al.,* 1991)

ramidal Si_3N_4 probes. Figure 14.39 shows briefly the fabrication steps in producing both the cantilever and the integrated probe.

Figure 14.40 shows SEM micrographs of the results of the above processing steps. Note that the tips are sharp to <300 Å; however, the cone angle of the structure is 55°. Such large cone angles would inhibit the penetration into high aspect features on the surface.

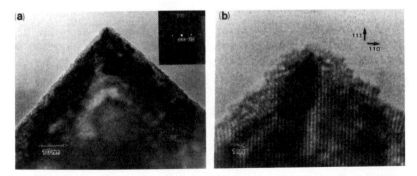

Figure 14.38. Phase-contrast, multibeam transmission electron micrographs of the same tungsten tip. (a) Lattice image showing (110) planes; inset shows (112) zone selected area diffraction: (b) magnified image of (a). (Biegelsen *et al.,* 1989)

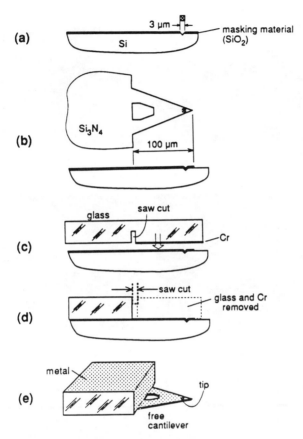

Figure 14.39. Fabrication of Si_3N_4 microcantilever with integrated pyramidal tips. (a) A pyramidal pit is etched in the surface of a (100) Si wafer using anisotropic etching. (b) A Si_3N_4 film is deposited over the surface and conforms to the shape of the pyramidal pit. (c) A glass plate is prepared with a saw cut and a Cr bond-inhibiting region. The glass is then anodically bonded to the annealed nitride surface. (d) A second saw cut releases the bond-inhibited part of the glass plate, exposing the cantilever. (e) All Si is etched away, leaving the Si_3N_4 microcantilever attached to the edge of a glass block. The back of the cantilever is coated with a metal to aid in deflection detection. (Albrecht, 1989)

To decrease the cone angle of integrated SFM probes, other microfabrication techniques have been developed. Si probes have been formed on Si_3N_4 cantilevers using an oxide mask and anisotropical etching of Si. This method produces tips having a radius of curvature better than about 400 Å, but the cone angle is still only about 46°. Figure 14.41 shows the steps used in this process.

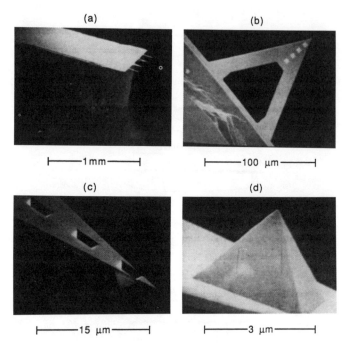

Figure 14.40. SEM micrographs of Si_3N_4 cantilevers with integrated pyramidal tips. (a) The Si_3N_4 film is attached to the surface of a glass block with dimensions of $2 \times 3 \times 0.7$ mm. Four cantilevers protrude from the edge of the block. (b) Four pyramidal tips can be seen at the end of this V-shaped cantilever. (c) The pyramidal tips are hollow when viewed from the back side. (d) Each tip has very smooth sidewalls, and the tip appears to terminate virtually at a point, with less than 300 Å radius. (Albrecht, 1989)

Several other microfabrication techniques exist for making different shaped probes. The formation of solid sharp cones by evaporation of material from a point source through an orifice onto a substrate has been demonstrated by Anbar and Aberth (1974) and Spindt *et al.* (1976). This allows the use of other materials such as Nb, W, and Ir.

14.6. STM Theory

One of the very appealing features of STM is that it gives an image which appears to be simply a direct topographic map of the surface structure. This view is quite adequate in most cases, although it is a naive view of what is actually occurring. With the advent of atomic resolution, it quickly became

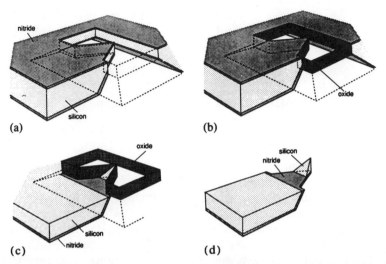

Figure 14.41. (a) The process for creating cantilevers with integral tips begins with fabrication of a freestanding Si and Si$_3$N$_4$ cantilever. (b) Exposed Si surfaces are selectively oxidized. (c) The nitride is removed from the front of the wafer and the exposed Si is anisotropically etched leaving a tetrahedral volume at the end of the cantilever. (d) Selective removal of the SiO$_2$ exposes the finished cantilever with integral tip. (Akamine *et al.*, 1990)

necessary to further understand what was occurring at the atomic level between the probe and sample surface. Many authors have studied the phenomena of metal–vacuum–metal (MVM) and metal–insulator–metal (MIM) tunneling well before the advent of the STM (Fowler and Nordheim, 1928; Simmons, 1963; Duke, 1969; Cohen *et al.*, 1962; Leipold and Feuchtwang, 1974; Feuchtwang, 1974a,b, 1975, 1976, 1979). The STM allowed, for the first time, the surface states of the tip and sample to be probed in a manner which would possibly allow the verification of various electron tunneling transport theories.

There exist three broad classes of tunneling theories to date:

1. Free electron models which consider the scattering and transmission of electrons incident on a potential barrier with corrugated boundaries. These methods are much more direct and transparent, and not restricted to low transmissivity.
2. Transfer Hamiltonian-type theories which can accommodate many-body effects and are more applicable to real surfaces.
3. More complete many-body theories based on Keldysh's infinite-order perturbation theory for nonequilibrium processes.

The reader is referred to several reviews on STM theory, including that of Stoll *et al.* (1984) and other work by Garcia *et al.* (1983), Baratoff (1984), Tersoff and Hamann (1983, 1985), and Lang (1985, 1986).

14.7. Scanning Force Microscope

14.7.1. Introduction

Interatomic and intermolecular forces acting between two bodies can be classified as attractive or repulsive, long range or short range. Some of the more common forces between bodies include van der Waals forces, repulsion, electrostatic, magnetic, friction, and adhesion. By measuring the forces between a sharpened probe and sample, the SFM can obtain images of the surface which depict contours of constant force. Furthermore, the SFM can qualitatively measure the long- and short-range interactions at any point within an image. Such measurements can be used to determine interfacial properties. In 1986, Binnig *et al.* developed the SFM having the capability of spatially measuring surface topography on an atomic scale. The strength of the SFM is its ability to operate on both conductive and nonconductive surfaces.

To understand the operation of the SFM it is necessary to understand the basic concepts of potential functions and surface forces. Figure 14.42 illustrates some of the types of commonly occurring intermolecular and intersurface potential functions and the different effects they have on molecule–molecule and particle–particle interactions.

The potential energy between two atoms can be simply described by the Lennard–Jones 6-12 potential for gases (Atkins, 1978):

$$U(r) = \frac{\alpha}{r^{12}} - \frac{\beta}{r^6} \tag{14.24}$$

where r is the distance between molecules and α and β are constants. The first term is attractive and is the instantaneous dipole–dipole interaction between molecules, i.e., the van der Waals energy. There are three contributions to the van der Waals energy: polarization, induction, and dispersion. The second term is repulsive and is the electrostatic repulsion felt by electrons in each molecule during orbital overlap. The balance between interatomic repulsion and attraction determines a variety of material properties. In SFM, it is the force, $\delta U/\delta D$, or the force gradient, $\delta^2 U/\delta D^2$, that is measured.

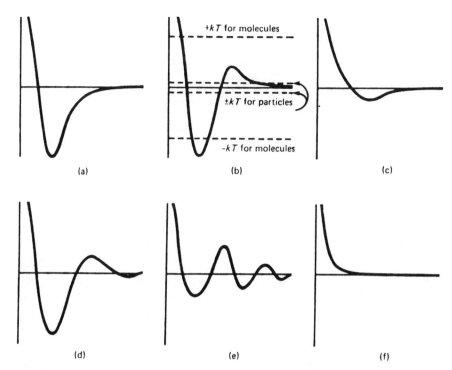

Figure 14.42. Typical interaction potentials encountered between molecules and particles in a medium. (a) This potential is typical of vacuum interactions but is also common in liquids. Both molecules and particles attract each other. (b) Molecules attract each other; particles repel each other. (c) Weak minimum. Molecules repel, particles attract. (d) Molecules attract strongly, particles attract weakly. (e) Molecules attract weakly, particles attract strongly. (f) Molecules repel, particles repel. (Israelachvili, 1985)

In the SFM, small forces near the surface of the sample act on the probe tip. These small forces are typically measured by the deflection of a small cantilever beam on which the probe is mounted. The cantilever beam has a known spring constant, and therefore, a measure of the deflection gives the force directly. Figure 14.43 shows the basic configuration of an SFM instrument used to measure the deflections of a cantilever beam. The method of sensing the deflection can be one of several different techniques. Variations in the methods to detect cantilever deflections have given rise to a number of SFM designs. Each new design has particular advantages over other techniques, such as cost, simplicity, resolution, or size. Several reviews (Burnham and Colton, 1990) and text (Sarid, 1991) have been written in the past few years on the applications of SFM. The reader should refer to these for a full description of the techniques.

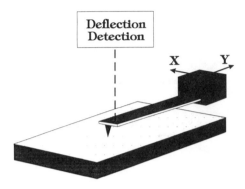

Figure 14.43. Block diagram of the SFM. Forces between the probe and sample surface cause minute deflections of the flexible cantilever beam. Deflections are measured with subangstrom sensitivity by a number of optical, capacitive, or tunneling detectors.

14.7.2. Designs

14.7.2.1. Tunneling Detection

The first SFM, developed by Binnig *et al.* in 1986, used an STM probe to measure cantilever deflections. Figure 14.44 shows the setup they used to obtain some of the first SFM images. In Fig. 14.44 an STM probe is brought to the back of a conductive cantilever beam. Deflections in the cantilever can be detected by a change in the tunneling current between the cantilever and the stationary STM probe. Since the tunneling current is exponentially dependent on the gap, deflection sensitivity < 0.01 Å can be measured. In this configuration, the sample is scanned in the x and y directions. Corrections in the z direction are used to maintain a constant bending force on the cantilever, i.e., a constant tunneling current. However, contamination and surface roughness on the cantilever make the tunneling measurement method non-ideal. Therefore, many authors have gone to a number of optical methods to measure deflections.

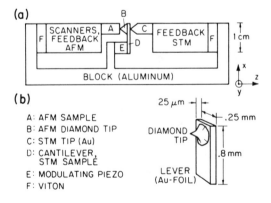

Figure 14.44. SFM experimental setup. The lever is not to scale in (a). Its dimensions are given in (b). The STM and SFM piezoelectric drives are facing each other, sandwiching the diamond tip that is glued to the lever. (Binnig *et al.*, 1986)

14.7.2.2. Deflection Detection

Because of the inherent problems with tunneling detection, various optical methods were developed. Optical methods are not affected by contaminants or roughness on the cantilever. The beam spot averages the surface roughness. The most popular optical method is the deflection detection method developed by Meyer and Amer (1988a,b) and Alexander *et al.* (1989). The deflection detection method is the primary system used in commercial SFM instruments. The mechanical and electrical system is simple to implement and low cost.

Figure 14.45 illustrates the deflection detection system. A collimated laser beam is focused on the back of a cantilever beam which supports the probe. The reflected path of light hits the junction of two closely spaced photodetectors, i.e., a position-sensitive detector (PSD). Small deflections of the cantilever cause one photodetector to collect more light than the other. The outputs of the photodetector are fed into a differential amplifier whose output is proportional to the deflection of the cantilever. Since the optical elements can be a large distance away from the cantilever system, mechanical crashes are avoided.

14.7.2.3. Homodyne Detection

Figure 14.46 illustrates the basic concepts in homodyne detection. A polarized beam of light passes through a beam splitter and is incident on a Fabry Perot which consists of two reflecting surfaces: an optical flat and the cantilever beam. The beam reflected from the Fabry Perot is incident on the same beam splitter which deflects the reflected beam onto a photodetector. The photocurrent from the photodetector indicates the cantilever deflection, i.e., the force. In a differential homodyne detection system, a reference beam

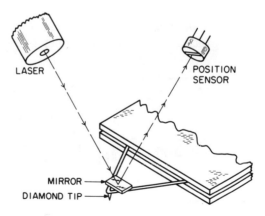

Figure 14.45. The cantilever deflection detection method. (Alexander *et al.*, 1989)

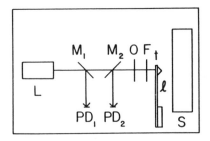

Figure 14.46. A schematic diagram of the homodyne detection system showing the laser (L), the beam splitters M_1 and M_2, the lever (l) supporting the force-sensing tip (t), the sample (S), the microscope objective (O), the optical flat (F), and the photodetectors PD_1 and PD_2. (Sarid, 1991)

is reflected by the first beam splitter onto a photodetector. The remaining portion of the beam is incident on the Fabry Perot and is reflected back onto the second beam splitter which deflects the beam onto a different photodetector. Several groups (McClelland *et al.*, 1987; Mate *et al.*, 1987; Erlandsson *et al.*, 1988) have employed the system shown in Fig. 14.46 without PD_1. Others (Stern *et al.*, 1988; Mamin *et al.*, 1988; Rugar, Mamin & Guethner, 1989) have used a fiber-coupled homodyne detection method similar to that shown in Fig. 14.46.

14.7.2.4. Heterodyne Detection

Martin *et al.* (1987) developed a heterodyne detection system to measure the deflection of the cantilever. Figure 14.47 shows the basic configuration for the heterodyne system. The latter system eliminates the effects of drift in the optical path length. The beam from a laser is split into two paths by the beam splitter. The first path passes through an acousto-optic modulator that shifts the beam frequency by Ω_m and the other is reflected onto a mirror as a reference beam. The beam with the shifted frequency is the signal beam and passes through a polarizing beam splitter, a quarter-wave plate, and finally a microscope objective that focuses the beam onto the cantilever. Upon reflection from the cantilever, the beam returns through the microscope objective and the quarter-wave plate which rotates the polarization. The reflected beam

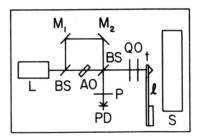

Figure 14.47. The heterodyne detection system, showing the laser (L), the two mirrors (M_1 and M_2), beam splitter (BS), the acousto-optic modulator (AO), the quarter-wave plate (Q), the lever (l) supporting the force-sensing tip (t), the sample (S), microscope objective (O), the polarizer, and the photodetector (PD). (Sarid, 1991)

is now rotated 90° relative to the reference beam after they are combined at the second beam splitter. The two beams are incident on a photodetector and interfere such that a photocurrent is generated. The photocurrent consists of a spectrum of frequencies. The photocurrent is fed into a single band receiver driving a phase-sensitive detector that provides the signal used to display the force on the cantilever. Several groups (Martin and Wickramasinghe, 1987; Martin *et al.,* 1987, 1988; Abraham *et al.,* 1988; Hobbs *et al.,* 1989) have employed a system similar to Fig. 14.47. However, instead of the two mirrors, they used a Dove prism to match the optical path length of the two arms of the interferometer.

14.7.2.5. Laser–Diode Detection

In 1988, Sarid *et al.* introduced a compact detection scheme using a laser diode. A laser is highly sensitive to optical feedback. Therefore, this sensitivity can be used to detect small variations in the feedback induced by changes in the position of a reflecting target, i.e., the cantilever beam. Figure 14.48 shows the basic configuration of the laser–diode feedback detection system. The cantilever beam supporting the force-sensing probe is positioned within several micrometers from the front facet of the laser. The lever and front facet combination act as a lossy Fabry Perot, whose reflectivity serves as the effective reflectivity of the front facet of the laser. Diffraction effects caused by successive reflections between the front facet and the cantilever decrease higher-order reflections. Therefore, only the first-order reflection is measured. The analysis of this technique is more complicated because of the nonlinear laser medium in which the optical and electronic processes occur. However, the technique is compact, inexpensive, easy to align, and can operate in vacuum.

14.7.2.6. Capacitive Detection

Capacitively sensing the displacement of a cantilever beam has similar advantages as the optical approaches in that the sensed area is averaged.

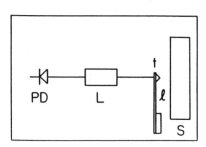

Figure 14.48. The laser–diode feedback detection system, showing the laser (L), the lever (l) supporting the force-sensing probe (t), the sample (S), and the photodetector (PD). (Sarid, 1991)

McClelland used capacitive sensing to measure lateral friction forces while scanning in STM and SFM. Goddenhenrich *et al.* (1990) developed a capacitance force sensor to measure magnetic bits and nonmagnetic topography with a resolution of greater than 1000 Å. Miller *et al.* (1991) and Joyce and Houston (1991) have developed methods to electrostatically balance a rocking beam cantilever while capacitively measuring its position. Figure 14.49 shows a schematic diagram of the operation of the force balance SFM developed by Miller *et al.* The feedback network is designed to maintain a constant capacitance on both sides of the rocker, therefore a constant balance position. However, as the probe encounters the surface, interaction forces try to unbalance the rocker one way or the other (attraction or repulsion). The feedback network responds to such interactions by applying a larger or smaller potential across one of the capacitor sides of the rocker. This response forces the rocker to maintain its position. The output of the feedback network used to maintain constant rocker position is linear to force. This output can then be used as the control signal for driving the z axis feedback used during scanning.

In this method, the sensitivity is decoupled from the stiffness of the cantilever, whereas they are directly coupled in all other methods. Also, this method allows the direct control of the position of the probe and cantilever with respect to the sample by the potentials applied to either side of the rocker. Therefore, in approaching the sample, the probe can be kept from jumping into the bottom of the potential well (Fig. 14.42) and getting stuck. This commonly occurs with many of the other techniques because the slope of the

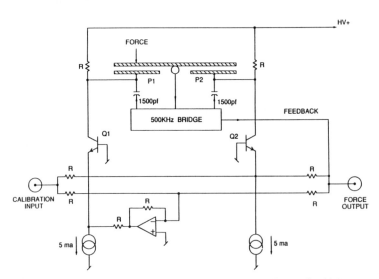

Figure 14.49. Force balance rocking beam SFM. (Miller *et al.*, 1991)

potential curve becomes larger than the stiffness of the cantilever and the probe jumps into the well. In the balanced rocking beam technique, the stiffness can be independently controlled allowing the probe to smoothly rise and fall through the full range of the potential curve.

14.7.3. Applications

The SFM has been used to image conductive and nonconductive samples from atomic resolution to 100-μm size fields. Samples vary from semiconductor device structures and polymers to biological samples such as DNA and blood cells.

Imaging is achieved similarly to that of STM; however, in SFM the control signal is that of force rather than current. Almost identical electronics can be used to control the probe height and the raster scanning of the probe. All of the same design considerations of stiffness and isolation apply.

After the invention of the first SFM in 1986, Binnig (1987) obtained the first atomic images of graphite using the SFM. Figure 14.50 shows the results of the first images of graphite.

A number of materials followed which could be imaged with atomic resolution: BN, NaCl, MoS_2, and Au. Later, interests shifted from imaging atomic features to measurement of the various surface properties and interactions with the SFM. This produced several papers involved in obtaining force profiles and measuring lateral forces during scanning in the STM and SFM. Figure 14.51 shows a typical force profile obtained on graphite by slowly moving the probe toward the sample until contact, then slowly moving the probe away from the surface. This curve is very similar in appearance and

Figure 14.50. First published atomic resolution images of graphite taken with an SFM. (Binnig, 1987)

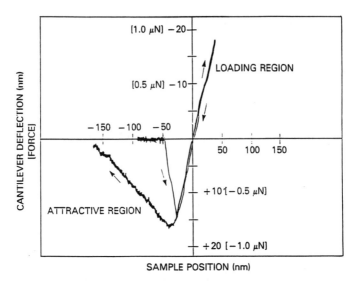

Figure 14.51. Force curves associated with the interaction of a tungsten probe with graphite. The probe is farthest from the sample on the left, and in contact on the right. (Burnham and Colton, 1990)

meaning to that seen in Fig. 14.42. The effect of the cantilever flopping into the potential well is illustrated in this figure by the abrupt slope while the probe is approaching the sample. In returning from contact, the potential well appears wider, deeper, and having a gentler return slope. This is believed to be an effect of adhesion. The properties of the force curves (i.e., slope, depth, width) have been studied to correlate their shapes to material properties such as surface energy, Young's modulus, adhesion, and contamination with nanometer resolution.

14.8. Magnetic Force Microscope

The formation and movement of domain walls and, in general, the domain structure of magnetic materials is a field of great interest because of its direct application in magnetic technology. The SFM can be employed to measure such magnetic domains by using a magnetic-sensitive probe (Saenz *et al.*, 1987, 1988; Abraham *et al.*, 1988; Rugar *et al.*, 1990; Schonenberg *et al.*, 1990; Grutter *et al.*, 1988). Probe materials such as Fe and CoPtCr can be fabricated similarly as STM and SFM probes to scan a surface and spatially measure the changes in the magnetic forces from domain to domain. This has enabled scientists to measure magnetic media such as recording tracks as shown in Fig. 14.52.

14.9. Scanning Capacitance Microscope

The scanning capacitance microscope (SCM) provides a mean for surface characterization through the measurement of local capacitance between probe and sample (Matey and Blanc, 1985; Bugg and King, 1988; Martin *et al.*, 1988; Williams *et al.*, 1989). It is useful for profiling both conductors and insulators. Perhaps more importantly, it provides a means for imaging variations in dielectric properties in or through insulating layers. This technique has demonstrated lateral resolutions of better than 25 nm (Williams *et al.*, 1989).

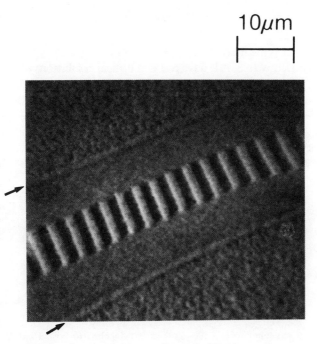

Figure 14.52. MFM image of an 8-μm-wide recording track. The bit transitions are spaced every 2 μm along the track. (Rugar *et al.*, 1990)

Several techniques have been developed to measure capacitances with the SCM; however, the most common is the use of high-sensitivity capacitance pickup circuitry developed by RCA (Radio Corporation of America). This circuitry was originally developed for a Video Disc capacitance sensor to read back a video signal stamped on a disk. An ultrahigh-frequency (UHF) oscillator (915 MHz) is coupled to detection circuitry through a resonant circuit. The resonant frequency of the circuit is largely determined by the total stray capacitance, but is modulated by a contribution introduced by the probe–sample capacitance. This probe–sample capacitance shifts the resonant frequency of the sensor, modifying the transmitted UHF signal from oscillator to detector. For small changes in capacitance, the output voltage changes linearly with the change in capacitance between probe and sample. Results indicate that the output sensitivity of a 10-nm radius tip is approximately 70 mV/fF at the tip. This measurement demonstrates a capacitive sensitivity in a 1-kHz bandwidth of $C = 1 \times 10^{-19}$ F. A demonstration of the resolution found in this technique is shown in Fig. 14.53.

14.10. Near-Field Scanning Optical Microscope

Near-field optical scanning (NFOS) microscopy or near-field scanning optical microscopy (NSOM) have reached optical resolutions of 20 nm (Durig *et al.,* 1986a), and resently $\lambda/50$ (Betzig and Chichester, 1994) on optically excited surfaces. The key element of the NSOM is a tiny aperture, typically

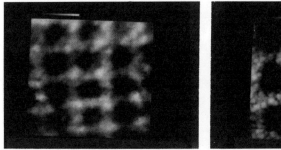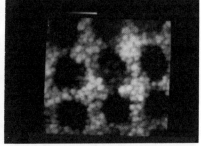

Figure 14.53. (Left) Gray-scale capacitance images of an array of 50-nm circular holes on 100-nm centers. The structures are in a 100-nm-thick PMMA film, overcoated with 20 nm of gold. The field of view is a 370 nm square. (Right) Gray-scale STM image of the same sample in a different region. The field of view is a 270 nm square. (Williams *et al.,* 1989)

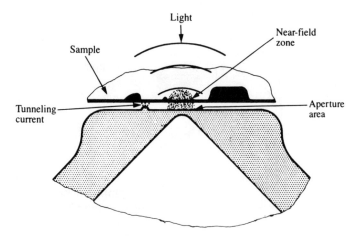

Figure 14.54. Aperture held at operating distance by means of tunneling. (Durig *et al.*, 1986b)

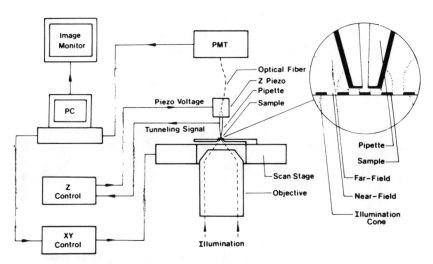

Figure 14.55. Instrumentation for the collection mode near-field scanning optical microscope. An expanded view is shown of the light transmitted through a sample being collected in the near field by an aperture. (Betzig *et al.*, 1987)

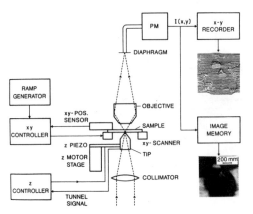

Figure 14.56. NFOS microscope, schematic. A high-resolution NFOS micrograph and the corresponding line scan of the border line of an opaque film on a transparent substrate are also shown. (Durig *et al.*, 1986a)

10 nm in diameter, positioned at the very apex of a conical or pyramidal probe. The probe is hollow such that the 10-nm aperture extends through the probe. Illumination of the aperture is handled through the hollow probe. The probe can then be scanned across a surface in close proximity of the surface, approximately 5 nm or less. Several modes of operation can be utilized: illumination of the aperture and transmission through an optically transparent sample (Durig *et al.*, 1986a), or illumination under the sample and measurement through the aperture (Bitzig *et al.*, 1987). The amount of radiation transmitted through the aperture and sample depends sensitively on the optical properties of the sample volume next to the aperture. The transmitted radiation

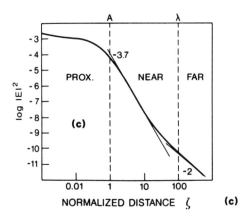

Figure 14.57. Calculated on-axis electrical-energy density versus distance from aperture, magnetic excitation. The distance from the aperture is normalized to the aperture radius a. (Durig *et al.*, 1986a)

can be determined by use of a photosensitive detector to measure transmitted intensity through the probe or sample. An optical microscope element is also necessary to focus the transmitted light onto the detector to receive maximum signal.

The distance that the probe travels above the sample is maintained by tunneling to a protruding point on the end of the probe. This is illustrated in Fig. 14.54.

A typical setup for the illuminated sample method is shown in Fig. 14.55. The setup for illumination of the aperture is shown in Fig. 14.56. There are several distinct advantages to receiving the signal through the probe instead of through the sample. First, owing to the presence of the incident wave, the geometry for the two modes is not symmetric, and indications suggest that collection through the probe may yield higher resolution. Second, luminescent samples could be readily imaged by collection through the probe. Finally, the

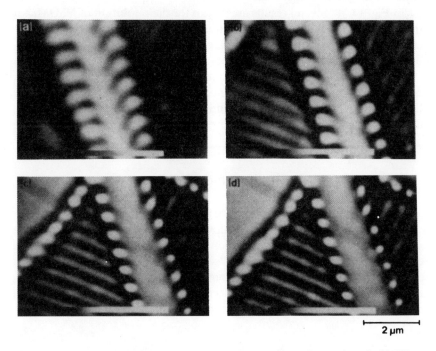

2 μm

Figure 14.58. Images generated at aperture-to-sample separations of approximately (a) 780 nm, (b) 330 nm, (c) 150 nm, and (d) in contact demonstrate the effect of the separation on the resolution. Differences in (c) and (d) are small, since the separation is comparable to the aperture size in (c). (Betzig *et al.*, 1987)

collection through the probe represents the first step toward a form of reflection mode NSOM in which the light from an external illuminator is reflected off the sample and collected by the aperture.

The diffraction limit is a very basic law of physics directly related to Heisenberg's uncertainty relation. In near-field microscopy, the incident radiation is forced through a subwavelength aperture. In terms of wave propagation theory, this is only possible by formation of wavelets with wavelength similar to the aperture diameter. The latter, known as evanescent waves, cannot propagate in free space. However, they can wind themselves around the aperture and therefore transmit radiative energy to the other side of the screen. Bethe's integrals can be easily calculated for the normal axis going through the center of the aperture. The resulting electric field energy density $|E|^2$, a major parameter with regard to absorption by sample material positioned nearby, is plotted as a function of distance in Fig. 14.57 for the case of purely magnetic excitation. This condition holds for an incident electric field polarized normal to the plane of incidence.

In Fig. 14.58, several results are shown for the resolution of NSOM as a function of sample-to-aperture separation. The sample was an electron lithographically produced pattern consisting of 250-nm-wide, 100-nm-high aluminum lines separated by 250 nm on a transparent, 100-nm-thick membrane of silicon nitride. One interesting feature of these images is the existence of strong intensity maxima present at the corners of the aluminum lines. These may result from high electric fields presumably caused by the high conductivity and sharp corners of the aluminum lines.

14.11. Future Trends

Scanning probe microscopies have become important tools for the evaluation of surfaces. The techniques allow subnanometer resolution imaging over hundreds of micrometers. A variety of methods allow SPMs to measure surface properties as well as some subsurface properties as well. Studies of surface chemistry, surface roughness, contamination growth, surface oxidation, corrosion, dopant density, surface dielectrics, nanometer-scale metrology and tribology are a few of the various areas actively studied using SPMs.

References

Abraham, D. W., Williams, C. C., and Wickramasinghe, H. K. (1988). *Appl. Phys. Lett.* **53**(15), 1447.

Akama, Y., Nishimura, E., Sakai, A., and Murakami, H. (1990). *J. Vac. Sci. Technol. A* **8**(1), 429.

Akamine, S., Barrett, R. C., and Quate, C. F. (1990). *Appl. Phys. Lett.* **57**(3), 316.

Albrecht, T. R. (1989). Ph.D. dissertation, G. L. No. 4529, Stanford University.

Alexander, S., Hellemans, L., Marti, O., Schneir, J., Elings, V., Hansma, P. K., Longmire, M., and Gurley, J. (1989). *J. Appl. Phys.* **65**, 164.

Anbar, M., and Aberth, W. H. (1974). *Anal. Chem.* **46**, 59.

Atkins, P. W. (1978). *Physical Chemistry,* Oxford University Press, London.

Baratoff, A. (1989). *Physica (Utrecht)* **127B**, 143.

Baro, A. M., Binnig, G., Gerber, C., Stoll, E., Baratoff, A., and Salvan, F. (1984). *Phys. Rev. Lett.* **52**, 1304.

Biegelsen, D. K., Ponce, F. A., Tramontana, J. C., and Koch, S. M. (1987). *Appl. Phys. Lett.* **50**(11), 696.

Biegelsen, D. K., Ponce, F. A., and Tramontana, J. C. (1989). *Appl. Phys. Lett.* **54**(13), 1223.

Binnig, G. (1987). *Phys. Scr.* **T19**.

Binnig, G., and Rohrer, H. (1982). *Helv. Phys.* **55**, 726.

Binnig, G., and Rohrer, H. (1987). *Rev. Mod. Phys.* **59**, 615.

Binnig, G., and Smith, D. P. E. (1986). *Rev. Sci. Instrum.* **57**(8).

Binnig, G., Rohrer, H., Gerber, C., and Weibel, E. (1982a). *Appl. Phys. Lett.* **40**, 178.

Binnig, G., Rohrer, H., Gerber, C., and Weibel, E. (1982b). *Physica* **109–110B**, 2075.

Binnig, G., Rohrer, H., Gerber, C., and Weibel, E. (1982c). *Phys. Rev. Lett.* **49**, 57.

Binnig, G., Rohrer, H., Gerber, C., and Weibel, E. (1983). *Phys. Rev. Lett.* **50**, 120.

Binnig, G., Quate, C. F., and Gerber, C. (1986). *Phys. Rev. Lett.* **56**(9), 930.

Betzig, E., and Chichester, R. J. (1994). *Science.* to be published.

Betzig, E., Isaacson, M., and Lewis, A. (1987). *Appl. Phys. Lett.* **51**(25), 2088.

Bryant, P. J., Kim, H. S., Zheng, Y. C., and Yang, R. (1987). *Rev. Sci. Instrum.* **58**, 1115.

Bugg, C. D., and King, P. J. (1988). *J. Phys. E* **21**, 147.

Burnham, N. A., and Colton, R. J. (1990). *Scanning Tunneling Microscopy: Theory and Application,* VCH Publishers, New York.

Chiang, S., and Wilson, R. J. (1986). *IBM J. Res. Dev.* **30**, 515.

Cohen, M. H., Falicov, L. M., and Phillips, J. C. (1962). *Phys. Rev. Lett.* **8**(8), 316.

Demuth, J. E., Koehler, U. K., and Hamers, R. J. (1988). *J. Microsc. (Oxford)* **152**, 299.

Duke, C. B. (1969). *Tunneling in Solids, Solid State Physics,* Suppl. 10, Academic Press, New York.

Durig, U., Pohl, D. W., and Rohner, F. (1986a). *J. Appl. Phys.* **59**(10), 3318.

Durig, U., Pohl, D. W., and Rohner, F. (1986b). *IBM J. Res. Dev.*

Erlandsson, R., McClelland, G. M., Mate, C. M., and Chiang, S. (1988). *J. Vac. Sci. Technol. A* **6**(2), 266.

Feuchtwang, T. E. (1974a). *Phys. Rev.* **B10**(10), 4121.

Feuchtwang, T. E. (1974b). *Phys. Rev.* **B10**(10), 4135.

Feuchtwang, T. E. (1975). *Phys. Rev.* **B12**(9), 3979.

Feuchtwang, T. E. (1976). *Phys. Rev.* **B13**(2), 517.

Feuchtwang, T. E. (1979). *Phys. Rev.* **B20**(2), 430.

Fowler, R. H., and Nordheim, L. (1928). *Proc. R. Soc. London* **A119**, 173.

Garcia, N., Ocal, C., and Flores, F. (1983). *Phys. Rev. Lett.* **50**(25), 2002.

Germano, C. P., (1959). *IRE Trans. Audio* **July–Aug.**

Goddenhenrich, T., Lemke, H., Hartmann, U., and Heiden, C. (1990). *J. Vac. Sci. Technol. A* **8**, 383.

Golovchenko, J. A. (1986). *Science* **232**, 48.

Griffith, J. E., and Kochanski, G. P. (1990). *Annu. Rev. Mater Sci.* **20**, 219.

Grutter, P., Meyer, E., Heinzelmann, H., Rosenthaler, L., Hidber, H. R., and Guntherodt, H. J. (1988). *J. Vac. Sci. Technol.* **A6**(2), 279.

Hansma, P. K., and Tersoff, J. (1987). *J. Appl. Phys.* **61**(2), R1.

Harrison, W. A. (1979). *Surf. Sci.* **55**, 1.

Hobbs, P. C. D., Abraham, D. W., and Wickramasinghe, H. K. (1989). *Appl. Phys. Lett.* **55**, 2357.

Israelachvili, J. N. (1985). *Intermolecular and Surface Forces,* Academic Press, New York.

Joyce, S. A., and Houston, J. E. (1991). *Rev. Sci. Instrum.* **62**(3), 710.

Kuk, Y., and Silverman, P. J. (1986). *Appl. Phys. Lett.* **48**(23), 1597.

Kuk, Y., and Silverman, P. K. (1989). *Rev. Sci. Instrum.* **60**(2).

Lang, N. D. (1985). *Phys. Rev. Lett.* **55**, 230.

Lang, N. D. (1986). *Phys. Rev. Lett.* **56**, 1164.

Leipold, W. C., and Feuchtwang, T. E. (1974). *Phys. Rev.* **B10**(6), 2195.

Lemke, H., Goddenhenrich, T., Bochem, H. P., Hartmann, U., and Heiden, C. (1990). *Rev. Sci. Instrum.* **61**(10), 2538.

Little, R. W. (1973). *Elasticity,* Prentice–Hall, Englewood Cliffs, N.J.

McClelland, G. M., Erlandsson, R., and Chiang, S. (1987). In *Review of Progress in Quantitative Nondestructive Evaluation* (D. O. Thompson and D. E. Chimenti, eds.), Vol. 6B, Plenum Press, New York, p. 307.

McCord, M. A., and Pease, R. F. (1987). *J. Vac. Sci. Technol.* Jan.–Feb.

Mamin, H. J., Rugar, D., Stern, J. E., Terris, B. D., and Lambert, S. E. (1988). *Appl. Phys. Lett.* **53**, 1563.

Martin, Y., and Wickramasinghe, H. K. (1987). *Appl. Phys. Lett.* **50**, 1455.

Martin, Y., Williams, C. C., and Wickramasinghe, H. K. (1987). *J. Appl. Phys.* **61**(10), 4723.

Martin, Y., Abraham, D. W., and Wickramasinghe, H. K. (1988). *Appl. Phys. Lett.* **52**, 1103.

Mate, C. M., McClelland, G. M., Erlandsson, R., and Chiang, S. (1987). *Phys. Rev. Lett.* **59**, 1942.

Matey, J. R., and Blanc, J. (1985). *J. Appl. Phys.* **47**, 1437.

Meyer, G., and Amer, N. M. (1988a). *Appl. Phys. Lett.* **53**(12), 1045.

Meyer, G., and Amer, N. M. (1988b). *Appl. Phys. Lett.* **53**(12), 2400.

Miller, G. L., Griffith, J. E., Wagner, E. R., and Grigg, D. A. (1991). *Rev. Sci. Instrum.* **62**(3), 705.

Muller, E. W., and Tsong, T. T. (1969). *Field Ion Microscopy,* American Elsevier, New York.

Musselman, I. H., and Russell, P. E. (1990). *J. Vac. Sci. Technol.* **A8**, 3558.

Pethica, J. B., and Sutton, A. P. (1988). *J. Vac. Sci. Technol.* A **6**(4), 249.

Phillips, C. L., and Nagle, H. T., Jr. (1984). *Digital Control System Analysis and Design,* Prentice–Hall, Englewood Cliffs, N.J.

Reiss, G., Schneider, F., Vancea, J., and Hoffmann, H. (1990). *Appl. Phys. Lett.* **57**(9), 867.

Rugar, D., Mamin, H. J., and Guethner, P. (1989). *Appl. Phys. Lett.* **55**, 2588.

Rugar, D., Mamin, H. J., Guethner, P., Lambert, S. E., Stern, J. E., and McFadyen, I. (1990). *J. Appl. Phys.* **68**(3), 1169.

Saenz, J. J., Garcia, N., Grutter, P., Meyer, E., Heinzelmann, H., Wiesendanger, R., Rosenthaler, L., and Hidber, H. R. (1987). *J. Appl. Phys.* **62**(10), 4293.

Saenz, J. J., Garcia, N., and Slonczewski, J. C. (1988). *Appl. Phys. Lett.* **53**(15), 1449.

Sarid, D. (1991). *Scanning Force Microscopy: With Applications to Electric, Magnetic, and Atomic Forces,* Oxford University Press, London.

Sarid, D., Iams, D., and Weissenberger, V. (1988). *Opt. Lett.* **13**, 1057.

Schonenberg, C., Alvarado, S. F., Lambert, S. E., and Sanders, I. L. (1990). *J. Appl. Phys.* **67**(12), 7278.

Simmons, J. G. (1963). *J. Appl. Phys.* **34**, 1793.

Smith, D. P. E., and Binnig, G. (1986). *Rev. Sci. Instrum.* **57**(10).

Spindt, C. A., Brodie, I., Humphrey, L., and Westerberg, E. R. (1976). *J. Appl. Phys.* **47**, 5248.

Stern, J. E., Terris, B. D., Mamin, H. J., and Rugar, D. (1988). *Appl. Phys. Lett.* **53**, 2717.

Stoll, E., Baratoff, A., Selloni, A., and Carnevalli, P. (1984). *J. Phys.* **C17**, 3073.

Tersoff, J., and Hamann, D. R. (1983). *Phys. Rev. Lett.* **50**, 25.

Tersoff, J., and Hamann, D. R. (1985). *Phys. Rev.* **B31**, 2.

Vasile, M. J., and Harriott, L. R. (1989). *J. Vac. Sci. Technol.* **B7**(6), 1954.

Vasile, M. J., Grigg, D. A., Griffith, J. E., Russell, P. E., and Fitzgerald, E. A. (1991). *Rev. Sci. Instrum.* **62**, 2167.

Williams, C. C., Hough, W. P., and Rishton, S. A. (1989). *Appl. Phys. Lett.* **55**(2), 203.

Ximen, H., and Russell, P. E. (1992). *Ultramicroscopy* **42–44**, 1526–1532.

Young, R. (1966). *Rev. Sci. Instrum.* **37**, 275.

Young, R. (1971). *Phys. Today* **24**, 42.

Young, R., Ward, J., and Scire, F. (1971). *Phys. Rev. Lett.* **27**, 922.

Young, R., Ward, J., and Scire, F. (1972). *Rev. Sci. Instrum.* **43**, 999.

Index